# Petroleum Engineering in Oil and Gas Production: Advances in Theory and Operation

# Petroleum Engineering in Oil and Gas Production: Advances in Theory and Operation

Editors

**Xingguang Xu**
**Kun Xie**
**Yang Yang**

Basel • Beijing • Wuhan • Barcelona • Belgrade • Novi Sad • Cluj • Manchester

*Editors*
Xingguang Xu
China University of Geosciences
Wuhan, China

Kun Xie
Northeast Petroleum University
Daqing, China

Yang Yang
Chengdu University of
Technology
Chengdu, China

*Editorial Office*
MDPI
St. Alban-Anlage 66
4052 Basel, Switzerland

This is a reprint of articles from the Special Issue published online in the open access journal *Energies* (ISSN 1996-1073) (available at: https://www.mdpi.com/journal/energies/special_issues/ Y3H698PD17).

For citation purposes, cite each article independently as indicated on the article page online and as indicated below:

Lastname, A.A.; Lastname, B.B. Article Title. *Journal Name* **Year**, *Volume Number*, Page Range.

**ISBN 978-3-0365-9332-6 (Hbk)**
**ISBN 978-3-0365-9333-3 (PDF)**
**doi.org/10.3390/books978-3-0365-9333-3**

# Contents

# About the Editors

**Xingguang Xu**

Dr. Xingguang Xu is a professor at the Department of Petroleum Engineering of the China University of Geosciences, Wuhan, China. Dr. Xu received his Ph.D in petroleum engineering from Curtin University, Australia, in 2016 before joining CSIRO as a postdoctoral research fellow (2016-2019). His research interests include enhanced oil and gas recovery for conventional and unconventional reservoirs, carbon capture storage and utilization (CCUS), and oilfield chemistry.

**Kun Xie**

Dr. Kun Xie is an associate professor at the College of Petroleum Engineering of Northeast Petroleum University. Dr. Kun received his Ph.D in petroleum and natural gas engineering from Northeast Petroleum University, China, in 2019. His research interests include enhanced oil recovery, oilfield chemistry, and physical stimulation.

**Yang Yang**

Dr. Yang Yang is a professor at the College of Energy of the Chengdu University of Technology, Chengdu, China. Dr. Yang received his Ph.D in petroleum engineering from Southwest Petroleum University, China, in 2019. His research interests include enhanced oil recovery, oilfield chemistry, and reservoir stimulation.

# Preface

Despite the increasing interest in renewable energy, oil and gas remain the predominant energy sources that spur economic growth. Along with escalating energy demand, the rapid decline in conventional oil and gas reservoirs has introduced additional challenges to sourcing oil and gas supplies. Over the past decade, significant research progress has been made in oil and gas industry, with the intention of stimulating oil and gas production from unconventional reservoirs that have either undesirable properties or harsh conditions.

Research papers included in this Reprint Book focus on recent advances in the basic theory and field practice of oil and gas production from (1) reservoirs that present unsatisfactory properties, such as low permeability and extreme reservoir heterogeneity with fractures, and (2) reservoirs that possess challenging conditions, including high temperature, high salinity, high oil viscosity, and deep water. Review papers focusing on state-of-the-art perspectives that provide enlightening support to the oil and gas production community are also included.

We hope that the published articles focusing on this specific topic will be helpful and enlightening for scientists, engineers, and governors who work in the field of oil and gas, and we sincerely thank all contributors to the successful publication of this Reprint Book.

**Xingguang Xu, Kun Xie, and Yang Yang**
*Editors*

*Article*

# Numerical Modeling of Shale Oil Considering the Influence of Micro- and Nanoscale Pore Structures

Qiquan Ran, Xin Zhou *, Dianxing Ren, Jiaxin Dong, Mengya Xu and Ruibo Li

Research Institute of Petroleum Exploration and Development, No. 20 Xueyuan Road, Haidian District, Beijing 100083, China; ranqq@petrochina.com.cn (Q.R.); rendianx@petrochina.com.cn (D.R.); djx1021@petrochina.com.cn (J.D.); xumengya@petrochina.com.cn (M.X.); liruibo01@petrochina.com.cn (R.L.)
* Correspondence: zhouxin510@petrochina.com.cn

**Abstract:** A shale reservoir is a complex system with lots of nanoscale pore throat structures and variable permeability. Even though shale reservoirs contain both organic and inorganic matter, the slip effect and phase behavior complicate the two-phase flow mechanism. As a result, understanding how microscale effects occur is critical to effectively developing shale reservoirs. In order to explain the experimental phenomena that are difficult to describe using classical two-phase flow theory, this paper proposes a new simulation method for two-phase shale oil reservoirs that takes into account the microscale effects, including the phase change properties of oil and gas in shale micro- and nanopores, as well as the processes of dissolved gas escape, nucleation, growth and aggregation. The presented numerical simulation framework, aimed at comprehending the dynamics of the two-phase flow within fractured horizontal wells situated in macroscale shale reservoirs, is subjected to validation against real-world field data. This endeavor serves the purpose of enhancing the theoretical foundation for predicting the production capacity of fractured horizontal wells within shale reservoirs. The impact of capillary forces on the fluid dynamics of shale oil within micro- and nanoscale pores is investigated in this study. The investigation reveals that capillary action within these micro- and nanoscale pores of shale formations results in a reduction in the actual bubble point pressure within the oil and gas system. Consequently, the reservoir fluid persists in a liquid monophasic state, implying a constrained mobility and diminished flow efficiency of shale oil within the reservoir. This constrained mobility is further characterized by a limited spatial extent of pressure perturbation and a decelerated pressure decline rate, which are concurrently associated with a relatively elevated oil saturation level.

**Keywords:** shale oil; two-phase flow simulation; microscale effect

**Citation:** Ran, Q.; Zhou, X.; Ren, D.; Dong, J.; Xu, M.; Li, R. Numerical Modeling of Shale Oil Considering the Influence of Micro- and Nanoscale Pore Structures. *Energies* **2023**, *16*, 6482. https://doi.org/10.3390/en16186482

Academic Editor: Marco Marengo

Received: 29 July 2023
Revised: 16 August 2023
Accepted: 26 August 2023
Published: 8 September 2023

## 1. Introduction

Shale reservoirs exhibit distinct features encompassing the proliferation of micro- and nanoscale pores along with elevated organic content, which collectively contribute to the manifestation of significant reservoir heterogeneity. The intricate interactions occurring at the liquid–solid interface under diverse occurrence states present substantial challenges when attempting to anticipate the flow dynamics of shale oil. Consequently, these challenges considerably hinder the effective advancement of unconventional shale oil development within our nation. The conventional permeability theory rooted in Darcy law falls short in precisely depicting the microscopic flow phenomena transpiring within shale formations under intricate conditions. As a result, the precise mechanism underlying the slippage of liquid hydrocarbons within nanoscale pores in shale remains elusive. The dominance of nanoscale pore velocity slip and other factors further complicates the understanding of the transport behavior of liquid hydrocarbons in different occurrence states within multiscale pore spaces, ultimately affecting the accuracy of predicting reservoir flow parameters [1–4].

Shale rocks exhibit an intricate and complex composition. Using X-ray diffraction mineral analysis, Ambrose and colleagues [5] found that shale reservoirs are made up of two separate components: organic and inorganic. The organic matter is classified as oil-wet kerogen, while the inorganic component includes hydrophilic minerals, such as quartz and clay minerals. These minerals exhibit different surface properties, leading to significant variations in their interaction mechanisms with fluids. Although many scholars have studied the flow mechanisms of gas in shale, the molecular diffusion of liquids is less pronounced compared to shale gas. Liquid–solid surface interactions are strong, and slip phenomena in microscale flows are more complex than in gas flows [6,7]. Song Fuquan [8] investigated the surface wetting and slip phenomena in liquid argon Poiseuille flow between parallel plates using molecular dynamics simulations as an example. Wang et al. [9–12] employed nonequilibrium molecular dynamics simulation methods within nanoscale pore throats to explore the microscopic flow behavior of methane, revealing significant differences in the flow characteristics of alkanes within pores with different mineral compositions. Attributable to disparities in frictional interactions between fluids and solids, the velocity of oil flow is most rapid within pores constituted by organic matter, followed by those composed of quartz, and notably slower within pores constituted by calcite. The slip length of saline water flowing over organic matter in shale was measured by Javadpour et al. [13] through atomic force microscopy. They developed a preliminary model for calculating the apparent permeability of liquids and found that the apparent permeability of liquids is significantly greater than the intrinsic matrix permeability. Most of the aforementioned studies are molecular simulation studies conducted in nanoscale channels. Arora et al. [14] constructed molecular dynamics structural models of curved single-layer carbon nanotubes using a splicing method and simulated diffusion mechanisms of pure $N_2$ and pure $O_2$ inside curved carbon nanotubes. The wettability and surface energy of reservoir rocks vary greatly with different mineral compositions. Zhang [15] conducted an extensive investigation on the wettability of clay minerals' surfaces, examining the impact of surface functional groups and mineralization on the wettability of reservoir surfaces. The study suggested that salt ions adsorbed on clay surfaces promote greater hydrophilicity. Wang et al. [9] used molecular simulations to show that the wetting angle of water in shale pores decreases with decreasing pore sizes, with a concomitant increase in the mercury contact angle. Chang et al. [16] systematically investigated the natural wettability of nanoscale rock surfaces, exploring the interactions between dodecane molecules and water molecules with rock surfaces. Graphene-dodecane bonding is more robust than water bonding, explaining why different rocks have different wettability properties at a microscopic scale. Research on fluid transport in nanopores forms the foundation of understanding microscale flow mechanisms. Many experts and scholars have investigated the flow behavior of different fluids in carbon nanotubes. Majumder et al. [13] investigated the augmented flow behavior of water, n-hexane, ethanol and alkanes within multilayered carbon nanotubes. The investigation revealed that the dimensions of the composite membrane were 4 to 5 orders of magnitude greater than values derived from calculations based on the slip-free Hagen–Poiseuille (HP). The determined slip length spanned from 3–70 μm, significantly surpassing the radius of the nanotubes (7 nm). Furthermore, owing to the robust interaction between liquid hydrocarbon molecules and the surface of carbon nanotubes, the slip length diminishes as the fluid exhibits increased hydrophobic characteristics. Whitby and Quirke [17,18] experimentally studied the water flow inside carbon nanotubes and discovered that the flow velocity of water in carbon nanotubes was 560 to 8400 times higher than the results calculated using the slip-free HP equation. Molecular dynamics simulations have been used by Falk et al. [19] to compute the coefficients of friction that govern the flow of fluids in carbon nanotubes with different radii and geometrical configurations. Their research also evaluated the augmentation factor of the Hagen–Poiseuille (HP) flow when slip was absent. In their work, Wu et al. [20] succinctly outlined the mechanisms governing water transport within nanopores, identifying noteworthy fluctuations in water flow dynamics contingent upon the relative intensity of interactions between liquid molecules and the boundaries of the nanopores. The aforementioned foundational theoretical studies mainly focused on the flow analysis and characterization of water or other fluids in carbon nanotubes. Wang et al. [9] introduced the concepts of "slip length" and "apparent viscosity" and developed mathematical models for alkane flow in nanoscale pore throats composed of different minerals. They established relationships between slip length and pore size, among other factors. Cui et al. [21] built an apparent permeability calculation model for liquids in organic nanopores based on the Mattia and Calabro [22] model. The study indicated the absence of initiation pressure gradients and nonlinear flow characteristics in oil within kerogen and that the physical adsorption of liquid hydrocarbons has a minimal impact on the permeability of organic matter nanopores. Zhang et al. [23] expanded the aforementioned model and mathematically represented a flow model considering the oil flow in both organic matter and nonorganic matter. In this manuscript, a numerical simulation framework is developed to address the intricate fluid flow dynamics within micro- and nanoscale pores of shale reservoirs. Specifically, the study focuses on mathematically characterizing the slip phenomenon prevalent in these pores. Leveraging the concept of apparent permeability, the slip effect is integrated into the foundational mathematical formulation of flow within porous media. Consequently, an advanced numerical simulation model is established, encapsulating the influence of micro- and nanoscale pores in shale oil dynamics. The construction of this model presents a novel avenue for the evaluation of shale oil reservoirs, thereby contributing a pioneering technological approach to reservoir assessment.

## 2. Microscale Effects in Shale Oil Reservoir

### 2.1. Stochastic Apparent Permeability Model

Fluids in porous media are usually transported in tortuous capillaries with a constant radius, according to researchers (Figure 1A). Natural porous media (like shale) have pores and throats, so fluid transport paths have variable radiuses. To clarify the oil transport process within shale formations, a modification has been made to the capillary model that previously relied on a constant radius assumption. This improved version of the model involves incorporating considerations for variable radii, as shown in Figure 1B. Capillaries with a constant radius are used to represent pores in this model (I, III and V in Figure 1B), and the throats are represented by the variable radius part between two pores (II and IV in Figure 1B).

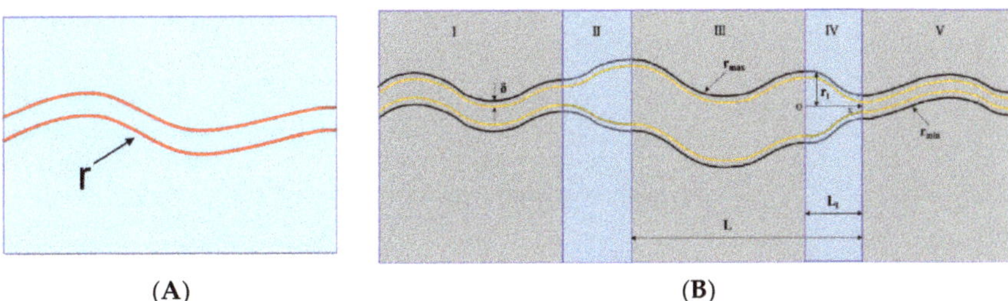
(A)          (B)

**Figure 1.** Schematic representation of the capillary model is depicted. (**A**) a tortuous capillary characterized by a constant radius is illustrated; (**B**) a tortuous capillary model is presented wherein the radius is variable. Parts I, III and V represent pores, while parts II and IV represent throats. Reproduced with permission from Jilong Xu (2020) [24].

Based on the no-slip H-P equation, the volumetric flow rate in a capillary with a constant radius is as follows:

$$Q_{no-slip} = \frac{\pi r^4}{8\mu_{bo}} \frac{\Delta p}{l} \tag{1}$$

where $Q_{no-slip}$ is the no-slip volumetric flow rate; $r$ is the capillary radius; $\mu_{bo}$ is the bulk oil viscosity; $\Delta p$ is the pressure drop across the capillary; and $l$ is the length of the tortuous capillary.

There are two different types of pores and throats in the shale: (1) pores and throats in organic matter and (2) pores and throats in the inorganic matrix. The sizes of the two types are mostly at a nanometer scale. Therefore, slip and viscosity correction of Equation (1) is needed, which can be expressed as

$$Q_{slip} = \left( \frac{\pi r^4}{8\mu_{eff}} + \frac{\pi l_s r^3}{2\mu_{io}} \right) \frac{\Delta p}{l} \tag{2}$$

where $Q_{slip}$ is the volumetric flow rate corrected by slip and viscosity; $\mu_{eff}$ is the adjusted viscosity that takes into account the influence of spatial variations in viscosity; $\mu_{io}$ is the oil viscosity near the wall; and $l_s$ is the slip length.

The slip length is a parameter characterized as the theoretical distance from the surface at which the extrapolation of the tangential velocity attains zero. For a smooth wall without dissolved gases, it is possible to calculate the slip length in terms of the contact angle.

$$l_s = \frac{C_s}{(1 + \cos \alpha)^2} \tag{3}$$

where $\alpha$ is the oil–wall contact angle; $C_s = 0.41$ is used in this work.

Based on the varying impact of wall molecular forces, the oil molecules within the nanotubes can be classified into two distinct regions: the interface region in proximity to the wall and the bulk region, as illustrated in Figure 1B. The viscosity of the oil within the interface region differs from that within the bulk region. Specifically, in nanopores embedded within organic matter, the substantial interactions between oil and the pore walls result in an elevated viscosity for the interface oil compared to the bulk oil viscosity [25–27]. In contrast, for nanopores in an inorganic matrix, the interface oil viscosity is smaller than the bulk oil viscosity. According to MDS results, the thickness of the interface region for oil in nanotubes is about 0.98 nm [28–30]. By employing experimental data and MDS data, it is possible to establish an expression governing the relationship of oil viscosity between these two distinct regions:

$$\mu_{io} = (-0.018\alpha + 3.25)\mu_{bo} \tag{4}$$

To account for the effect of the viscosity difference between two regions, the effective viscosity is introduced.

$$\mu_{eff} = \mu_{io} \frac{A_{io}}{A_t} + \mu_{bo} \frac{A_{bo}}{A_t} \tag{5}$$

where $A_{io}$, $A_{bo}$ and $A_t$ represent the area of the interface, the bulk and the entire tube. The relationship between these parameters can be expressed as follows: $A_t = A_{io} + A_{bo}$.

By analogy with the definition of resistance in electrical theory, the hydrodynamic resistance of oil transport in a nanotube can be expressed as follows:

$$R_{slip} = \frac{\Delta p}{Q_{slip}} = l \left( \frac{\pi r^4}{8\mu_{eff}} + \frac{\pi l_s r^3}{2\mu_{io}} \right)^{-1} \tag{6}$$

The mass flow rate through organic micro- and nanoscale pores can be expressed as follows:

$$J_{org} = Q_{slip}\rho_o = \left( \frac{\pi r^4}{8\mu_{eff}} + \frac{\pi l_s r^3}{2\mu_{io}} \right) \frac{\Delta p}{l} \rho_o \tag{7}$$

where $J_{org}$ is the organic matter mass flow rate and $\rho_o$ is the density of shale oil.

The mass flow rate of inorganic micro- and nanoscale pores is as follows:

$$J_{iorg} = Q_{slip}\rho_o = \left( \frac{\pi r^4}{8\mu_{eff}} \right) \frac{\Delta p}{l} \rho_o \tag{8}$$

where $J_{iorg}$ is the inorganic matter mass flow rate.

The main types of porous media in shale include organic porous media within the kerogen and inorganic porous media within the matrix. Therefore, considering the absence of mass exchange between the inorganic and organic porous media, the introduction of the organic porosity content $\beta$ allows us to obtain the coupled mass flow rate within the shale porous media. Among them, the porosity content of organic matter $\beta$ is approximately replaced by the TOC in shale reservoirs.

$$J_t = (1 - \beta)J_{iorg} + \beta J_{org} = \frac{\Delta p}{l}\rho_o\left[(1-\beta)\left(\frac{\pi r^4}{8\mu_{eff}}\right) + \beta\left(\frac{\pi r^4}{8\mu_{eff}} + \frac{\pi l_s r^3}{2\mu_{io}}\right)\right] \tag{9}$$

where $J_t$ is the coupled mass flow rate within the porous medium and $\beta$ is the dimensionless organic porosity content of shale.

The expression for mass flow rate in Darcy's law is given as follows:

$$J_v = -\frac{\rho_o K_o}{\mu_{eff}}\Delta p \tag{10}$$

Equations (9) and (10) are used to express the apparent permeability of porous media in shale:

$$k_\alpha = -\left[(1-\beta)\left(\frac{\pi r^4}{8\mu_{eff}}\right) + \beta\left(\frac{\pi r^4}{8\mu_{eff}} + \frac{\pi l_s r^3}{2\mu_{io}}\right)\right]\frac{\mu_{eff}}{l} \tag{11}$$

Building upon the aforementioned equation, it becomes feasible to construct a mathematical model that elucidates the dynamics of two-phase flow involving oil and water within the porous medium of a shale reservoir.

$$\frac{\partial_t}{\partial}\left[\phi^m\frac{\rho_o}{\rho_{osc}}S_o\right] + \nabla\cdot\left(-\frac{\rho_o}{\rho_{osc}}\frac{K_\alpha k_{ro}}{\mu_o}(\nabla p_o - \rho_o g\nabla z)\right) = \nabla\cdot\left(q_o^m - \sum_{i=1}^{N_f}q_o^{m,if}\right) \tag{12}$$

$$\frac{\partial_t}{\partial}\left[\phi^m\frac{\rho_w}{\rho_{wsc}}S_w\right] + \nabla\cdot\left(-\frac{\rho_w}{\rho_{wsc}}\frac{K_\alpha k_{rw}}{\mu_w}(\nabla p_w - \rho_w g\nabla z)\right) = \nabla\cdot\left(q_w^m - \sum_{i=1}^{N_f}q_w^{im,if}\right) \tag{13}$$

where $\phi^m$ is the pore volume, $\rho_{osc}, \rho_{wsc}$ are the densities of the oil–water two-phase system under standard surface conditions, $\nabla z$ is the altitude variation, $S_o, S_w$ are the saturation of oil and saturation of water, $k_{ro}, k_{rw}$ are the relative permeability values of oil and water, $N_f$ is the number of fractures, $q_o^{m,if}, q_w^{im,if}$ are the fluid flux between the $i$-th fracture and the matrix for oil and water, and $q_o^m, q_w^m$ are the source and sink terms of oil and water in the matrix.

For the problem of the channeling flow between fracture and matrix. The model developed in this paper uses the embedded discrete fracture model (EDFM), where the expression of the channeling flow between the matrix and the fracture is shown in Formula (14).

$$T_{ik}^{NNC} = \frac{\frac{K}{\mu}A_{ik}}{d_{ik}} \tag{14}$$

where $K$ is the fracture permeability, $A_{ik}$ is the area of intersection between the fracture grid and matrix grid, and $d_{ik}$ is the average normal distance from the matrix mesh to the fracture mesh.

### 2.2. Phase Change Characteristics in Micro- and Nanopores

Shale reservoirs have developed nanoscale pores, and the influence of capillary forces on the flow of oil and gas in both phases is not negligible. Combining the capillary force model with the thermodynamic equation of state can effectively describe the characteristics of oil and gas phase change in shale micro- and nanopores. The Peng–Robinson equation demonstrates a remarkable level of accuracy when applied to computations involving saturated vapor pressure, the molar volume of the liquid phase and similar properties. When the oil and gas system contains n components, taking the oil phase as an example, the fugacity coefficient of each component can be expressed as follows:

$$ln\phi_i = \frac{b_i}{b}(Z-1) - \ln\frac{P(V-b)}{RT} - \frac{a}{2\sqrt{2}bRT}\left(\frac{2\sum x_j a_{ij}}{a} - \frac{b_i}{b}\right)\ln\left(\frac{V+2.414b}{V-0.414b}\right) \tag{15}$$

According to the mixing rule, the expressions $a$ and $b$ can be obtained as follows:

$$a = \sum_{i=1}^{n}\sum_{j=1}^{n} x_i x_j \sqrt{a_i a_j}\left(1-k_{ij}\right) \tag{16}$$

$$b = \sum_{i=1}^{n} x_i b_i \tag{17}$$

where $x_i$ is molar proportion of component $i$ in the oil phase, with no factorization; $x_j$ is themolar proportion of component $j$ in the oil phase, with no factorization; and $k_{ij}$ signifies the binary interaction coefficient between components $i$ and $j$, with no factorization.

The expression of the capillary force in shale nanopores is given by the following:

$$f_{il} = x_i\phi_{il}p_l \tag{18}$$

where $p_g$ is the gas-phase pressure, $10^{-6}$ MPa; $p_l$ is the oil-phase pressure, $10^{-6}$ MPa; $\sigma$ is the oil–gas interfacial tension, N·m$^{-1}$; $\theta$ is the wetting angle, °; and $r$ is the pore radius, m.

The oil–gas interfacial tension can be expressed as follows:

$$\sigma = \left[\sum_i^n K_i\left(x_i\overline{\rho_l} - y_i\overline{\rho_g}\right)\right]^4 \tag{19}$$

where $K_i$ is the isotropic specific volume of component $i$, cm$^3$·mol$^{-1}$·mJ$^{-1/4}$·m$^{-1/2}$; $y_i$ is the molar fraction of component $i$ in the gas phase, which is factorless; $\overline{\rho_l}$ is the molar density of the oil phase, mol·cm$^{-3}$; and $\overline{\rho_g}$ is the molar density of the gas phase, mol·cm$^{-3}$.

The pressures and chemical potentials of the oil and gas phases become equal as soon as the fluid reaches the two-phase equilibrium between the gas and liquid inside the nanopore:

$$f_{ig} = f_{il} \tag{20}$$

$$\sigma = \left[\sum_i^n K_i\left(x_i\overline{\rho_l} - y_i\overline{\rho_g}\right)\right]^4 \tag{21}$$

Then,

$$f_{ig} = y_i\phi_{ig}p_g \tag{22}$$

$$f_{il} = x_i\phi_{il}p_l \tag{23}$$

The following relationship exists at the bubble point (saturation pressure):

$$\sum_{i=1}^{n} y_i = 1 \tag{24}$$

The following relationship exists at the dew point:

$$\sum_{i=1}^{n} x_i = 1 \tag{25}$$

In the scenario of diminutive pore radii, the capillary force exerts a heightened influence, leading to a discernible deviation between the actual bubble pressure and the bulk phase's bubble pressure. The impact of the pore radius on the bubble point pressure diminishes with an elevation in the methane molar fraction. Given that nanoscale pores dominate the spatial composition of shale reservoirs, it becomes imperative to incorporate the influence of capillary forces in simulating the phase change attributes of oil and gas during two-phase flow within shale oil systems.

### 2.3. Shale Oil Reservoir Two-Phase Flow Mechanism

After the pressure drops to the saturation pressure, the dissolved gas in the reservoir undergoes a process of gas bubble nucleation, growth and coalescence, ultimately forming a continuous gas phase as the reservoir pressure decreases. The pore throat size of shale reservoirs is much smaller than that of conventional reservoirs. The continuous gas phase formed after gas bubble coalescence is subjected to the shear effect of the shale nanopore structure during flow and is then divided into dispersed gas bubbles.

When the reservoir pressure is high, the interfacial tension between oil and gas is small, and gas bubbles are not prone to coalesce. The coalescence rate of gas bubbles is smaller than the breakup rate, so gas bubbles mainly exist in a dispersed state in the pore space and cannot form a two-phase fluid flow of oil and gas. As the reservoir pressure further decreases, the interfacial tension between oil and gas gradually increases, and the coalescence rate of gas bubbles increases. When the coalescence rate of gas bubbles surpasses the breakup rate, the dispersed gas bubbles undergo swift coalescence, culminating in the establishment of a two-phase fluid flow encompassing oil and gas. Considering the theory of heavy oil dissolved in the gas drive flow, the pressure at which gas bubbles begin to coalesce rapidly in the rock pores is defined as the pseudo-bubble point pressure. According to the comprehensive analysis presented above, the progression of the shale oil two-phase flow can be categorized into four distinct stages, contingent upon variations in the reservoir pressure. During the initial stage, characterized by a reservoir pressure surpassing the saturation pressure (determined within the PVT chamber), both oil and gas constituents within the reservoir persist in a liquid state. In the second stage, the reservoir pressure is lower than the saturation pressure but higher than the actual bubble point pressure. If the effect of capillary force is not considered, bubbles start to nucleate and grow at this time, but the effect of the capillary force of shale micro–nanopores causes the actual bubble point pressure of the oil and gas system to decrease, and the reservoir fluid still exists in a liquid form; in stage 3, the reservoir pressure is lower than the actual bubble point pressure but higher than proposed bubble point pressure, and, at this time, bubbles nucleate, grow and merge in the pore, and the continuous gas formed after merging is affected by the shearing effect of the shale micro–nanopore throat. At this time, bubbles nucleate, grow and merge in the pore space and are affected by the shearing effect of the shale micro–nanopore throat, and the continuous gas formed after merging is divided into small bubbles. The gas bubbles become a continuous gas phase, forming oil–gas two-phase flow, and the gas phase flow rate exceeds that of the oil phase, leading to a rapid increase in the production gas–oil ratio and a decrease in the oil drive efficiency.

## 3. Two-Phase Flow Shale Simulation and Model Validation

### 3.1. Model Validation

Historical fitting was executed employing empirical data extracted from a volumetric fractured horizontal well situated within the Changqing shale-oil-producing region in China (referred to as Changqing). The reservoir and fluid parameters pertinent to this analysis are outlined in Table 1. The adsorbed state oil per unit mass of shale to TOC ratio, adsorbed state oil recoverable per unit pressure difference to TOC ratio, and reaction frequency coefficients were determined by fitting production data as shown in Figure 2. Other model parameters were calculated following the numerical simulation characterization method of shale oil microscale flow in the reaction model. The overall permeability of the reservoir modification area is provided in the literature, so a single media model represents two-phase flow characteristics of this fractured horizontal well, assuming that the hydraulic fractures are equally spaced along the horizontal well, and a quarter of a single section of fractures is taken for simulation considering the symmetry of the model.

**Table 1.** Reservoir and fluid parameters for oil–water two-phase flow in shale formation.

| Parameter | Value | Parameter | Value |
|---|---|---|---|
| Reservoir thickness/m | 45.72 | TOC/% | 3.00 |
| Length of horizontal section/m | 1188.72 | Initial water saturation/% | 30 |
| Number of fracturing | 14 | Initial oil saturation/% | 70 |
| Hydraulic fracture length/m | 152.4 | Initial gas saturation/% | 0 |
| Matrix porosity | 0.1 | Temperature/°C | 154.00 |
| Permeability/$10^{-3}$ $\mu m^2$ | 0.13 | Initial pressures/MPa | 51.71 |
| Composite compressibility/MPa$^{-1}$ | 0.00018 | Theoretical bubble point pressure/MPa | 27.58 |
| Parameter | Value | Parameter | Value |
| Reservoir thickness/m | 45.72 | fracture widths | 0.004 |

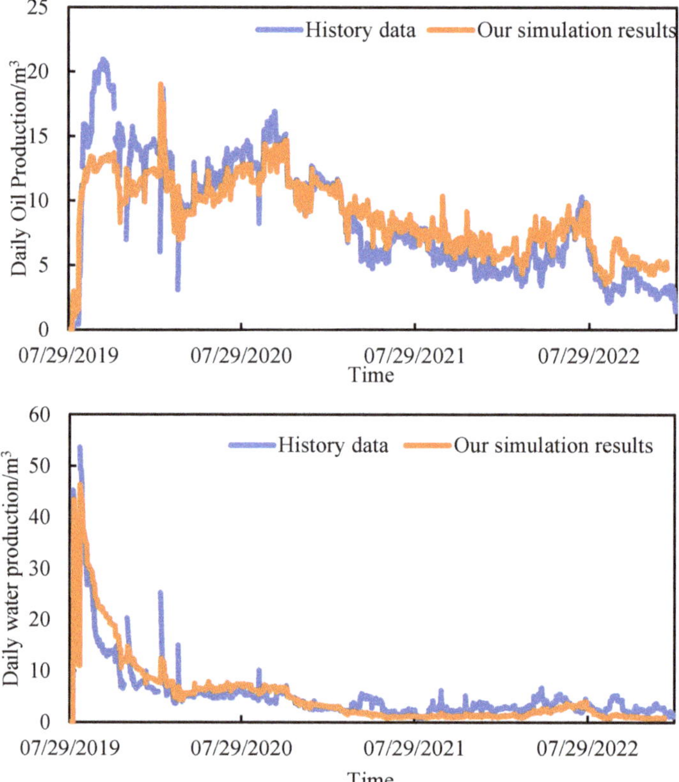


**Figure 2.** History match of oil and water production.

The changes in oil and gas production were simulated. They were derived from the real bottomhole pressure of the well, and the changes in the oil and gas production of the fractured horizontal well without considering the influence of shale nanopores were also simulated and compared with the production data. In scenarios where the influence of shale nanopores is disregarded, the theoretical calculation yields an approximate bubble point pressure of 28 MPa, extrapolated from the fluid component data available in the literature. Given that the bottomhole pressure registers below the bubble point pressure, the process of production leads to the liberation of dissolved gas, subsequently fostering the development of oil and gas two-phase flow. The simulated oil and gas production is therefore larger than the actual value. When factoring in the impact of shale nanopores, the bubble point pressure for both oil and gas phases decreases, leading to a higher bottomhole pressure. The fluid in the reservoir always remains in a single-phase liquid flow, resulting in a lower oil drive efficiency and reduced oil and gas production. Therefore, the simulation results are more in line with the real production.

*3.2. Influence of Capillary Forces on Reservoirs Considering the Effects of Micro- and Nanoscale Pores*

To conduct a more in-depth exploration of the impact exerted by micro- and nanoscale pores on the production and recovery scope of shale oil reservoirs, we extend the model by integrating capillary forces, as delineated in Equation (24). This augmentation enables us to capture the intricate influence stemming from the presence of micro- and nanoscale pores within shale oil reservoirs. Subsequently, we conduct a sensitivity analysis on the impact of introducing capillary forces on the mobilization of reservoir oil.

$$p_{cow}^{m} = p_{o}^{m} - p_{w}^{m} \tag{26}$$

where $p_{cow}^{m}$ is the capillary forces between oil and water phases; $p_{o}^{m}$ is the oil phase pressure; and $p_{w}^{m}$ is the water phase pressure.

Building upon the reservoir and fracture data elucidated in Table 1, we have developed a numerical conceptual model for shale oil reservoirs. We conducted a five-year production simulation considering the effects of capillary forces due to micro- and nanoscale pore considerations and compared it with a simulation without considering these effects. The cumulative oil production comparison curve is shown in Figure 3, which illustrates the impact of capillary forces on production over a five-year period when considering micro- and nanoscale pore effects. From Figure 3, it can be observed that incorporating capillary forces results in an approximately 6% lower oil production in shale oil reservoirs considering micro- and nanoscale pores over the five-year period.

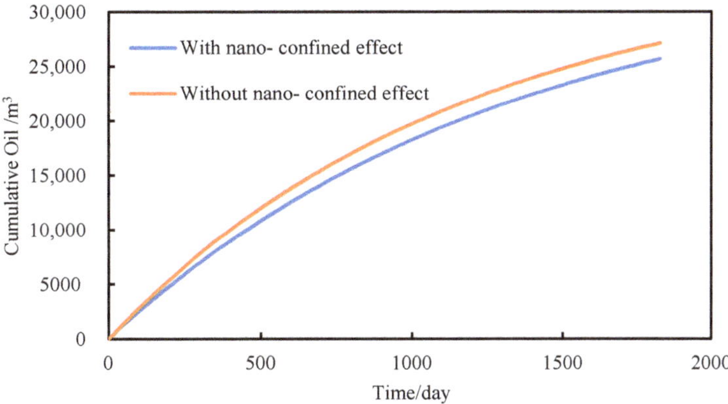


**Figure 3.** Cumulative oil production curves considering the effects of capillary forces with and without the consideration of micro- and nanoscale pores.

The underlying rationale for this phenomenon stems from the intricate interplay of capillary forces, which are intrinsically influenced by the pore radius within the shale oil reservoir. Notably, the shale reservoir exhibits an exceptional prevalence of micro- and nanoscale pores, thereby inducing an intensified manifestation of capillary forces within the system. The consequential impact of these capillary forces is disproportionately pronounced in shaping the fluid migration behavior within the reservoir. As a result of the dominance of capillary effects, the cumulative oil production from the shale oil reservoir, when factoring in the influence of micro- and nanoscale pores, experiences a reduction of approximately 6%.

Through the aforementioned simulation, we obtained the variations in the oil saturation field and pressure field considering the effects of capillary forces with and without the consideration of micro- and nanoscale pores, as shown in Figures 4 and 5. We analyzed the impact of including capillary forces and considering micro- and nanoscale pores on the pressure and flow field in shale oil reservoirs undergoing dynamic variations. From the figures, it can be observed that considering the effects of micro- and nanoscale pores leads to a smaller decline in the reservoir pressure and higher pressure values compared to the case where micro- and nanoscale pore effects are not considered. Moreover, analyzing the saturation field reveals that the oil saturation is generally higher when considering the effects of micro- and nanoscale pores compared to not considering them. This is attributed to the capillary forces acting on the micro- and nanoscale pores, which lower the actual bubble point pressure of the oil–gas system. As a result, the reservoir fluids exist predominantly in a liquid phase, limiting the mobility and flow efficiency of shale oil. Consequently, the pressure propagation range is smaller, and the pressure decline rate is reduced, resulting in a higher oil saturation.

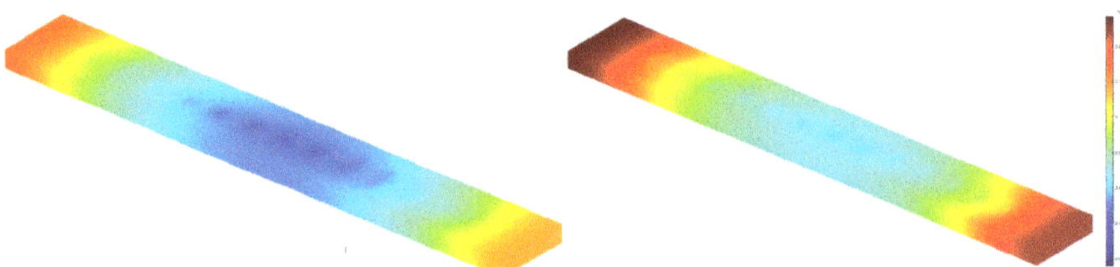

**Figure 4.** Pressure field variations when capillary forces are included, comparing the scenario without considering the effects of micro- and nanoscale pores (**left**) and the scenario considering these effects (**right**).

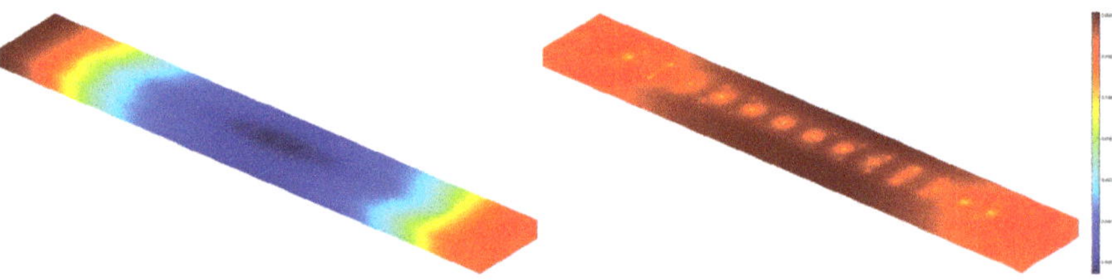

**Figure 5.** Variation in oil saturation field when capillary forces are included, comparing the scenario without considering the effects of micro- and nanoscale pores (**left**) and the scenario considering these effects (**right**).

### 3.3. Influence of Artificial Fractures on Productivity of Fractured Horizontal Wells in Shale Reservoirs

#### 3.3.1. Fracture Number

The productivity of staged fractured horizontal wells within shale reservoirs is evaluated across various scenarios involving 6, 8, 10, 12 and 14 artificial fractures, respectively. The impact of different fracture numbers on productivity is scrutinized, as depicted in Figure 6. From the figure, it is observable that the number of artificial fractures mainly affects the initial phase of production from the horizontal well. The higher the number of artificial fractures, the greater the daily oil production and cumulative oil production, but, with the increase in the number of fractures, the increase in horizontal well productivity is smaller and smaller. In addition, the production of fracturing and nonfracturing in horizontal wells is quite different, so the shale reservoir has a better production effect after hydraulic fracturing.

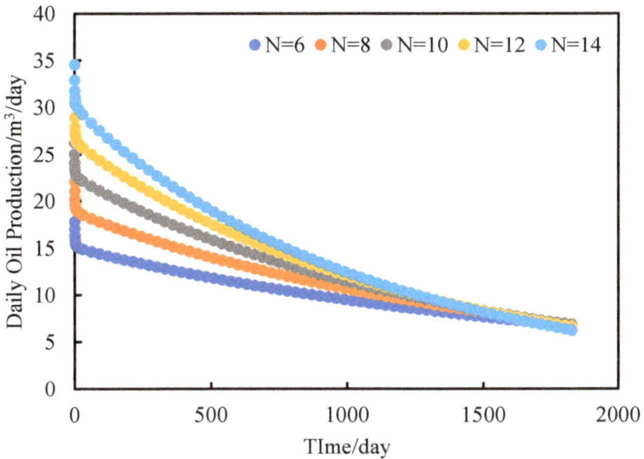

(**a**) Daily oil production

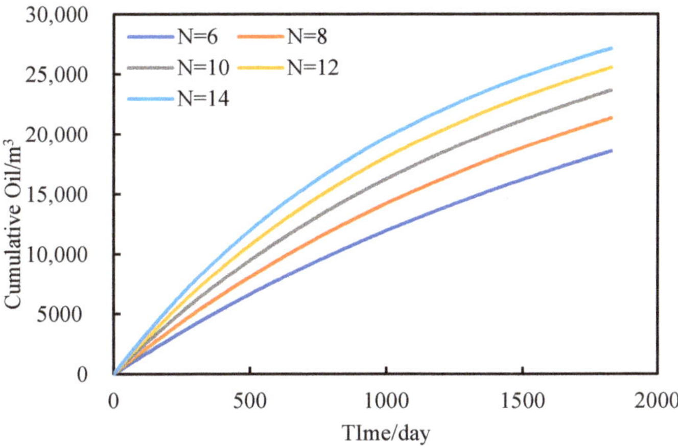

(**b**) Cumulative oil production

**Figure 6.** Figures of daily oil production (**a**) and cumulative production (**b**) of staged fracturing horizontal wells with different numbers of artificial fractures.

### 3.3.2. Fracture Spacing

Figure 7 illustrates the dynamic change in the productivity of fractured horizontal wells in shale reservoirs with micro–nano pores when the number of artificial fractures is fixed at 14 and the spacing of artificial fractures is 60 m, 80 m and 100 m. The graphical representation highlights that, under a constant number of fractures, an observable trend emerges: as the fracture spacing increases, the daily oil production of horizontal wells registers an elevation, accompanied by an augmented cumulative oil production. This is because, when the number of fractures is fixed, the larger the fracture spacing, the larger the area of fracturing transformation and the larger the range of artificial fractures, so the better the fracturing effect.

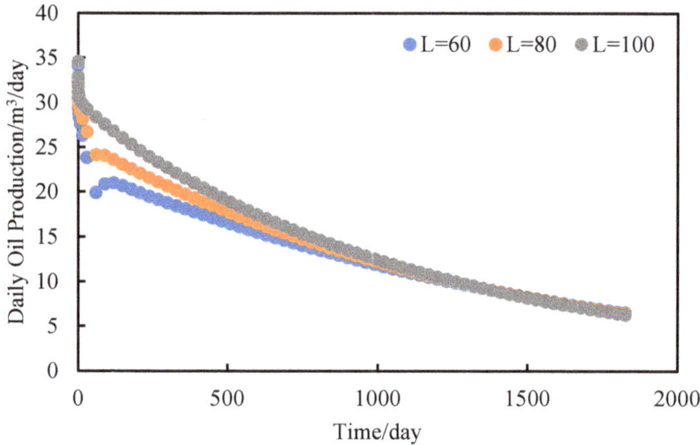

(**a**) Daily oil production

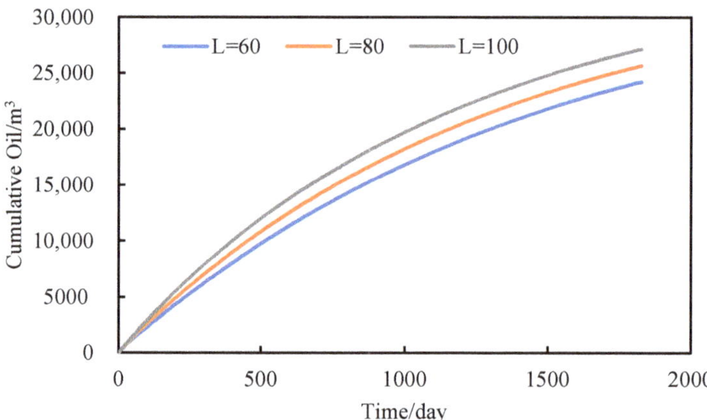

(**b**) Cumulative oil production

**Figure 7.** The daily oil production (**a**) and cumulative production (**b**) change diagram of staged fracturing horizontal wells with different artificial fracture spacing.

## 4. Conclusions

In this work, we derived the formula for the volumetric flow rate in a single nanotube with a variable radius and then formulated an apparent permeability model of a shale oil reservoir. Our study fully considers the impact of micro/nanopores on the flow of shale oil. The outcomes of our study elucidate the following:

(1) By combining the capillary force model with the thermodynamic state equation and modifying the critical properties of each component, a reaction model can describe the flow mechanism of shale oil in different storage states and flow of dissolved gas, effectively characterizing two-phase flow characteristics of shale oil.

(2) The oil and gas production rates of horizontal wells increase as a function of the enhancement of the vertical flow, and the impact of the vertical flow on shale oil production in a two-phase flow is more significant compared to a single-phase liquid flow. When the bottomhole pressure of the horizontal well is high, the oil and gas

production rates are low. However, as the bottomhole pressure decreases, the elastic energy used for oil displacement increases, and more dissolved gas escapes from the crude oil, resulting in an increase in oil and gas production rates, with a greater increase in gas production compared to oil production.

(3) Upon analyzing the dynamic evolution of reservoir pressure fields and permeability fields with and without accounting for capillary forces, it is observed that the cumulative oil yield from shale oil, when considering the influence of micro- and nanoscale pores, is diminished by approximately 6% in comparison to the scenario without such consideration. Under the influence of micro- and nanoscale pores, the reservoir experiences heightened pressure fields and an elevated overall oil saturation distribution. This outcome is attributed to the capillary forces exerted by micro- and nanoscale pores within the shale. These capillary forces contribute to the reduction in the actual bubble point pressure within the oil and gas system. Consequently, the fluid within the reservoir remains in a monophasic liquid state, thus constraining the mobility of oil within the shale and subsequently leading to a reduced flow efficiency.

(4) The number of artificial fractures mainly affects the early stage of horizontal well production. A discernible correlation emerges: an increase in the count of artificial fractures corresponds to a heightened daily oil production and an augmented cumulative oil production. However, with the increase in the number of fractures, the increase in the horizontal well productivity becomes smaller and smaller. Shale gas reservoirs have good production effects after hydraulic fracturing; when the number of fractures is fixed, the greater the fracture spacing, the higher the daily oil production of horizontal wells. This trend is further underscored by the observation that a higher cumulative oil production accompanies a more pronounced fracturing effect.

**Author Contributions:** Conceptualization, Q.R. and X.Z.; methodology, Q.R.; software, Q.R.; validation, X.Z., D.R. and Q.R.; formal analysis, D.R.; investigation, J.D.; resources, M.X.; data curation, R.L.; writing—original draft preparation, R.L.; writing—review and editing, M.X.; visualization, M.X.; supervision, J.D.; project administration, J.D.; funding acquisition, Q.R. All authors have read and agreed to the published version of the manuscript.

**Funding:** This research was funded by [Key Core Technology Research Projects of PetroChina Company Limited] grant number [2020B-4911]. And the APC was funded by [Research Institute of Petroleum Exploration and Development].

**Data Availability Statement:** Data available on request due to restrictions eg privacy or ethical. The data presented in this study are available on request from the corresponding author. The data are not publicly available due to the data in this paper is required to be confidential.

**Acknowledgments:** All individuals included in this section have consented to the acknowledgement.

**Conflicts of Interest:** The authors declare no conflict of interest.

## References

1. Hu, D.; Wei, Z.; Liu, R.; Wei, X.; Chen, F.; Liu, Z. Enrichment control factors and explo-ration potential of lacustrine shale oil and gas: A case study of Jurassic in the Fuling area of the Sichuan Basin. *Nat. Gas Ind. B* **2022**, *9*, 1–8. [CrossRef]
2. Ju, W.; Niu, X.B.; Feng, S.B.; You, Y.; Xu, K.; Wang, G.; Xu, H.R. Predicting the present-day in situ stress distribution within the Yanchang Formation Chang 7 shale oil reservoir of Ordos Basin, central China. *Pet. Sci.* **2020**, *17*, 912–924. [CrossRef]
3. Hu, T.; Pang, X.Q.; Jiang, F.J.; Wang, Q.F.; Wu, G.Y.; Liu, X.H.; Jiang, S.; Li, C.R.; Xu, T.W.; Chen, Y.Y. Key factors controlling shale oil enrichment in saline lacustrine rift basin: Implications from two shale oil wells in Dongpu Depression. *Bohai Bay Basin. Pet. Sci.* **2021**, *18*, 687–711. [CrossRef]
4. Tang, Y.; Cao, J.; He, W.J.; Guo, X.G.; Zhao, K.B.; Li, W.W. Discovery of shale oil in alkaline lacustrine basins: The Late Paleozoic Fengcheng Formation, Mahu Sag, Junggar Basin, China. *Pet. Sci.* **2021**, *18*, 1281–1293. [CrossRef]
5. Ambrose, R.J.; Hartman, R.C.; Diaz Campos, M.; Akkutlu, I.Y.; Sondergeld, C. New pore-scale considerations for shale gas in place calculations. In Proceedings of the SPE Unconventional Gas Conference, Pittsburgh, PA, USA, 23–25 February 2010. SPE-131772-MS.
6. Wu, H.A.; Chen, J.; Liu, H. Molecular dynamics simulations about adsorption and displacement of methane in carbon nanochannels. *J. Phys. Chem. C* **2015**, *119*, 13652–13657. [CrossRef]

7.   Botan, A.; Vermorel, R.; Ulm, F.J.; Pellenq, R.J.M. Molecular simulations of supercritical fluid permeation through disordered microporous carbons. *Langmuir* **2013**, *29*, 9985–9990. [CrossRef]

8.   Song, F.; Chen, X. Molecular dynamics study of liquid wall slip. *Hydrodyn. Res. Prog.* **2012**, *27*, 80–86.

9.   Wang, S.; Javadpour, F.; Feng, Q. Molecular dynamics simulations of oil transport through inorganic nanopores in shale. *Fuel* **2016**, *171*, 74–86. [CrossRef]

10.   Wang, S.; Javadpour, F.; Feng, Q. Fast mass transport of oil and supercritical carbon dioxide through organic nanopores in shale. *Fuel* **2016**, *181*, 741–758. [CrossRef]

11.   Wang, S.; Feng, Q.; Javadpour, F.; Xia, T.; Li, Z. Oil adsorption in shale nanopores and its effect on recoverable oil-in-place. *Int. J. Coal Geol.* **2015**, *147–148*, 9–24. [CrossRef]

12.   Wang, S.; Feng, Q.; Cha, M.; Lu, S.; Qin, Y.; Xia, T.; Zhang, C. Molecular dynamics simulation of liquid alkane fugacity in shale organic matter pore joints. *Pet. Explor. Dev.* **2015**, *42*, 772–778. [CrossRef]

13.   Javadpour, F.; McClure, M.; Naraghi, M.E. Slip-corrected liquid permeability and its effect on hydraulic fracturing and fluid loss in shale. *Fuel* **2015**, *160*, 549–559. [CrossRef]

14.   Arora, G.; Sandler, S.I. Nanoporous carbon membranes for separation of nitrogen and oxygen: Insight from molecular simulations. *Fluid Phase Equilibria* **2007**, *259*, 3–8. [CrossRef]

15.   Zhang, L.; Lu, X.; Liu, X.; Yang, K.; Zhou, H. Surface Wettability of Basal Surfaces of Clay Minerals: Insights from Molecular Dynamics Simulation. *Energy Fuels* **2016**, *30*, 149–160. [CrossRef]

16.   Chang, X.; Xue, Q.; Li, X.; Zhang, J.; Zhu, L.; He, D.; Zheng, H.; Lu, S.; Liu, Z. Inherent wettability of different rock surfaces at nanoscale: A theoretical study. *Appl. Surf. Sci.* **2018**, *434*, 73–81. [CrossRef]

17.   Majumder, M.; Chopra, N.; Andrews, R.; Hinds, B. Nanoscale hydrodynamics: Enhanced flow in carbon nanotubes. *Nature* **2005**, *438*, 44. [CrossRef] [PubMed]

18.   Whitby, M.; Quirke, N. Fluid flow in carbon nanotubes and nanopipes. *Nat. Nanotechnol.* **2007**, *2*, 87–94. [CrossRef] [PubMed]

19.   Falk, K.; Sedlmeier, F.; Joly, L.; Netz, R.R.; Bocquet, L. Ultralow liquid/solid friction in carbon nanotubes: Comprehensive theory for alcohols, alkanes, OMCTS, and water. *Langmuir* **2012**, *8*, 14261–14272. [CrossRef]

20.   Wu, K.; Chen, Z.; Li, J.; Li, X.; Xu, J.; Dong, X. Wettability effect on nanoconfined water flow. *Proc. Natl. Acad. Sci. USA* **2017**, *114*, 3358–3363. [CrossRef]

21.   Cui, J.; Sang, Q.; Li, Y.; Yin, C.; Li, Y.; Dong, M. Liquid permeability of organic nanopores in shale: Calculation and analysis. *Fuel* **2017**, *15*, 426–434. [CrossRef]

22.   Mattia, D.; Calabro, F. Explaining high flow rate of water in carbon nanotubes via solid-liquid molecular intercations. *Microfluid Nanofluid* **2012**, *13*, 125–130. [CrossRef]

23.   Qi, Z.; Yuliang, S.; Wendong, W.; Mingjing, L.; Guanglong, S. Apparent permeability for liquid transport in nanopores of shale reservoirs: Coupling flow enhancement and near wall flow. *Int. J. Heat Mass Transf.* **2017**, *115*, 224–234.

24.   Xu, J.; Su, Y.; Wang, W.; Wang, H. Stochastic Apparent Permeability Model of Shale Oil Considering Geological Control. In Proceedings of the Unconventional Resources Technology Conference, Virtual, 20–22 July 2020; pp. 2443–2460.

25.   Wang, X.; Li, J.; Jiang, W.; Zhang, H.; Feng, Y.; Yang, Z. Characteristics, current exploration practices, and prospects of continental shale oil in China. *Adv. Geo-Energy Res.* **2022**, *6*, 454–459. [CrossRef]

26.   Sharifi, M.; Kelkar, M.; Karkevandi-Talkhooncheh, A. A workflow for flow simulation in shale oil reservoirs: A case study in woodford shale. *Adv. Geo-Energy Res.* **2021**, *5*, 365–375. [CrossRef]

27.   Feng, Q.; Xu, S.; Xing, X.; Zhang, W.; Wang, S. Advances and challenges in shale oil development: A critical review. *Adv. Geo-Energy Res.* **2020**, *4*, 406–418. [CrossRef]

28.   Su, Y.; Sun, Q.; Wang, W.; Guo, X.; Xu, J.; Li, G.; Pu, X.; Han, W.; Shi, Z. Spontaneous imbibition characteristics of shale oil reservoir under the influence of osmosis. *Int. J. Coal Sci. Technol.* **2022**, *9*, 69. [CrossRef]

29.   Bai, L.; Wang, Y.; Zhang, K.; Yang, Y.; Bao, K.; Zhao, J.; Li, X. Spatial variability of soil moisture in a mining subsidence area of northwest China. *Int. J. Coal Sci. Technol.* **2022**, *9*, 64. [CrossRef]

30.   Hamanaka, A.; Sasaoka, T.; Shimada, H.; Matsumoto, S. Amelioration of acidic soil using fly ash for mine revegetation in post-mining land. *Int. J. Coal Sci. Technol.* **2022**, *9*, 33. [CrossRef]

 *energies*

*Article*

# Estimation of Free and Adsorbed Gas Volumes in Shale Gas Reservoirs under a Poro-Elastic Environment

Reda Abdel Azim [1,*], Abdulrahman Aljehani [2] and Saad Alatefi [3]

1. Petroleum Engineering Department, American University of Kurdistan, Sumel 42003, Iraq
2. Faculty of Earth Sciences, King Abdulaziz University, Jeddah 21589, Saudi Arabia
3. Department of Petroleum Engineering Technology, College of Technological Studies, The Public Authority for Applied Education and Training (PAAET), P.O. Box 42325, Kuwait City 70654, Kuwait; se.alatefi@paaet.edu.kw
* Correspondence: reda.abdulrasoul@auk.edu.krd

**Abstract:** Unlike conventional gas reservoirs, fluid flow in shale gas reservoirs is characterized by complex interactions between various factors, such as stress sensitivity, matrix shrinkage, and critical desorption pressure. These factors play a crucial role in determining the behavior and productivity of shale gas reservoirs. Stress sensitivity refers to the stress changes caused by formation pressure decline during production, where the shale gas formation becomes more compressed and its porosity decreases. Matrix shrinkage, on the other hand, refers to the deformation of the shale matrix due to the gas desorption process once the reservoir pressure reaches the critical desorption pressure where absorbed gas molecules start to leave the matrix surface, causing an increase in shale matrix porosity. Therefore, the accurate estimation of gas reserves requires careful consideration of such unique and complex interactions of shale gas flow behavior when using a material balance equation (MBE). However, the existing MBEs either neglect some of these important parameters in shale gas reserve analysis or employ an iterative approach to incorporate them. Accordingly, this study introduces a straightforward modification to the material balance equation. This modification will enable more accurate estimation of shale gas reserves by considering stress sensitivity and variations in porosity during shale gas production and will also account for the effect of critical desorption pressure, water production, and water influx. By establishing a linear relationship between reservoir expansion and production terms, we eliminate the need for complex and iterative calculations. As a result, this approach offers a simpler yet effective means of estimating shale gas reserves without compromising accuracy. The proposed MBE was validated using an in-house finite element poro-elastic model which accounts for stress re-distribution and deformation effects during shale gas production. Moreover, the proposed MBE was tested using real-field data of a shale gas reservoir obtained from the literature. The results of this study demonstrate the reliability and usefulness of the modified MBE as a tool for accurately assessing free and adsorbed shale gas volumes.

**Keywords:** shale gas reservoirs; poro-elasticity is shale gas; gas absorption–desorption process

**Citation:** Azim, R.A.; Aljehani, A.; Alatefi, S. Estimation of Free and Adsorbed Gas Volumes in Shale Gas Reservoirs under a Poro-Elastic Environment. *Energies* **2023**, *16*, 5798. https://doi.org/10.3390/en16155798

Academic Editors: Xingguang Xu, Kun Xie, Yang Yang and Shu Tao

Received: 15 May 2023
Revised: 18 July 2023
Accepted: 27 July 2023
Published: 4 August 2023

## 1. Introduction

The complexities involved in shale gas systems present challenges when it comes to estimating the volume and production of shale gas. This is primarily due to factors such as low permeability and highly heterogeneous porous media, which make conventional estimation methods ineffective. Therefore, a crucial step in determining shale gas production is selecting the appropriate material balance equation. Gas not only exists in small pores and open fractures but is also adsorbed into the grains of shale. Previous research and field data indicate that over 50% of the total gas is adsorbed within the matrix of the shale formation. The Langmuir pressure and volume parameters are used to estimate this volume of adsorbed gas [1–8].

Research by Darishchev et al. 2013 [9] demonstrates how the gas desorption volume is measured using shale samples at different temperatures. The study findings provide clear insights into the requirement of maintaining reservoir pressure below the Langmuir pressure to effectively release adsorbed gas. In rich shale formations, the overall recovery of gas heavily relies on the quantity of adsorbed gas that can be liberated. Despite hydraulic fracturing being employed to enhance permeability in shale matrices, there remains a portion of residual absorbed gas that cannot be extracted without utilizing additional high-value stimulation techniques like thermal stimulation [10–12].

As the main gas production contribution from the adsorbed gas, neglecting the desorption process may lead to unrealistic estimations of well production rates [13–18]. The desorption of gas is a vital factor in the deformation of the solid matrix of a reservoir. Therefore, this study takes this process into account using the poro-elasticity theory based on finite element modeling. Unlike conventional reservoirs, shale gas reservoirs require extraordinary technologies to extract the gas at an economical rate due to their extremely low matrix permeability and porosity. Accordingly, modeling shale gas well production and estimating accurate gas in place volumes, both free and adsorbed, are essential to predict the economic viability of the gas production process.

Shale gas reservoirs are different than the conventional gas reservoirs, in that they contain not only free gas, but also adsorbed gas and dissolved gas [19–21]. King (1990, 1993) [22,23] developed an MBE for a shale gas reservoir with the condition of limited water influx. The author considered the adsorption/desorption characteristics of the adsorbed gas and free gas. In addition, the author constructed a linear relationship through a pseudo-deviation factor to estimate the original gas in place (OGIP). However, the work of King required an intensive iteration process. Clarkson and Smith [24] (1997) considered gas in the adsorbed state only when deriving an MBE for a coalbed methane (CBM) reservoir. Liu et al. (2020) [25] considered free gas, formation compressibility, gas adsorption/desorption, and water expansion when proposing an MBE for a CBM reservoir. On the other hand, Moghadam et. al. (2011) [26] ignored water and formation compressibility when deriving the MBE for CBM and shale gas reservoirs. Meng et al. (2017) [27] used the same principle as King's to derive an MBE and presented a new definition of the Z factor. The original adsorbed gas in place and original free gas in place were estimated [28–37] using a developed MBE for a CBM reservoir. Moreover, Shi et al. (2018a) proposed a modified material balance equation that incorporates the effects of gas desorption and matrix shrinkage, using a pseudo-deviation factor. However, determining this pseudo-deviation factor is complex and requires an iterative process. More recently, Meng et al., Shi et al., and Yang et al. [38–40] also developed modified MBEs for shale gas reservoirs that account for critical desorption pressure, stress sensitivity, and matrix shrinkage effects. Nevertheless, these equations still rely on the use of a pseudo-deviation factor which requires an iteration process. In summary, it can be concluded that current MBE formulations either neglect important parameters in shale gas reservoir analysis (such as stress sensitivity, matrix shrinkage, and critical desorption pressure), or they include these factors but require iterative calculations. Therefore, the primary objective of this study was to introduce a straightforward adjustment to the material balance equation. This modification will enable more accurate estimation of free and adsorbed shale gas volumes by considering stress sensitivity and variations in porosity during shale production, along with accounting for the effect of critical desorption pressure. By establishing a linear relationship between reservoir expansion and production terms, we eliminate the need for complex and iterative calculations. As a result, this approach offers a simpler yet effective means of estimating shale gas reserves without compromising accuracy. The structure of this study is as follows: the derivation of the proposed modified shale gas material balance equation is presented in Section 2. Then, an introduction to the in-house finite element simulator which accounts for the coupled poro-elasticity (deformation-diffusion process) usually encountered in shale gas reservoirs flow behavior is presented in Section 3. Finally, verification of the modified MBE using the in-house poro-elastic model and real-field cases are presented in Section 4.

## 2. General Material Balance Equation of the Shale Gas Reservoir

The derivation of the modified shale gas material balance equation (MBE) is presented in this section. The proposed MBE takes into consideration three distinct forms in which gas is stored: free, absorbed, and dissolved gases. However, the amount of gas dissolved in formation water can be ignored as it is negligible when compared to the volumes of free and absorbed gases. When the pressure decreases below a certain point, known as the critical desorption pressure, the organic matter releases (desorbs) its stored gas into matrix pores causing deformation in the shale due to stress changes and matrix shrinkage. Additionally, its assumed that during production, there is an isothermal process occurring and natural fractures act primarily as pathways rather than storage spaces.

The material balance equation can be expressed as

$$\text{Production = Expansion (including expansion in both Organic and Inorganic matters)} \tag{1}$$

The production term in Equation (1) is stated as

$$G_p \beta_g + W_p \beta_w - W_i \tag{2}$$

where $G_p$ and $W_p$ are the cumulative gas and water production respectively, while $W_i$ is the injection volume. In Equation (1), all production and injection volumes are in cubic meters unit. $B_g$ and $B_w$ are the volume factors of gas and water respectively, in ($m^3/m^3$) unit.

Moreover, the expansion term can be stated as:

$$\text{Expansion = volume change of inorganic matrix + volume change of organic matrix + desorbed gas volume + pore volume occupied by free gas—produced water.} \tag{3}$$

All the volume changes included in the expansion term in Equation (3) are presented in the next sections (Sections 2.1–2.3).

### 2.1. The Inorganic Matrix Volume Changes

As the pressure in shale gas reservoirs decreases, there is a change in the volume of water and the pore spaces within the inorganic matrix. This change can be attributed to two factors: elastic variation in pore space within the reservoir, as well as expansion of formation water. The combined effect of these changes accounts for alterations observed in the inorganic matrix volume.

First, the variation in the inorganic matrix pore volume is equal to:

$$c_p = \frac{1}{v_p} \frac{\Delta v_p}{\Delta p} \tag{4}$$

where $c_p$ is the pore space compressibility in $MPa^{-1}$ and $v_p$ is the pore volume in $m^3$.

$$c_w = \frac{1}{v_w} \frac{\Delta v_w}{\Delta p} \tag{5}$$

Similar to Equation (4) $c_w$ and $v_w$ are the water compressibility volumes, respectively.

Re-arranging Equations (4) and (5) will give the change in reservoir pore and water volumes as follows:

$$\Delta v_p = c_p \, \Delta p \, v_p \tag{6}$$

$$\Delta v_w = c_w \Delta p \, v_w \tag{7}$$

Adding Equations (6) and (7), will present the total inorganic matrix volume change due to expansion as stated in Equation (8):

$$\Delta v_{exp} = c_w \Delta p \, v_w + c_p \, \Delta p \, v_p \tag{8}$$

Using the definition of water saturation ($s_w$), the water volume can be expressed in terms of pore volume, as follows:

$$v_w = s_w v_p \tag{9}$$

Thus, Equation (8) can be re-written as follows:

$$\Delta v_{exp} = c_w \Delta p\, s_w v_p + c_p\, \Delta p\, v_p = \Delta p\, v_p \left( c_w\, s_w + c_p \right) \tag{10}$$

Moreover, the pore volume ($v_p$) can be written in terms of the inorganic matrix free gas in place ($G_{inorg}$) as follows:

$$v_p = \frac{v_g}{s_{gi}} = \frac{G_{inorg}\, \beta_{gi}}{1 - s_{wi}} \tag{11}$$

where ($G_{inorg}$) is the inorganic matrix free gas in-place (m$^3$) and $B_{gi}$ is the initial gas volume factor, m$^3$/m$^3$. Therefore, the total expansion of pore space and formation is equal to:

$$\Delta v_{exp} = c_w \Delta p\, s_w v_p + c_p\, \Delta p\, v_p = \Delta p\, \frac{G_{inorg}\, \beta_{gi}}{1 - s_{wi}} \left( c_w\, s_w + c_p \right) \tag{12}$$

In addition, the free gas expansion in the inorganic matrix pores can be stated in terms of the inorganic matrix free gas and gas volume factor as follows:

$$G_{inorg} \left( \beta_g - \beta_{gi} \right) \tag{13}$$

where $B_{gi}$ is the initial gas volume factor and $B_g$ is the gas volume factor at any pressure P and both are measured in (m$^3$/m$^3$) units.

### 2.2. The Organic Matrix Volume Changes

Shale deformation can be characterized by considering the impact of compression and matrix shrinkage, as explained by Plamer et al. (1998) [41], as follows:

$$\frac{\phi_{sh}}{\phi_{ish}} = 1 + \frac{c_m}{\phi_{sh}}(P - P_i) + \frac{\xi_l}{\phi_{sh}} \left( \frac{K}{M} - 1 \right) \left( \frac{P}{P_L + P} - \frac{P_i}{P_L + P_i} \right) \tag{14}$$

where $\Phi_{sh}$ and $\Phi_{ish}$ are the initial shale porosity at the initial reservoir pressure and the shale porosity at any reservoir pressure, both dimensionless. $C_m$ is the expansion factor of organic matter, $K$ is the bulk modulus (MPa), $\xi_l$ is the volumetric strain change (dimensionless), $M$ is the axial strain modulus (MPa), $P_L$ is Langmuir's pressure (MPa), and $V_L$ is the maximum amount of adsorbed gas (Langmuir's volume, m$^3$).

$$c_m = \frac{1}{M} - \left( \frac{K}{M} + 0.5 \right) \gamma \tag{15}$$

Here, $\gamma$ is the solid compressibility of shale; $K/M$ in Equation (15) is a function of Poisson's ratio ($\nu$) as follows:

$$\frac{K}{M} = \frac{1}{3} \left( \frac{1 + \nu}{1 - \nu} \right) \tag{16}$$

The solid compressibility of shale ($\gamma$), can be neglected, so Equation (15) is re-written as:

$$C_m = \frac{1}{M} \tag{17}$$

Equation (14) could be further simplified as follows:

$$\varnothing_{sh} - \varnothing_{shi} = \frac{1}{M}(P - P_i) + \xi_l \left( \frac{1}{3} \left( \frac{1 + \nu}{1 - \nu} \right) - 1 \right) \left( \frac{P}{P_L + P} - \frac{P_i}{P_L + P_i} \right) \tag{18}$$

Therefore, the organic matrix volume change is written as:

$$\Delta V_{sh} = \frac{G_{forg}\beta_{gi}}{\varnothing_{sh}}\left\{\frac{1}{M}(P - P_i) + \xi_l\left(\frac{1}{3}\left(\frac{1+\upsilon}{1-\upsilon}\right) - 1\right)\left(\frac{P}{P_L + P} - \frac{P_i}{P_L + P_i}\right)\right\} \tag{19}$$

where $G_{forg}$ is the organic matrix free gas volume in m$^3$ and $B_{gi}$ is the initial gas volume factor.

### 2.3. Volume Change of the Adsorbed Shale Gas

The desorbed gas volume in the organic matrix can be calculated as follows:

$$V_{des} = V_L V_b \rho_{sh} \frac{P}{P + P_L} \tag{20}$$

where $V_L$ and $P_L$ are Langmuir's volume and pressure respectively; $P$ is the reservoir pressure $V_b$ is the reservoir bulk volume (m$^3$), $\rho_{sh}$ is the shale density.

Using Equation (20), the gas desorption rate into the matrix pore space can be written as follows:

$$-\frac{\partial V_{des}}{\partial t} = -V_L V_b \rho_{sh} \frac{1}{(P + P_L)^2}\frac{\partial P}{\partial t} \tag{21}$$

The volume of desorbed gas is written in term of the organic matrix free gas volume as follows:

$$V_{des} = \frac{G_{forg}\beta_{gi}}{\phi_{sh}}\rho_{sh}\beta_g\left(\frac{V_L P_i}{P_i + P_L} - \frac{V_L P}{P + P_L}\right) \tag{22}$$

### 2.4. The Shale Gas Material Balance Equation

The general form of the shale gas material balance equation can be written by summing all the derived volume changes in organic and non-organic matrix (i.e.; the sum of Equations (12), (13), (19) and (22)), as follows:

$$
\begin{aligned}
G_p\beta_g + W_p\beta_w - w_i \\
= G_{inorg}&\left[(\beta_g - \beta_{gi}) + \Delta p\frac{\beta_{gi}}{1-S_{wi}}(c_w s_w + c_p)\right] \\
+ G_{forg}&\left\{
\begin{aligned}
&(\beta_g - \beta_{gi}) - \frac{\beta_{gi}}{\phi_{sh}}\left\{
\begin{aligned}
&\frac{1}{M}(P - P_i)+ \\
&\xi_l\left(\frac{1}{3}\left(\frac{1+\upsilon}{1-\upsilon}\right) - 1\right)\left(\frac{P}{P_L+P} - \frac{P_i}{P_L+P_i}\right)
\end{aligned}\right\} \\
&+ \frac{\beta_{gi}}{\phi_{sh}}\rho_{sh}\beta_g\left(\frac{V_L P_i}{P_i+P_L} - \frac{V_L P}{P+P_L}\right)
\end{aligned}
\right\}
\end{aligned} \tag{23}
$$

To simplify the presentation of Equation (23), the new parameters ($F$, $E_g$, and $X$) are introduced. The parameter $F$ represents the production term in Equation (23). The inorganic matrix expansion term in Equation (23) is represented by parameter $E_g$, while the organic matrix expansion term is represented by the parameter $X$. The three parameters are expressed as follows:

$$F = G_p\beta_g + W_p\beta_w - W_i \tag{24}$$

$$E_g = \left[(\beta_g - \beta_{gi}) + \Delta p\frac{\beta_{gi}}{1 - S_{wi}}(c_w s_w + c_p)\right] \tag{25}$$

$$X = \left\{
\begin{aligned}
&(\beta_g - \beta_{gi}) - \frac{\beta_{gi}}{\varnothing_{sh}}\left\{\frac{1}{M}(P - P_i) + \xi_l\left(\frac{1}{3}\left(\frac{1+\upsilon}{1-\upsilon}\right) - 1\right)\left(\frac{P}{P_L+P} - \frac{P_i}{P_L+P_i}\right)\right\} \\
&+ \frac{\beta_{gi}}{\varnothing_{sh}}\rho_{sh}\beta_g\left(\frac{V_L P_i}{P_i+P_L} - \frac{V_L P}{P+P_L}\right)
\end{aligned}
\right\} \tag{26}$$

The parameter $X$ can be written in a shorter form as follows:

$$X = \{x_1 - x_2 + x_3\} \tag{27}$$

where $x_1$ is the organic matrix volume change due to free gas expansion (the first term in Equation (26)), $x_2$ is the organic matrix volume change due to deformation from matrix shrinkage and stress sensitivity (the second term in Equation (26)), and $x_3$ is the organic matrix volume change due to gas desorption process (the third term in Equation (26)).

In conclusion, Equation (23) can be re-arranged in terms of the new parameters ($F$, $E_g$, and $X$) as follow:

$$\frac{F}{E_g} = G_{inorg} + G_{forg}\frac{X}{E_g} \tag{28}$$

Equation (28) is a linear relationship between the production and expansion terms of the shale gas material balance equation; where the slope of the straight line in Equation (28) presents the free gas volume in the organic matrix $G_{forg}$, and the intercept is the free gas volume of the inorganic matrix $G_{inorg}$. Both volumes can be determined through linear fitting using the plot of $F/E_g$ vs. $X/E_g$.

Finally, the volume of adsorbed gas ($G_{Absd}$) as a function of the free gas volume in the organic matrix $G_{forg}$ and the Langmuir's parameters is as follow:

$$G_{Absd} = \frac{G_{forg}\beta_{gi}}{\phi_{sh}}\rho_{sh}\left(\frac{V_L P_i}{P_i + P_L}\right) \tag{29}$$

where, $V_L$ is the Langmuir's volume and $P_L$ is the Langmuir's pressure; $\rho_{sh}$ and $\varphi_{sh}$ are the shale density and porosity, respectively.

The calculation procedures of the proposed MBE are summarized as follows:

1. Obtain the rock/fluid properties needed for the calculation;
2. Obtain the dynamic production data including: the reservoir pressure profile, cumulative gas and water production, and water influx if available;
3. Calculate steps 4 to 6 for each cumulative production record;
4. Calculate the production term (i.e., parameter $F$ from Equation (24));
5. Calculate the inorganic matrix expansion term (i.e., parameter $E_g$ from Equation (25));
6. Calculate the organic matrix expansion term (i.e., parameter $X$ from Equation (26));
7. Plot ($F/E_g$) vs. ($X/E_g$). Based on Equation (28), the plot is a straight-line relationship with a slope of free gas volume of organic matrix ($G_{forg}$), and an intercept of the free gas volume of the inorganic matrix ($G_{inorg}$);
8. Calculate the adsorbed gas volume using Equation (29).

In the following section, a brief description of the in-house poro-elastic finite element simulator which will be used to validate the proposed material balance equation is given.

### 3. Poro-Elastic In-House Numerical Simulation Model

The flow of fluid in shale gas reservoirs was modeled in this study using an in-house poro-elastic simulator. The finite element method was utilized to discretize the equations that govern fluid flow in two-dimensional space.

The 2-D continuity equation of gas flow is written as follows:

$$\frac{\partial}{\partial x}(\rho_g u_g) - \alpha\frac{\partial}{\partial t}\left(\frac{\partial}{\partial x}\frac{\partial}{\partial y}\right)\vec{u} = -\frac{\partial}{\partial t}(\phi\rho_g) \tag{30}$$

where $\rho_g$ *and* $u_g$ are the gas density and velocity, respectively.

By incorporating the gas formation volume factor and the source/sink term ($q_g$) into Equation (30), we obtain the following equation:

$$\frac{\partial}{\partial x}(\beta_g u_g) - \alpha\frac{\partial}{\partial t}\left(\frac{\partial}{\partial x}\frac{\partial}{\partial y}\right) - q_g = -\frac{\partial}{\partial t}(\phi\beta_g) \tag{31}$$

Assuming off-diagonal components of the permeability tensor are zero $\overrightarrow{k} = \begin{pmatrix} k_x & 0 \\ 0 & k_y \end{pmatrix}$, Equation (31) can be reformulated as Darcy's velocity and utilized as follows:

$$\frac{\partial}{\partial x}\left(\frac{ck_x\beta_g}{\mu_g}\frac{\partial p_g}{\partial x}\right) + q_g = \phi\frac{\partial p_g}{\partial t}\frac{\partial \beta_g}{\partial p_g} \tag{32}$$

Writing Equation (32) in 2D form will give:

$$\frac{\partial}{\partial x}\left(\frac{ck_x\beta_g}{\mu_g}\frac{\partial p_g}{\partial x}\right) + \frac{\partial}{\partial z}\left(\frac{ck_y\beta_g}{\mu_g}\frac{\partial p_g}{\partial z}\right) + q_g = \phi\frac{\partial p_g}{\partial t}\frac{\partial \beta_g}{\partial p_g} \tag{33}$$

Modifying Equation (33) to include the amount of adsorbed gas as a source term (i.e., injection well) will give:

$$\frac{\partial}{\partial x}\left(\frac{ck_x\beta_g}{\mu_g}\frac{\partial p_g}{\partial x}\right) + \frac{\partial}{\partial y}\left(\frac{ck_y\beta_g}{\mu_g}\frac{\partial p_g}{\partial y}\right) + V_L\rho_R\frac{1}{(p+p_L)^2}\frac{\partial p}{\partial t} - \alpha\frac{\partial}{\partial t}\left(\frac{\partial}{\partial x}\frac{\partial}{\partial y}\right)\overrightarrow{u} = \phi\frac{\partial p_g}{\partial t}\frac{\partial \beta_g}{\partial p_g} \tag{34}$$

Multiplying both sides of Equation (34) by a trial function $w$ and integrating over the domain, $\Omega$ yields:

$$\int_\Omega\left(wc_t\phi\frac{\partial p_g}{\partial t}\frac{\partial \beta_g}{\partial p_g}\right)d\Omega - V_L\rho_R\frac{1}{(p+p_L)^2}\int_\Omega\left(w\frac{\partial p_g}{\partial t}\right)d\Omega - $$
$$\int_\Omega\left[w\alpha\frac{\partial}{\partial t}\left(\frac{\partial}{\partial x}\frac{\partial}{\partial y}\right)\overrightarrow{u}\right]d\Omega = \int_\Omega w\left(\begin{array}{c}\frac{\partial}{\partial x}\left(\frac{ck_x\beta_g}{\mu_g}\frac{\partial p_g}{\partial x}\right)+ \\ \frac{\partial}{\partial y}\left(\frac{ck_y\beta_g}{\mu_g}\frac{\partial p_g}{\partial y}\right)\end{array}\right)d\Omega \tag{35}$$

Applying Green Formula, Equation (35) is written as:

$$\int_\Omega\left(wc_t\phi\frac{\partial p_g}{\partial t}\frac{\partial \beta_g}{\partial p_g}\right)d\Omega - V_L\rho_R\frac{1}{(p+p_L)^2}\int_\Omega\left(w\frac{\partial p_g}{\partial t}\right)d\Omega = $$
$$\int_\Omega\frac{\partial w}{\partial x}\left(\frac{ck_x\beta_g}{\mu_g}\frac{\partial p_g}{\partial x}\right)d\Omega + \frac{\partial w}{\partial y}\left(\frac{ck_y\beta_g}{\mu_g}\frac{\partial p_g}{\partial y}\right)d\Omega$$
$$- \int_\Omega\oint_r w\left(\frac{ck_x\beta_g}{\mu_g}\frac{\partial p_g}{\partial x}n_x + \frac{ck_y\beta_g}{\mu_g}\frac{\partial p_g}{\partial y}n_y\right)d\Gamma + \int_\Omega\left[w\alpha\frac{\partial}{\partial t}\left(\frac{\partial}{\partial x}\frac{\partial}{\partial y}\right)\overrightarrow{u}\right]d\Omega \tag{36}$$

where $(\Gamma)$ is the boundary and $(n)$ is the outward normal to the boundary. The finite difference method is used to discretize the terms including differentiation with respect to time. Permeability and porosity remain constant with changes in time. Re-arranging Equation (36) gives:

$$\int_\Omega\left(wc_t{}^{i-1}\phi^{i-1}\frac{p^i-p^{i-1}}{\Delta t^i}\frac{p^i-p^{i-1}}{\beta_g{}^i-\beta_g{}^{i-1}}\right)d\Omega + V_L\rho_R\frac{1}{(p+p_L)^2}\int_\Omega\left(w\frac{p^i-p^{i-1}}{\Delta t^i}\right)d\Omega$$
$$+\int_\Omega\frac{\partial w}{\partial x}\left(\frac{ck_x{}^{i-1}\beta_g}{\mu_g}\frac{\partial p^i}{\partial x}\right)d\Omega+$$
$$\int_\Omega\frac{\partial w}{\partial y}\left(\frac{ck_y{}^{i-1}\beta_g}{\mu_g}\frac{\partial p^i}{\partial y}\right)d\Omega = \int_\Omega\left[w\alpha\left(\frac{\partial}{\partial x}\frac{\partial}{\partial y}\right)\frac{\overrightarrow{u}^i-\overrightarrow{u}^{i-1}}{\Delta t^i}\right]d\Omega \tag{37}$$

In which $(i)$ and $(i-1)$ are current and previous times respectively. Using Galerkin approach [42], Equation (37) is re-written as follow:

$$\left[\int_\Omega c_t\phi^i\overrightarrow{N_p}^T\overrightarrow{N_p}d\Omega\right]\frac{\overrightarrow{P}^i-\overrightarrow{P}^{i-1}}{\Delta t^i}\frac{p^i-p^{i-1}}{\beta_g{}^i-\beta_g{}^{i-1}} + V_L\rho_R\frac{1}{(p+p_L)^2}\left[\int_\Omega\overrightarrow{N_p}^T\overrightarrow{N_p}d\Omega\right]\frac{\overrightarrow{P}^i-\overrightarrow{P}^{i-1}}{\Delta t^i}$$
$$+\left[\int_\Omega\left(\frac{ck_x{}^{i-1}\beta_g}{\mu_g{}^{i-1}}\frac{\partial \overrightarrow{N_p}}{\partial x}^T\frac{\partial \overrightarrow{N_p}}{\partial x} + c\beta_g\frac{k_y{}^{i-1}}{\mu^{i-1}}\frac{\partial \overrightarrow{N_p}}{\partial y}^T\frac{\partial \overrightarrow{N_p}}{\partial y}\right)d\Omega\right]\overrightarrow{P}^i + \left(\int_\Omega\overrightarrow{N_p}^T\alpha\left(\frac{\partial}{\partial x}\frac{\partial}{\partial y}\right)d\Omega\right)\frac{\overrightarrow{u}^i-\overrightarrow{u}^{i-1}}{\Delta t^i} = 0 \tag{38}$$

where

$$
\begin{aligned}
\vec{P}^{T} &= (p_1 p_2 \cdots p_n) \\
\vec{N_p}^{T} &= (N_1 N_2 \cdots Nn) \\
\vec{N_u} &= \begin{bmatrix} N_1 0 N_2 \cdots 0 \\ 0 N_1 0 \cdots N_n \end{bmatrix} \\
\vec{U}^{T} &= (u_{x1} u_{y1} u_{x2} \cdots u_{yn})
\end{aligned}
\tag{39}
$$

$(n)$ is the total number of nodes.

Equation (37) can be further re-arranged as follow:

$$
\vec{\vec{L}}\left(\vec{p}^{i} - p^{i-1}\right) + \Delta t^{i} \vec{\vec{H}} \vec{p}^{i} - \vec{\vec{Q}}\left(\vec{U}^{i} - U^{i-1}\right) = \vec{0}
\tag{40}
$$

where $\vec{\vec{L}}$ is as follow:

$$
\vec{\vec{L}} = \sum_{e=1}^{ne} \vec{\vec{L^{e}}}
\tag{41}
$$

Here, $(e)$ denotes to element and $(ne)$ is the number of elements.

$$
\vec{\vec{L^{e}}} = \int_{\Omega}\left(\frac{P^{i} - P^{i-1}}{\beta_g^{i} - \beta_g^{i-1}} c_t^{i-1} \phi^{ei-1} \vec{N_p}^{T} \vec{N_p}\right) d\Omega + V_{L}\rho_{R}\frac{1}{(p+p_L)^2}\int_{\Omega}\left(\vec{N_p}^{T}\vec{N_p}\right)
\tag{42}
$$

In which

$$
\vec{\vec{H}} = \sum^{\substack{e=1 \\ ne}} \vec{\vec{H^{e}}}
\tag{43}
$$

$$
\vec{\vec{H^{e}}} = \int_{\Omega} e\left(c\beta_g \frac{k_x^{ei-1}}{\mu^{ei-1}} \frac{\partial \vec{N_p^{e}}^{T}}{\partial x} \frac{\partial \vec{N_p^{e}}}{\partial x} + c\beta_g \frac{k_y^{ei-1}}{\mu^{ei-1}} \frac{\partial \vec{N_p^{e}}^{T}}{\partial y} \frac{\partial \vec{N_p^{e}}}{\partial y}\right) d\Omega
$$

$$
\vec{\vec{Q}} = \sum_{e=1}^{ne} \vec{\vec{Q^{e}}}
\tag{44}
$$

where

$$
Qe = \alpha \int_{\Omega} e\vec{N_p}^{eT}\left(\frac{\partial N_u^{e}}{\partial x} + \frac{\partial N_u^{e}}{\partial x}\right) d\Omega
\tag{45}
$$

Equation (40) can be written in an incremental form as follows:

$$
\left(\vec{\vec{L}} + \Delta t^{i} \vec{\vec{H}}\right) \Delta P^{i} - \vec{\vec{Q}}\Delta \vec{U}^{i} = \vec{f_1}
$$

$$
\Delta \vec{P} = \vec{P}^{i} - \vec{P}^{i-1}, \Delta \vec{U} = \vec{U}^{i} - \vec{U}^{i-1}
\tag{46}
$$

$$
\vec{f}_1^{e} = -\Delta t^{i} \vec{\vec{H}} \vec{P}^{i-1}
$$

$$
\begin{aligned}
\vec{N_p}^{T} &= (N_1 N_2 \cdots Nn) \\
\vec{N_u} &= \begin{bmatrix} N_1 & 0 & N_1 & \cdots & 0 \\ 0 & N_1 & 0 & \cdots & N_1 \end{bmatrix}
\end{aligned}
\tag{47}
$$

$\vec{N_p}^{T}$ is the pressure shape function, $n$ is the number of nodes, $\vec{N_u}$ is the displacement function, and P represents the pressure nodal values.

Gaussian quadrature is applied for integration. Also, for the coupled equilibrium equation multiplying Equation (46) by trial function and integrating over the domain $\Omega$ leads to:

$$\int_\Omega \left( \vec{W}^T \vec{S}^T \Delta\vec{\sigma} \right) d\Omega = \vec{0}$$

$$\vec{W} = \begin{bmatrix} W_1(x,y) \\ W_2(x,y) \end{bmatrix}$$

$$(48)$$

and $W_1$ and $W_2$ are trial functions. Using Green's identity, one obtains:

$$\int_\Gamma \left( \vec{W}^T \vec{M}^T \Delta\vec{\sigma} \right) d\Gamma - \int_\Omega \Delta\vec{\sigma}^T \left( \vec{S}\vec{W} \right) d\Omega = \vec{0}$$

$$\vec{M}^T = \begin{bmatrix} n_x & 0 & n_y \\ 0 & n_y & n_x \end{bmatrix}$$

$$(49)$$

substituting Equation (49) into Equation (48) and rearranging, we will obtain:

$$\int_\Omega \left( \vec{S}\vec{W} \right)^T \vec{D}_e \vec{S} \Delta\vec{U}^i \, d\Omega + \alpha \int_\Omega \left( \vec{S}\vec{W} \right) \Delta\vec{P}^i \, d\Omega = \int_\Gamma \left( W^T \vec{M}^T \Delta\vec{\sigma}^i \right) d\Gamma$$

$$(50)$$

Using Galerkin method Equation (50) is re-written as follows:

$$\alpha \left( \int_\Omega \left( \left( \frac{\partial \vec{N}_u}{\partial X} \frac{\partial \vec{N}_u}{\partial X} \right) \vec{N}_P \right)^T d\Omega \right) \Delta\vec{P} + V_L \rho_R \frac{1}{(p+p_L)^2} \int_\Omega \left( \int_\Omega \left( \left( \frac{\partial \vec{N}_u}{\partial X} \frac{\partial \vec{N}_u}{\partial X} \right) \vec{N}_P \right)^T d\Omega \right)^i$$

$$= \left( \int_\Omega \left( \vec{S}\vec{N}_u \right)^T \vec{D}_e \vec{S}\vec{N}_u d\Omega \right) \Delta\vec{U}^i$$

$$(51)$$

Equation (51) in compact form is as follow:

$$\vec{K}\Delta\vec{U} + \vec{Q}^T \Delta\vec{P} = \vec{f}_2$$

$$(52)$$

Finally, Equations (53) to (57) are the final finite element equations simultaneously solved as a system of linear equations as follows:

$$\vec{K} = \sum_{e=1}^{ne} \vec{K}^e$$

$$(53)$$

$$\vec{K}^e = \int_{\Omega^e} \left( \vec{S}\vec{N}_u^e \right)^T \vec{D}_e \vec{S}\vec{N}_u^e \, d\Omega$$

$$(54)$$

$$\vec{f}_2 = \sum_{e=1}^{ne} \vec{f}_2^{\,e}$$

$$(55)$$

$$\vec{f}_2 = \int_{\Gamma^e} \left( \vec{N}_u \vec{M}^T \Delta\vec{\sigma}^i \right) d\Gamma$$

$$(56)$$

$$\begin{bmatrix} \vec{L} + \Delta t \vec{H} & -\vec{Q} \\ \vec{Q}^T & \vec{K} \end{bmatrix} \begin{bmatrix} \Delta\vec{P}^i \\ \Delta\vec{U}^i \end{bmatrix} = \begin{bmatrix} \vec{f}_1 \\ \vec{f}_2 \end{bmatrix}$$

$$(57)$$

A finite element poro-elastic model is presented in this section which accounts for the deformation effects in shale gas reservoirs resulting from the combined matrix shrinkage, stress sensitivity, and gas desorption processes. The model is then used to validate the proposed material balance equation presented earlier in Section 1 of this study. Moreover, two published real-field cases are also used to further validate the material balance equation using real production data, as will be shown in the next section. Finally, it should be noted that a brief demonstration of the poro-elastic model's ability to capture stress and deformation changes around production wells is presented in Appendix A.

## 4. Results and Discussion

### 4.1. Verification of the Modified MBE Using the In-House Numerical Poro-Elastic Model

The in-house simulator was used to verify the proposed modified material balance equation by conducting several numerical cases to simulate the production of a shale gas reservoir under a poro-elastic environment that accounts for the coupled effects of stress sensitivity, matrix shrinkage, and critical desorption pressure. Figure 1 presents a demonstration of one of the in-house simulator cases used during the verification process; in this figure the in-house simulator was run for a shale gas reservoir with a drainage radius of 100 m and a matrix permeability of 0.001 md. Other rock and fluid properties used in this run are shown in Table 1. Figure 1A presents the pressure decline rate for two elements (element 1 and element 2): element 1 is close to the wellbore while the other is far from the wellbore. It can be seen from this figure that the mesh element which is close to the well vicinity goes through a faster rate of pressure decline while the far field element shows a slower decline rate due to the low matrix permeability. Moreover, Figure 1B presents the amount of desorbed gas volumes in element 1 and element 2 discussed in Figure 1A; it can be seen from Figure 1B that the element which is close to the wellbore vicinity (element 1) starts producing a desorbed gas volume at a rate faster than that seen in the element far from the wellbore due to the fact that the pressure decline rate is faster near the wellbore where the pressure will quickly reach the critical desorption pressure and the absorbed gas layer in the matrix wall will start to desorb gas into the matrix pore spaces and then to the wellbore. Moreover, the well was run to an abonnement pressure of 5 MPa; then, the proposed modified material balance was used to estimate the expansion/production parameters (i.e., the F, X, and $E_g$ parameters from Equations 24,25, and 26) for the determination of total gas in place. Figure 2A presents the cumulative gas production versus the average reservoir pressure, while Figure 2B presents the change in the gas volume factor as a function of the average reservoir pressure; data in Figure 2A, B were used as an input to the MBE calculation. Moreover, Table 2 presents the calculation procedures of the proposed MBE, and Figure 3 presents the relationship between the production and reservoir expansion terms. Based on Figure 3, it is evident that a strong correlation exists between the production and reservoir expansion terms of the studied reservoir. The relationship between these two terms can be described as a linear relationship, with a coefficient of determination of 0.99 ($R^2 = 0.99$), a slope of ($1.22 \times 10^6$), and an intercept of ($0.16 \times 10^4$). Thus, according to Equation (28) the total free gas in place is equal to $1.2216 \times 10^6$ m$^3$, which includes a free gas volume in shale (organic matrix) of $1.22 \times 10^6$ m$^3$ and a volume of $0.16 \times 10^4$ m$^3$ of free gas in the inorganic matrix. Furthermore, the total free gas in place from the in-house simulator was equal to $1.16 \times 10^6$ m$^3$, hence the error between the total free gas volume from the modified MBE and that from the in-house poro-elastic numerical model is around 5%, which highlights the reliability of the presented MBE in calculating shale gas in place. In the following section, additional validation tests of the proposed MBE will be presented using real-field cases.

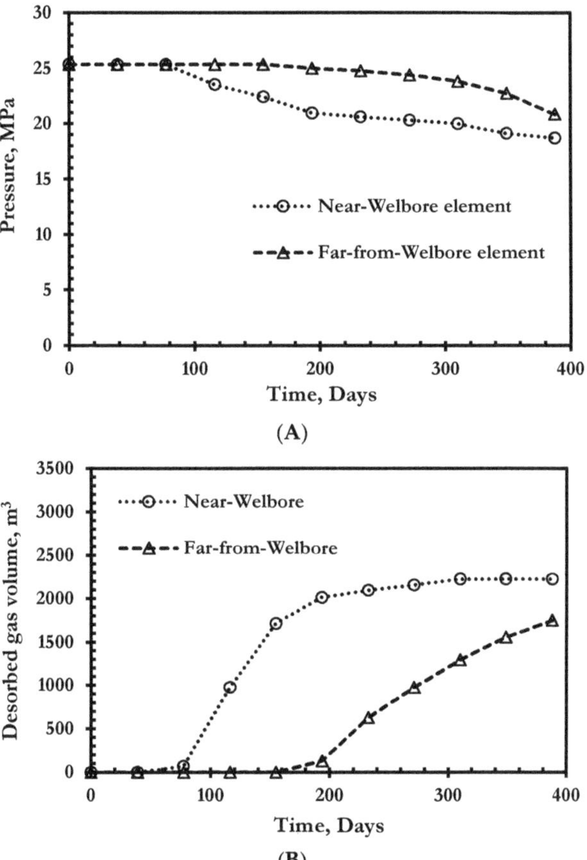

**Figure 1.** (**A**) Pressure decline rate in two elements (element-1 near-wellbore and element-2 far- from wellbore); and (**B**) desorbed gas volume in two elements (element-1 near-wellbore and element-2 far-from wellbore), after one year of production.

**Table 1.** Reservoir and rock properties for the in-house numerical model case.

| Property | Unit |
| --- | --- |
| Initial reservoir Pressure Pi | 25.5 MPa |
| Gas volume factor @ Pi | $0.00523 \text{ m}^3/\text{m}^3$ |
| Initial water saturation, $S_{wi}$ | 0.23 |
| Shale density, $\rho_{sh}$ | $1.43 \text{ t}/\text{m}^3$ |
| Langmuir volume $V_L$, | $23 \text{ m}^3/\text{t}$ |
| Shale porosity, $\varphi_{sh}$ | 0.14 |
| Langmuir pressure PL | 21 MPa |
| Modulus of elasticity M | 5800 |
| Poisson's ratio of shale $\nu$ | 0.31 |
| Langmuir curve parameter $\xi$ | 0.0158 |

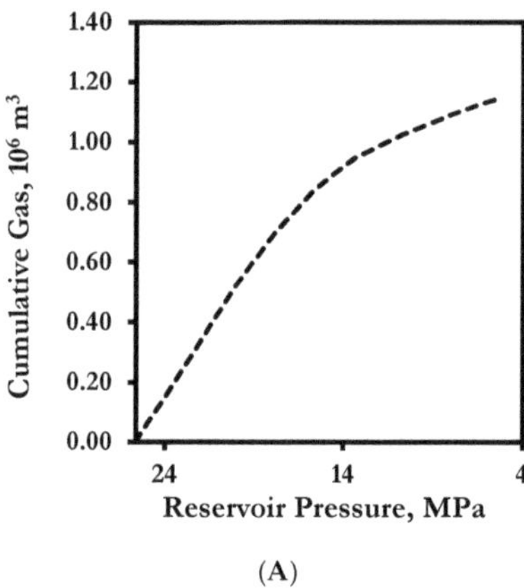

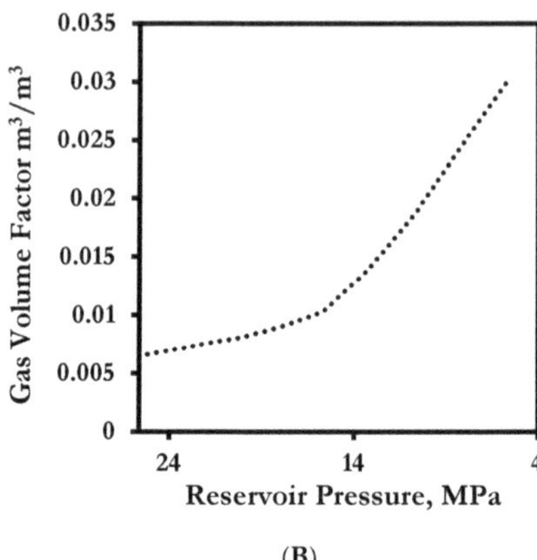

**Figure 2.** (**A**) Cumulative gas production versus average reservoir pressure; and (**B**) gas volume factor versus average reservoir pressure.

**Table 2.** MBE equation calculation procedures for the numerical in-house model run.

| Pressure (MPa) | $F$ | $x_1$ | $x_2$ | $x_3$ | $X$ | $E_g$ |
|---|---|---|---|---|---|---|
| 25.03 | 1721.17 | 0.00127 | $-4.7205 \times 10^{-5}$ | $1.78995 \times 10^{-7}$ | 0.001317384 | 0.06051544 |
| 19 | 4193.28 | 0.00287 | $-2.0503 \times 10^{-4}$ | $2.44282 \times 10^{-6}$ | 0.003077472 | 0.56670584 |
| 17.58 | 6335.44 | 0.00387 | $-2.6574 \times 10^{-4}$ | $3.44507 \times 10^{-6}$ | 0.004139181 | 0.68653144 |
| 15.48 | 8641.83 | 0.00507 | $-3.5674 \times 10^{-4}$ | $5.18537 \times 10^{-6}$ | 0.005431929 | 0.86345944 |
| 13.2 | 12961.85 | 0.00847 | $-5.4843 \times 10^{-4}$ | $8.99202 \times 10^{-6}$ | 0.009027424 | 1.05764984 |
| 9.87 | 19189.08 | 0.01347 | $-8.7398 \times 10^{-4}$ | $1.72092 \times 10^{-5}$ | 0.014361185 | 1.34130424 |
| 7.85 | 27349.73 | 0.01977 | $-1.2486 \times 10^{-3}$ | $2.77518 \times 10^{-5}$ | 0.021046344 | 1.51663784 |
| 5.45 | 34794.86 | 0.02517 | $-1.6009 \times 10^{-3}$ | $4.17467 \times 10^{-5}$ | 0.026812674 | 1.72286984 |

$x_1$ = volume changes due to free gas expansion in Organic matrix, $x_2$ = volume changes due to shale matrix shrinkage and stress sensitivity, $x_3$ = volume changes due to desorbed gas volume. Note: $X$ = is the summation of ($x_1$, $x_2$, and $x_3$).

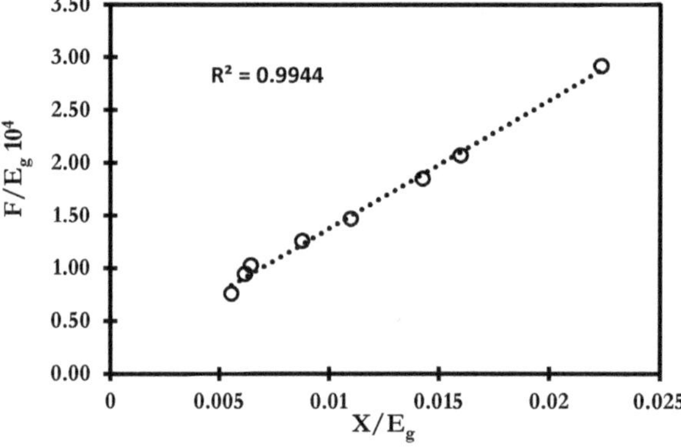


**Figure 3.** Linear relationship between the production and expansion parameters proposed in the modified material balance equation presented in this study (using the production data from the numerical case of the in-house model).

### 4.2. Verification of the Modified MBE Using Real-Field Case Study

To further validate the proposed modified material balance equation, two real-field cases of shale gas reservoirs located in China were used from the literature, Hu et al.

2019 [37]. Tables 3 and 4 presents the cumulative gas and water production versus the average reservoir pressure for two wells, Well-A and Well-B. Moreover, Table 5 presents the rock and fluid properties of the reservoir.

**Table 3.** Shale gas production history of Well-A. Reprinted with permission from reference [37] Copyright 2019 Elsevier.

| Well-A | | | |
|---|---|---|---|
| Pressure MPa | Cumulative Gas Production m$^3$ | Cumulative Water Production m$^3$ | Gas Volume Factor m$^3$/m$^3$ |
| 17.2 | $7.162 \times 10^7$ | $1.142 \times 10^4$ | 0.0065 |
| 13.8 | $9.133 \times 10^7$ | $1.361 \times 10^4$ | 0.0081 |
| 12.4 | $9.605 \times 10^7$ | $2.424 \times 10^4$ | 0.0091 |
| 11 | $1.046 \times 10^8$ | $2.765 \times 10^4$ | 0.0103 |
| 8.5 | $1.222 \times 10^8$ | $3.365 \times 10^4$ | 0.0137 |
| 6.3 | $1.326 \times 10^8$ | $4.631 \times 10^4$ | 0.0187 |
| 4.8 | $1.464 \times 10^8$ | $6.279 \times 10^4$ | 0.025 |
| 4 | $1.533 \times 10^8$ | $7.833 \times 10^4$ | 0.0304 |

**Table 4.** Shale gas production history of Well-B. Reprinted with permission from reference [37]. Copyright 2019 Elsevier.

| Well-B | | | |
|---|---|---|---|
| Pressure MPa | Cumulative Gas Production m$^3$ | Cumulative Water Production m$^3$ | Gas Volume Factor m$^3$/m$^3$ |
| 17.2 | $3.485 \times 10^7$ | $1.504 \times 10^4$ | 0.0065 |
| 13.8 | $5.286 \times 10^7$ | $1.614 \times 10^4$ | 0.0081 |
| 12.4 | $6.028 \times 10^7$ | $2.875 \times 10^4$ | 0.0091 |
| 11 | $6.783 \times 10^7$ | $3.164 \times 10^4$ | 0.0103 |
| 8.5 | $8.214 \times 10^7$ | $4.202 \times 10^4$ | 0.0137 |
| 6.3 | $9.573 \times 10^7$ | $5.437 \times 10^4$ | 0.0187 |
| 4.8 | $1.065 \times 10^8$ | $6.731 \times 10^4$ | 0.025 |
| 4 | $1.133 \times 10^8$ | $8.435 \times 10^4$ | 0.0304 |

**Table 5.** Reservoir and rock properties for the real-field case study. Reprinted with permission from reference [37]. Copyright 2019 Elsevier.

| Property | Unit |
|---|---|
| Initial reservoir Pressure Pi | 25.738 MPa |
| Gas volume factor @ Pi | 0.00516 m$^3$/m$^3$ |
| Initial water saturation, $S_{wi}$ | 0.23 |
| Shale density, $\rho_{sh}$ | 1.325 t/m$^3$ |
| Langmuir volume $V_L$, | 2.34 m$^3$/t |
| Shale porosity, $\varphi_{sh}$ | 0.05 |
| Langmuir pressure PL | 2.63 MPa |
| Modulus of elasticity M | 5800 |
| Poisson's ratio of shale $\nu$ | 0.31 |
| Langmuir curve parameter $\xi$ | 0.0158 |

Table 6 presents the proposed MBE calculation procedure for Well-A, and Figure 4 presents the linear relationship between the production and expansion terms of Well-A dynamic production data. In Figure 4, the slope of the straight line (i.e., the free gas volume in the organic matrix) is equal to $1.57 \times 10^8$ m$^3$ while the intercept (i.e., the free gas volume in the inorganic matter) is equal to $2.25 \times 10^5$ m$^3$; moreover, the absorbed gas volume is equal to $4.56 \times 10^7$ m$^3$ using equation 29. Accordingly, the sum of these three volumes will give a total gas in place volume of $2.03 \times 10^8$ m$^3$ which is in close agreement with the total gas in place of $2.10 \times 10^8$, as reported by Hu et al. 2018 [37], (i.e., the difference between the reported total gas in place and the one calculated using the modified MBE presented herein is around 3%). Furthermore, it should be noted that if the matrix shrinkage and stress sensitivity effects as well as the change in volume due to gas adsorption are ignored (i.e., ignoring the values of columns ($x_2$ and $x_3$) in Table 6) then the estimated total gas volume will be equal to $2.13 \times 10^8$ m$^3$, which gives an over-estimation of 5% compared to the case where all effects are included.

**Table 6.** MBE equation calculation procedures for Well-A production data.

| Pressure (MPa) | $F$ | $x_1$ | $x_2$ | $x_3$ | $X$ | $E_g$ |
|---|---|---|---|---|---|---|
| 17.2 | $4.770 \times 10^5$ | 0.00134 | $-1.9345 \times 10^{-4}$ | $8.30205 \times 10^{-8}$ | 0.001533529 | 0.711386692 |
| 13.8 | $7.534 \times 10^5$ | 0.00294 | $-3.1549 \times 10^{-4}$ | $1.74589 \times 10^{-7}$ | 0.003255668 | 0.995741289 |
| 12.4 | $8.984 \times 10^5$ | 0.00394 | $-3.8584 \times 10^{-4}$ | $2.39559 \times 10^{-7}$ | 0.00432608 | 1.113169653 |
| 11 | $1.105 \times 10^6$ | 0.00514 | $-4.6934 \times 10^{-4}$ | $3.30384 \times 10^{-7}$ | 0.005609674 | 1.230798017 |
| 8.5 | $1.708 \times 10^6$ | 0.00854 | $-6.8753 \times 10^{-4}$ | $6.29435 \times 10^{-7}$ | 0.009228155 | 1.442105809 |
| 6.3 | $2.526 \times 10^6$ | 0.01354 | $-9.7975 \times 10^{-4}$ | $1.20748 \times 10^{-6}$ | 0.014520956 | 1.630064666 |
| 4.8 | $3.722 \times 10^6$ | 0.01984 | $-1.3018 \times 10^{-3}$ | $2.08990 \times 10^{-6}$ | 0.021143915 | 1.761109341 |
| 4 | $4.738 \times 10^6$ | 0.02524 | $-1.5469 \times 10^{-3}$ | $2.95677 \times 10^{-6}$ | 0.026789896 | 1.833039835 |

$x_1$ = volume changes due to free gas expansion in Organic matrix, $x_2$ = volume changes due to shale matrix shrinkage and stress sensitivity, $x_3$ = volume changes due to desorbed gas volume. Note: $X$ = is the summation of ($x_1$, $x_2$, and $x_3$).

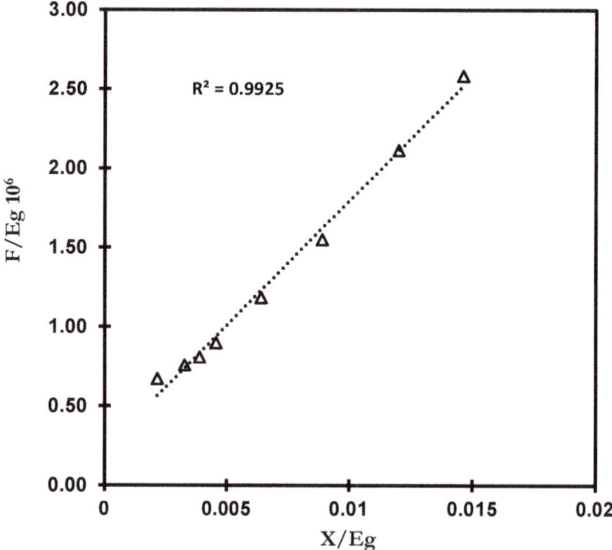


**Figure 4.** The linear relationship between the production and expansion parameters proposed in the modified material balance equation presented in this study (using the production data from the real-field production data of Well-A).

Table 7 presents the proposed MBE calculation procedure for Well-B. The total gas in place of Well-B is equal to $1.24 \times 10^8$ as reported by Hu et al. 2018 [37]. Figure 5 presents the linear relationship between the production and expansion terms of Well-B production data. In Figure 5, the slope of the straight line (i.e., the free gas volume in organic matrix) is equal to $1.28 \times 10^8$ m$^3$, while the intercept (i.e., the free gas volume in the inorganic matter) is equal to $2.38 \times 10^4$ m$^3$; moreover, the absorbed gas volume is equal to $3.72 \times 10^6$ m$^3$ using equation 29. Therefore, the MBE calculated total gas in place volume of Well-B is equal to $1.32 \times 10^8$ m$^3$ which is in close agreement with the total gas in place of $1.24 \times 10^8$ as reported by Hu et al. 2018 [37], (i.e., the difference between the reported total gas in place and the one calculated using the modified MBE presented herein is around 6%). Such encouraging results from the real-field data of Well-A and Well-B highlight the reliability of the presented modified material balance equation in estimating shale gas in place, accounting for the combined effects of stress change, matrix shrinkage, water production, and critical desorption pressure.

**Table 7.** MBE equation calculation procedures for Well-B production data.

| Pressure (MPa) | $F$ | $x_1$ | $x_2$ | $x_3$ | $X$ | $E_g$ |
|---|---|---|---|---|---|---|
| 17.2 | $2.416 \times 10^5$ | 0.00134 | $-1.9345 \times 10^{-4}$ | $8.30205 \times 10^{-8}$ | 0.001533529 | 0.710815 |
| 13.8 | $4.444 \times 10^5$ | 0.00294 | $-3.1549 \times 10^{-4}$ | $1.74589 \times 10^{-7}$ | 0.003255668 | 0.994941 |
| 12.4 | $5.774 \times 10^5$ | 0.00394 | $-3.8584 \times 10^{-4}$ | $2.39559 \times 10^{-7}$ | 0.00432608 | 1.112276 |
| 11 | $7.304 \times 10^5$ | 0.00514 | $-4.6934 \times 10^{-4}$ | $3.30384 \times 10^{-7}$ | 0.005609674 | 1.229810 |
| 8.5 | $1.168 \times 10^6$ | 0.00854 | $-6.8753 \times 10^{-4}$ | $6.29435 \times 10^{-7}$ | 0.009228155 | 1.440951 |
| 6.3 | $1.845 \times 10^6$ | 0.01354 | $-9.7975 \times 10^{-4}$ | $1.20748 \times 10^{-6}$ | 0.014520956 | 1.628762 |
| 4.8 | $2.730 \times 10^6$ | 0.01984 | $-1.3018 \times 10^{-3}$ | $2.08990 \times 10^{-6}$ | 0.021143915 | 1.759706 |
| 4 | $3.529 \times 10^6$ | 0.02524 | $-1.5469 \times 10^{-3}$ | $2.95677 \times 10^{-6}$ | 0.026789896 | 1.831583 |

$x_1$ = volume changes due to free gas expansion in Organic matrix, $x_2$ = volume changes due to shale matrix shrinkage and stress sensitivity, $x_3$ = volume changes due to desorbed gas volume. Note: $X$ = is the summation of ($x_1$, $x_2$, and $x_3$).

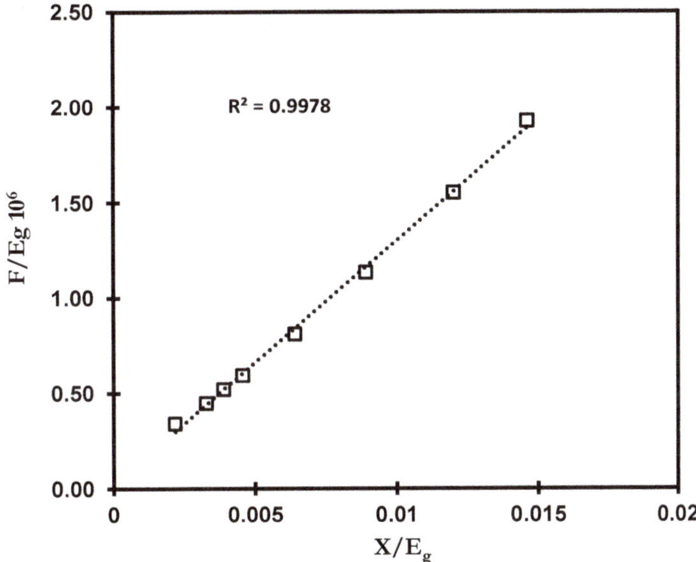


**Figure 5.** The linear relationship between the production and expansion parameters proposed in the modified material balance equation presented in this study (using the production data from the real-field production data of Well-B).

## 5. Conclusions

The accurate estimation of recoverable reserves and production forecasting is incredibly important in shale gas reservoirs. For unconventional reservoirs like shale gas, the material balance equation is a critical tool used to estimate these reserves and forecast production. With the global demand for energy increasing and the potential for shale gas resources to meet this demand, it becomes even more crucial to have precise estimates of recoverable reserves and accurate production forecasts. To address this need, our study presents a comprehensive derivation of a modified material balance equation for estimating free and adsorbed gas in place within shale gas reservoirs. This modification will enable more accurate estimation of shale gas reserves by considering stress sensitivity and variations in porosity during shale gas production, along with accounting for the effect of critical desorption pressure, water production, and water influx. By establishing a linear relationship between reservoir expansion and production terms, the proposed MBE eliminates the need for complex and iterative calculations usually encountered in existing shale gas MBEs. As a result, this approach offers a simpler yet effective means of estimating shale gas reserves without compromising accuracy. The proposed MBE was validated using an in-house finite element poro-elastic model which accounts for stress redistribution and deformation effects during shale gas production. Moreover, the proposed MBE was tested using real-field data of a shale gas reservoir obtained from the literature. The results of this study demonstrate the reliability and usefulness of the modified MBE as a tool for accurately assessing free and adsorbed shale gas volumes under a coupled poro-elastic environment.

**Author Contributions:** Conceptualization, R.A.A. and S.A.; methodology, R.A.A., S.A. and A.A.; software, R.A.A.; validation, R.A.A., S.A. and A.A.; formal analysis, S.A.; investigation, R.A.A. and S.A.; resources, A.A.; data curation, S.A.; writing—original draft preparation, R.A.A.; writing—review and editing, S.A.; visualization, A.A.; supervision, R.A.A.; project administration, A.A.; funding acquisition, A.A. All authors have read and agreed to the published version of the manuscript.

**Funding:** This research work was funded by Institutional Fund Projects under grant no. (IFPIP: 423-145-1443).

**Data Availability Statement:** The datasets used and/or analyzed during the current study are available from the corresponding author upon reasonable request.

**Acknowledgments:** This research work was funded by Institutional Fund Projects under grant no. (IFPIP: 423-145-1443). The authors gratefully acknowledge technical and financial support provided by the Ministry of Education and King Abdulaziz University, DSR, Jeddah, Saudi Arabia.

**Conflicts of Interest:** The authors declare no conflict of interest.

## Appendix A

Figure A1 shows the changes of *X*-component of the effective stresses around the wellbore. It can be seen from Figure A1 that the values of the effective stress simulated numerically are in a good agreement with the analytical solution at different simulation time steps. Hence, the poro-elastic effect led to changes in pore pressure due to fluid production and, consequently, altered the state of the stresses.

Figure A2 shows the pressure distribution around the wellbore due to fluid production after one hour of starting the production process at $P_{wf}$ = 1000 psi, while Figure A3 shows the contour map of the shear component of effective stress after 1 h of the production process. It can be seen from Figure A3 that the maximum shear stress is around the wellbore and increases away from the wellbore.

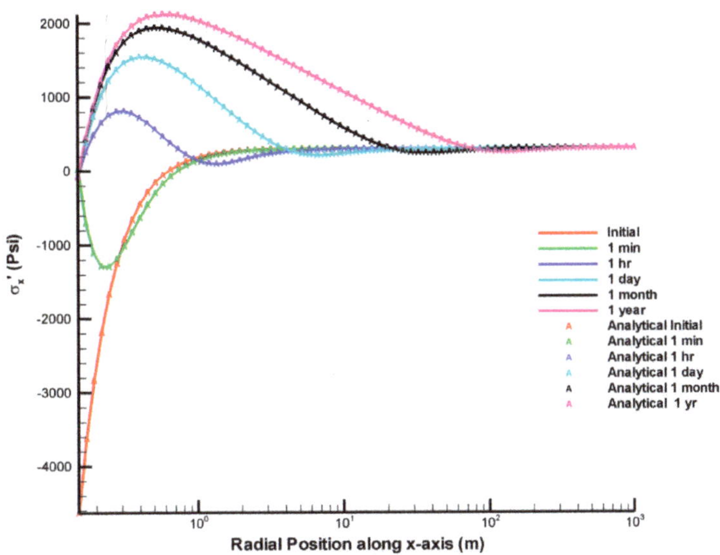

**Figure A1.** X component of effective stress analytical versus numerical as a function of radius and time along the x-axis, for $\sigma_H$ = 5800 psi, $\sigma_h$ = 5500 psi, Pi = 5500 psi, $P_w$ = 1000 psi, $k_x$ = 0.01 md, $k_y$ = 0.01 md (solid lines are the results obtained from the in-house numerical simulator).

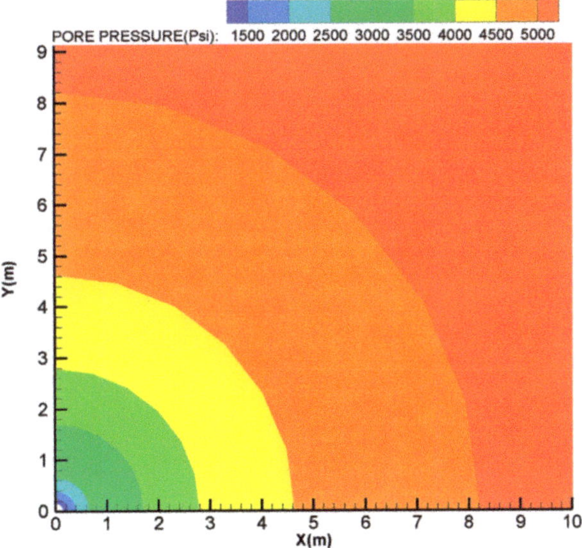

**Figure A2.** Pore pressure contour map after 1 h (numerical results, $\sigma_H$ = 5800 psi, $\sigma_h$ = 5500 psi, Pi = 5500 psi, $P_w$ = 1000 psi, $k_x$ = 0.01 md, $k_y$ = 0.01 md).

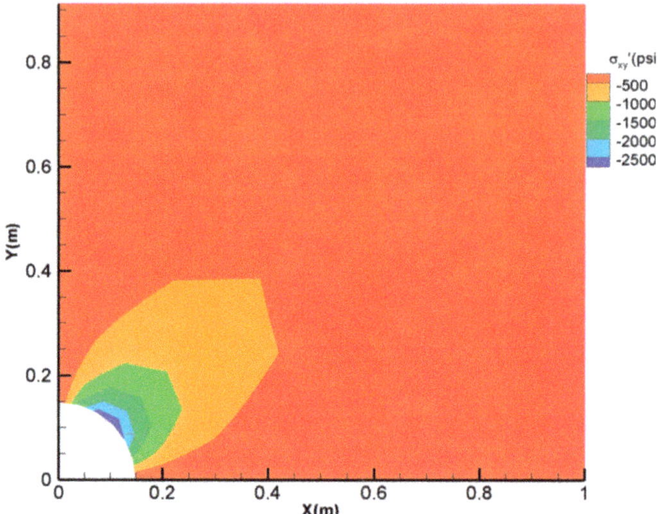

**Figure A3.** Contour map of shear component of effective stress after 1 h (Numerical results, $\sigma_H$ = 5800 psi, $\sigma_h$ = 5500 psi, Pi = 5500 psi, $P_w$ = 1000 psi, $k_x$ = 0.01 md, $k_y$ = 0.01 md).

## References


1.  Ahmed, T.H.; Centilmen, A.; Roux, B.P. A generalized material balance equation for coalbed methane reservoirs. In Proceedings of the SPE Annual Technical Conference and Exhibition, San Antonio, TX, USA, 24–27 September 2006. [CrossRef]
2.  Akkutlu, I.; Efendiev, Y.; Savatorova, V. Multi-scale asymptotic analysis of gas transport in shale matrix. *Transp. Porous Media* **2015**, *107*, 235–260. [CrossRef]
3.  Chen, Y.; Hu, J. Derivation of methods for estimating OGIP and recoverable reserves and recovery ratio of saturated coal-seam gas reservoirs. *Oil Gas Geol.* **2008**, *29*, 151–156. (In Chinese)
4.  Clarkson, C.R.; McGovern, J.M. Study of the potential impact of matrix free gas storage upon coalbed gas reservoirs and production using a new material balance equation. In Proceedings of the 2001 International Coalbed Methane Symposium, Tuscaloosa, AL, USA, 14–18 May 2001.
5.  Fuentes-Cruz, G.; Vasquez-Cruz, M. Reservoir performance analysis through the material balance equation: An integrated review based on field examples. *J. Petrol. Sci. Eng.* **2022**, *208*, 109377. [CrossRef]
6.  Feng, D.; Li, X.; Wang, X.; Li, J.; Sun, F.; Sun, Z.; Zhang, T.; Li, P.; Chen, Y.; Zhang, X. Water adsorption and its impact on the pore structure characteristics of shale clay. *Appl. Clay Sci.* **2018**, *155*, 126–138. [CrossRef]
7.  Jensen, D.; Smith, L. A practical approach to coalbed methane reserve prediction using a modified material balance technique. In Proceedings of the International Coalbed Methane Symposium, Tusaloosa, AL, USA, 12–17 May 1997.
8.  Kazemi, N.; Ghaedi, M. Production data analysis of gas reservoirs with edge aquifer drive: A semi-analytical approach. *J. Nat. Gas Sci. Eng.* **2020**, *80*, 103382. [CrossRef]
9.  Darishchev, A.; de Nancy, E.N.; Lemouzy, P.; Rouvroy, P. On simulation of flow in tight and shale gas reservoirs. In SPE Unconventional Gas Conference and Exhibition. In Proceedings of the SPE Unconventional Gas Conference and Exhibition, Muscat, Oman, 28–30 January 2013.
10. Wang, H.; Merry, H.; Amorer, G.; Kong, B. Enhance hydraulic fractured coalbed methane recovery by thermal stimulation. In Proceedings of the SPE Canada Unconventional Resources Conference, Calgary, AL, Canada, 20–22 October 2015.
11. Wang, J.; Liu, H.; Wang, L.; Zhang, H.; Luo, H.; Gao, Y. Apparent permeability for gas transport in nanopores of organic shale reservoirs including multiple effects. *Int. J. Coal Geol.* **2015**, *152*, 50–62. [CrossRef]
12. Weng, X. Modeling of complex hydraulic fractures in naturally fractured formation. *J. Unconv. Oil Gas Resour.* **2015**, *9*, 114–135. [CrossRef]
13. Li, J.; Zhang, T.; Li, Y.; Liang, X.; Wang, X.; Zhang, J.; Zhang, Z.; Shu, H.; Rao, D. Geochemical characteristics and genetic mechanism of the high-N2 shale gas reservoir in the Longmaxi Formation, Dianqianbei Area. *China. Petrol. Sci.* **2020**, *17*, 939–953. [CrossRef]
14. Lane, H.S.; Lancaster, D.E.; Watson, A.T. Estimating gas desorption parameters from Devonian shale well test data. In Proceedings of the SPE Eastern Regional Meeting, Columbus, OH, USA, 31 October–2 November 1990.
15. Lu, X.C.; Li, F.C.; Watson, A.T. Adsorption measurements in Devonian shales. *Fuel* **1995**, *74*, 599–603. [CrossRef]
16. Zhao, Y.; Hu, Y.; Zhao, B.; Yang, D. Nonlinear coupled mathematical model for solid deformation and gas seepage in fractured media. *Transp. Porous Media* **2004**, *55*, 119–136. [CrossRef]

17. Zhang, H.; Liu, J.; Elsworth, D. How sorption-induced matrix deformation affects gas flow in coal seams: A new FE model. *Int. J. Rock Mech. Min. Sci.* **2008**, *45*, 1226–1236. [CrossRef]
18. Lewis, A.M.; Hughes, R.G. Production data analysis of shale gas reservoirs. In Proceedings of the SPE Annual Technical Conference and Exhibition, Denver, CO, USA, 21–24 September 2008.
19. Montgomery, S.L.; Jarvie, D.M.; Bowker, K.A.; Pollastro, R.M. Mississippian Barnett Shale, Fort Worth basin, north-central Texas: Gas-shale play with multi–trillion cubic foot potential. *AAPG Bull.* **2005**, *89*, 155–175. [CrossRef]
20. Ross, D.J.; Bustin, R.M. The importance of shale composition and pore structure upon gas storage potential of shale gas reservoirs. *Mar. Pet. Geol.* **2009**, *26*, 916–927. [CrossRef]
21. Schepers, K.C.; Gonzalez, R.J.; Koperna, G.J.; Oudinot, A.Y. Reservoir modeling in support of shale gas exploration. In Proceedings of the SPE Latin America and Caribbean Petroleum Engineering Conference, Cartagena de Indias, CO, USA, 31 May–3 June 2009.
22. King, G. Material balance techniques for coal seam and Devonian shale gas reservoirs. In Proceedings of the SPE Annual Technical Conference and Exhibition, New Orleans, LA, USA, 23–26 September 1990. [CrossRef]
23. King, G. Material-balance techniques for coal-seam and Devonian shale gas reservoirs with limited water influx. *SPE Reserv. Eng.* **1993**, *8*, 67–72. [CrossRef]
24. Clarkson, C.R.; Jensen, J.L.; Blasingame, T.A. Reservoir engineering for unconventional gas reservoirs: What do we have to consider? In Proceedings of the SPE Unconventional Resources Conference/Gas Technology Symposium. The Woodlands, TX, USA, 14–16 June 2011.
25. Liu, J.; Xie, L.; He, B.; Zhao, P.; Ding, H. Performance of free gases during the recovery enhancement of shale gas by CO2 injection: A case study on the depleted Wufenge Longmaxi shale in northeastern Sichuan Basin. *China. Petrol. Sci.* **2021**, *18*, 530–545. [CrossRef]
26. Moghadam, S.; Jeje, O.; Mattar, L. Advanced gas material balance in simplified format. *J. Can. Petrol. Technol.* **2011**, *50*, 90–98. [CrossRef]
27. Meng, Y.; Li, X.; He, M.; Jiang, M. A semi-analytical model research of liquid collar shape and coalescence in pore throat during snap-off. *Arab. J. Geosci.* **2019**, *12*, 468. [CrossRef]
28. Nie, H.; Chen, Q.; Zhang, G.; Sun, C.; Wang, P.; Lu, Z. 2021. An overview of the characteristic of typical Wufeng-Longmaxi shale gas fields in the Sichuan Basin, China. *Nat. Gas Ind. B* **2021**, *8*, 217–230. [CrossRef]
29. Pan, X.; Zhang, G.; Chen, J. The construction of shale rock physics model and brittleness prediction for high-porosity shale gas-bearing reservoir. *Petrol. Sci.* **2020**, *17*, 658–670. [CrossRef]
30. Qiao, X.; Zhao, C.; Shao, Q.; Hassan, M. Structural characterization of corn stover lignin after hydrogen peroxide presoaking prior to ammonia fiber expansion pretreatment. *Energy Fuel* **2018**, *32*, 6022–6030. [CrossRef]
31. Sun, Z.; Li, X.; Liu, W.; Zhang, T.; He, M.; Nasrabadi, H. Molecular dynamics of methane flow behavior through realistic organic nanopores under geologic shale condition: Pore size and kerogen types. *Chem. Eng. J.* **2020**, *398*, 124341. [CrossRef]
32. Yang, B.; Zhang, H.; Kang, Y.; She, J.; Wang, K.; Chen, Z. In situ sequestration of a hydraulic fracturing fluid in Longmaxi Shale gas formation in the Sichuan Basin. *Energy Fuels* **2020**, *33*, 6983–6994. [CrossRef]
33. Zhang, T.; Javadpour, F.; Yin, Y.; Li, X. Upscaling water flow in composite nanoporous shalematrix using lattice Boltzmann method. *Water Resour. Res.* **2020**, *56*, e2019WR026007. [CrossRef]
34. Zhao, C.; Qiao, X.; Cao, Y.; Shao, Q. Application of hydrogen peroxide presoaking prior to ammonia fiber expansion pretreatment of energy crops. *Fuel* **2017**, *205*, 184–191. [CrossRef]
35. Zhang, L.; Chen, G.; Zhao, Y.; Liu, Q.; Zhang, H. A modified material balance equation for shale gas reservoirs and a calculation method of shale gas reserves. *Nat. Gas. Ind.* **2013**, *33*, 66–70. (In Chinese)
36. Williams-Kovacs, J.; Clarkson, C.; Nobakht, M. Impact of material balance equation selection on rate-transient analysis of shale gas. In Proceedings of the SPE Annual Technical Conference and Exhibition, San Antonio, TX, USA, 8–10 October 2012. [CrossRef]
37. Hu, S.; Hu, X.; He, L.; Chen, W. A new material balance equation for dual-porosity media shale gas reservoir. *Energy Procedia* **2019**, *158*, 5994–6002. [CrossRef]
38. Meng, Y.; Li, M.; Xiong, X.; Liu, J.; Zhang, J.; Hua, J.; Zhang, Y. Material balance equation of shale gas reservoir considering stress sensitivity and matrix shrinkage. *Arab. J. Geosci.* **2020**, *13*, 535. [CrossRef]
39. Shi, J.-T.; Jia, Y.-R.; Zhang, L.-L.; Ji, C.-J.; Li, G.-F.; Xiong, X.-Y.; Huang, H.-X.; Li, X.-F.; Zhang, S.-A. The generalized method for estimating reserves of shale gas and coalbed methane reservoirs based on material balance equation. *Pet. Sci.* **2022**, *19*, 2867–2878. [CrossRef]
40. Yang, L.; Zhang, Y.; Zhang, M.; Liu, Y.; Bai, Z.; Ju, B. Modified flowing material balance equation for shale gas reservoirs. *ACS Omega* **2022**, *7*, 20927–20944. [CrossRef]
41. Palmer, I.; Mansoori, J. How permeability depends on stress and pore pressure in coalbeds: A new model. *SPE Reserv. Eval. Eng.* **1998**, *1*, 539–544. [CrossRef]
42. Zienkiewicz, O.C.; Taylor, R.L. *The Finite Element Method: Solid Mechanics Vol 2*, 5th ed.; Butterworth-Heinemann: Oxford, UK, 2000.

*Article*

# Study of the Wellbore Instability Mechanism of Shale in the Jidong Oilfield under the Action of Fluid

Xiaofeng Xu [1], Chunlai Chen [1], Yan Zhou [1], Junying Pan [1], Wei Song [1], Kuanliang Zhu [1], Changhao Wang [2,*] and Shibin Li [2]

[1]  Institute of Drilling and Production Technology, Petro China Jidong Oilfield Company, Tangshan 063000, China
[2]  School of Petroleum Engineering, Northeast Petroleum University, Daqing 163318, China
*  Correspondence: chwang88@nepu.edu.cn

**Abstract:** Wellbore instability is the primary technical problem that restricts the low-cost drilling of long-interval horizontal wells in the shale formation of the Jidong Oilfield. Based on the evaluation of the mineral composition, structure and physicochemical properties of shale, this paper investigates the mechanical behavior and instability characteristics of shale under fluid action by combining theoretical analysis, experimental evaluation and numerical simulation. Due to the existence of shale bedding and microcracks, the strength of shale deteriorates after soaking in drilling fluid. The conductivity of the weak surface of shale is much higher than that of the rock matrix. The penetration of drilling fluid into the formation along the weak surface directly reduces the strength of the structural surface of shale, which is prone to wellbore collapse. The collapse pressure of the shale formation in the Nanpu block of the Jidong oilfield was calculated. The well inclination angle, azimuth angle and drilling fluid soaking time were substituted in the deterioration model of rock mechanics parameters, and the safe drilling fluid density of the target layer was given. This work has important guiding significance for realizing wellbore stability and safe drilling of hard brittle shale in the Jidong Oilfield.

**Keywords:** wellbore stability; rock strength deterioration; shale formation; oil-based drilling fluid

**Citation:** Xu, X.; Chen, C.; Zhou, Y.; Pan, J.; Song, W.; Zhu, K.; Wang, C.; Li, S. Study of the Wellbore Instability Mechanism of Shale in the Jidong Oilfield under the Action of Fluid. *Energies* **2022**, *16*, 2989. https://doi.org/10.3390/en16072989

Academic Editors: Reza Rezaee, Jalel Azaiez and Paul Stewart

Received: 28 August 2022
Accepted: 25 November 2022
Published: 24 March 2023

## 1. Introduction

The wellbore instability of shale has always been a complex problem in drilling engineering worldwide. Borehole wall instability causes great difficulties in drilling engineering, mainly manifested as diameter reduction and collapse sticking. These problems not only prolong the drilling cycle but also increase the drilling cost [1–3]. Borehole instability is due to the bedding development and strong water sensitivity of shale. The invasion of drilling fluid causes fractures to continuously extend and expand along the formation, eventually forming a complex fracture network. This greatly reduces rock strength [4,5]. The integrity of near-wellbore water content and cementation changes the formation strength and stress field around a borehole, causing stress concentration and failure to establish a new equilibrium in the borehole, resulting in borehole instability [6,7].

Current research on the wellbore stability of shale focuses on the mechanical–chemical–thermal coupling of multiple fields and is basically in the preliminary research stage [8,9]. However, whether the strength of layered shale under the action of fluid changes with the action time and the influence of strength deterioration on the collapse pressure of well walls remains ambiguous [10,11]. The outstanding feature of the shale formation in the Jidong Oilfield is that it has approximately parallel bedding planes, that is, the dividing planes between rock layers. Bedding is different from microcrack with no obvious displacement, and the strength of the rock affected by the bedding plane has stronger anisotropy. In this paper, laboratory experiments are carried out on the bedded shale in the Nanpu block of Jidong

Oilfield. The hydration of shale is analyzed by the finite difference method and numerical simulation, and the degradation mechanism of the shale in the lower layer is revealed. This research has guiding significance in the design of drilling fluids for shale formations and for realizing high-quality, safe, efficient and low-cost drilling.

## 2. Experiment on Mineral Composition and Mechanical Parameters of Bedded Shale

In order to study the degradation law of shale under the action of different fluids, cores from Well NP2-46 in the Jidong Oilfield were selected for laboratory experiments. The core was taken by pressure maintaining coring on site and sealed with heat-shrinkable tube to prevent it from drying. Because the bedding and fracture of the shale are obviously developed, the wire-cutting method was adopted to conduct secondary sampling for the core, which ensures that the rock sample does not contact any fluid during processing. This ensures that the experimental results accurately reflect the real situation in the well.

### 2.1. Mineral Composition and Fracture Development Characteristics

X-ray diffraction (X-RD) and scanning electron microscopy experiments were carried out on the shale in the Nanpu block of Jidong Oilfield. Figure 1 shows the X-RD experimental test results. It can be seen from the figure that the main mineral components of the shale are clay and quartz minerals. Minerals with extremely poor expansiveness, such as quartz, are attached to the surrounding of the clay. After the clay is hydrated and expanded under the action of the drilling fluid, a pressure difference is generated, which leads to the instability of the shale. In addition, hard and brittle minerals such as plagioclase, calcite and dolomite are developed to different degrees. The average content of clay minerals is 34.85%. The clay minerals mainly consist of illite and illite-mixed layers, followed by kaolinite and chlorite without montmorillonite. The illite content is higher in the illite-mixed layer, which also shows the typical brittleness characteristics of deep shale in the structure. This kind of lithology is not easy to expand, but the illite in it also undergoes rapid microexpansion. This is due to the adsorption of water molecules on illite, which leads to the surface hydration of rocks, and the main driving force is the surface hydration energy. This reduces the strength of the shale and causes its spalling.

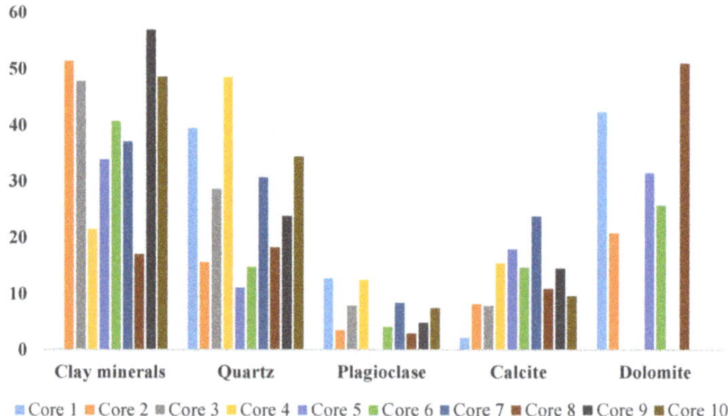


**Figure 1.** Mineral composition of shale in the Nanpu block.

The shale in the Nanpu block is mainly argillaceous, with a compact and hard structure. Figure 2 shows the observation results of scanning electron microscopy. It can be found that the pores in the shale are not developed, and the micro pores are scattered, but the microfractures and micropores are relatively developed. There is no typical occurrence of minerals and clay minerals in the core. The fracture width in the SEM image is 2.48 μm. When shale is placed in water, a large number of small air bubbles are attached to the surface of the shale. Over time, obvious cracks appeared in the rock sample, which gradually opened and finally collapsed.

**Figure 2.** SEM image of deep shale in Nanpu block.

### 2.2. Triaxial Compressive Strength Test of Shale

Due to the significant anisotropy of bedded shale, coring was carried out along different bedding angles. The samples were then machined into standard core columns with a diameter of 25 mm and a length of 50 mm for triaxial compressive strength testing of the shale in an unimmersed state. Figure 3 shows the fracture patterns of the core after loading in the parallel and vertical bedding directions.

(**a**) Loading in the parallel direction.  (**b**) Loading in the vertical direction.

**Figure 3.** The failure form of shale under different loading directions.

According to the photos of rock breakage, the fracture morphology of parallel and vertical stratified rocks is different after loading. When loading is parallel to the direction of stratification, multiple openings occur along the bedding cracks, forming multiple groups of fragments along the direction of the bedding plane. However, shear cracks are generated when the vertical bedding direction is loaded, and large volume fractures are formed after the shear cracks connect with the microcracks along the bedding. Rock loaded in the parallel bedding direction produces tensile failure along the structural plane, and rock loaded in the vertical bedding direction needs to overcome the shear strength of the shale body.

Table 1 shows the experimental test results. The strength of the shale under parallel bedding and vertical bedding loading is very different, and the anisotropy is strong. When the confining pressure is set to 60 MPa, the elastic modulus of the shale loaded by vertical bedding is high and the Poisson ratio is low, indicating that the shale in the Nanpu block is relatively brittle.

**Table 1.** Triaxial stress test results.

| Loading Direction | Lithology | Confining Pressure (MPa) | Compressive Strength (MPa) | Young's Modulus (MPa) | Poisson's Ratio |
|---|---|---|---|---|---|
| Vertical bedding direction | Light gray fine shale | 60 | 99.817 | 14,924.7 | 0.16 |
| Parallel bedding direction | Dark gray mud shale | 60 | 67.146 | 9979.2 | 0.20 |

Due to the well-developed bedding and significant anisotropy of the shale, the compressive strength experiments under loading at angles of 30°, 45° and 60° with the bedding were continued. Combined with the above triaxial compressive strength test results loaded in the parallel and vertical bedding directions, the cohesive force and internal friction angle of the rock were calculated as shown in Table 2.

**Table 2.** Experimental results of rock mechanics parameters in different loading directions.

| Load Direction | Cohesion (MPa) | Angle of Internal Friction (°) | Coefficient of Internal Friction |
|---|---|---|---|
| Parallel bedding direction | 11.386 | 12.95 | 0.23 |
| Vertical bedding di-rection | 22.464 | 13.06 | 0.232 |
| 30° bedding direction | 15.698 | 7.57 | 0.133 |
| 45° bedding direction | 8.338 | 6.39 | 0.112 |
| 60° bedding direction | 4.775 | 4.4 | 0.077 |

According to the analysis of the triaxial stress experimental results in different loading directions, the failure of the shale can be divided into three regions, as shown in Figure 4. When the loading direction is perpendicular to the shale bedding, the strength is the highest, and with the increase in the included angle, the strength gradually decreases. According to Mohr–Coulomb criterion, when the angle is 45° + $\varphi/2$ ($\varphi$ is the angle of internal friction), the intensity reaches the lowest value and then gradually increases.

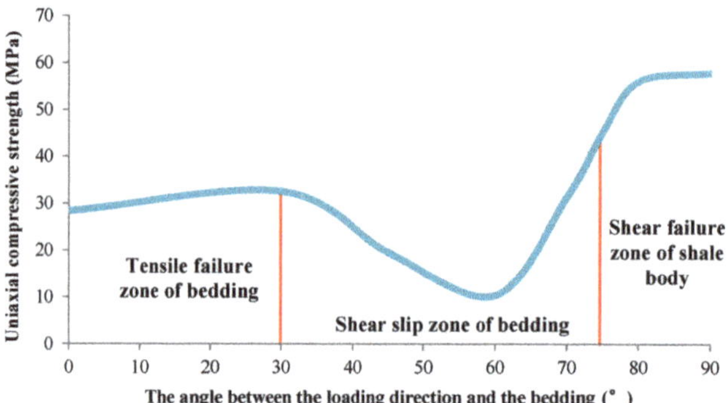


**Figure 4.** Core anisotropic failure law.

According to the different loading directions, shale failure forms can be expressed as follows:

(1) When the loading direction is 0°~30° to the bedding, splitting failure mainly occurs along the bedding plane.

(2) Shear slip mainly occurs along the bedding plane when the loading direction is between 30° and 75°, and the strength is the lowest between 50° and 60°.

(3) When the loading direction is in the range of 75°~90° with bedding, shear failure of the shale body mainly occurs, and the strength is the largest.

### 2.3. Triaxial Compressive Strength Test of Shale under Different Fluid Immersions

The vertical coring shale and the horizontal coring shale were immersed in pure water and diesel oil, respectively. After different times, the triaxial compressive strength test was conducted again, and the cohesion and internal friction angle were calculated. Experimental photos are shown in Figures 5 and 6.

(a) Before soaking.

(b) After soaking.

**Figure 5.** Soaking in water for 48 h.

(a) Before soaking.

(b) After soaking.

**Figure 6.** Soaking in diesel oil for 48 h.

When the shale was soaked in water, three cores were directly broken within 1 h, and two cores had obvious cracks. After soaking in oil for 48 h, the surface of one core peeled off, and the other cores were not significantly broken. A comparison of the amount and time of core damage in water and in oil revealed that the damage rate of water to shale cores is much higher than that of oil to shale cores [12].

After soaking for 48 h, the unbroken cores were subjected to triaxial testing. According to the experimental results in Table 3, it can be seen that the shale cored parallel to the bedding direction was completely broken after soaking in water. The triaxial strength was reduced to 0, and the strength was reduced by 100%. The average triaxial strength of the shale after being soaked in oil was 45.728 MPa, and the strength decreased by 31.9%. The average triaxial strength of the shale in the vertical bedding direction was 38.907 MPa after soaking in water, and the strength decreased by 61%. The average triaxial strength after oil soaking was 63.874 MPa, which was almost unchanged compared with the original rock sample.

**Table 3.** 48 h triaxial stress test results of shale soaking.

| Types of Experiments | Confining Pressure (MPa) | Compressive Strength (MPa) | Young's Modulus (MPa) | Poisson's Ratio |
|---|---|---|---|---|
| Soaking in water (parallel) | broken | 0<br>0<br>0 | 0<br>0<br>0 | 0<br>0<br>0 |
| Soaking in oil (parallel) | 60<br>broken<br>60 | 49.677<br>0<br>41.778 | 9537.9<br>0<br>13,057.2 | 0.21<br>0<br>0.2 |
| Soaking in water (vertical) | 60<br>60 | 38.944<br>38.869 | 15,396.5<br>8608.5 | 0.21<br>0.23 |
| Soaking in oil (vertical) | 60<br>60 | 64.225<br>63.523 | 13,322.7<br>10,173.9 | 0.19<br>0.22 |

*2.4. Tensile Strength Test of Shale*

The tensile strength of the shale was tested by the Brazilian splitting method. It can be seen from the test results in Table 4 that the tensile strength of the shale varies greatly in the vertical and horizontal directions.

**Table 4.** Shale tensile strength test results in vertical/parallel bedding directions.

| Coring Direction | Stratum | Tensile Strength (MPa) | Average Tensile Strength (MPa) |
|---|---|---|---|
| Vertical | Es1 | 7.99<br>8.15 | 8.07 |
| Parallel | Es1 | 3.39<br>3.05 | 3.22 |

According to the fracture morphology of the rock in Figure 7, tensile failure is prone to occur along the bedding plane of the rock, which is an important factor that affects the stability of the well wall [13].

(a) Vertical bedding loading.    (b) Parallel bedding loading.

**Figure 7.** Fragmentation morphology of loaded shale in different bedding directions.

### 3. Study on the Deterioration Law of Shale under the Action of Fluid

Shale is a rock that is composed of water-sensitive clay minerals whose interaction with drilling fluids is inevitable [14]. Due to the characteristics of shale's structure and composition, the effect of different drilling fluid systems is also very different. Most shale is strong enough to withstand borehole stress for some time as the bit cuts through the shale formation [15]. However, over time, the strength of the shale decreases significantly, regardless of the type of drilling fluid used. This is due to hydration, which occurs even with very little water uptake by the shale. This creates a zone of softening around the well, leading to the delayed failure of the shale [16].

The main components of shale are clay minerals. The composition, content and microstructure of the clay minerals determine the basic physical and chemical properties and hydration mechanism of the shale. Understanding the hydration mechanism of shale is the basis of analyzing the stability of shale wellbore. When the shale is in contact with the drilling fluid, the transport of water and ions is induced, driven by hydraulic and chemical potential gradients. These include the Darcy flow, which is driven by the pressure difference between the drilling fluid column pressure and the pore pressure, and ion diffusion, which is driven by the chemical potential difference between the drilling fluid and the shale [17].

According to C.H. Yew [18], hydration of the shale formation around the well can be described by a diffusion equation of water molecules. The water absorption diffusion equation of shale wellbore can be established by the law of conservation of mass. Let $q$ be the mass flow of water adsorption, and let $W(r, t)$ be the weight percentage of adsorbed water at point $r$ away from the well axis and time $t$. In the conservation of mass, the following expression is needed:

$$\nabla q = \frac{\partial W}{\partial t} \tag{1}$$

Assumptions:

$$q = C_f \nabla W \tag{2}$$

where $\nabla$ is the gradient operator, and $C_f$ represents the adsorption constants of materials, which are related to the properties of the shale and the drilling fluid and can be measured with the water adsorption test.

Substitute Equation (2) into Equation (1). In the cylindrical coordinate system, the basic equation of water adsorption can be obtained as follows:

$$C_f \frac{1}{r} \frac{\partial}{\partial r}\left( r \frac{\partial W}{\partial r} \right) = \frac{\partial W}{\partial t} \tag{3}$$

Under boundary conditions at infinity, the formation water content is the original formation water content. At the well wall, the formation water content is considered saturated and does not change with time:

$$W|_{r=r_w} = W_s, \quad W|_{r\to\infty} = W_0, \ 0 < t < \infty \tag{4}$$

Initial conditions:

$$W|_{t=0} = W_0 \tag{5}$$

Using the finite difference method to solve the above equation, the water content of the formation around the borehole at different locations at different times can be obtained. Let the time step be $\Delta t$ and the radial distance step be $\Delta h$. Therefore, the radial distance is $r_i = i\Delta h$, and the time is $t_i = j\Delta t (i = (0,1,2,\ldots n); j = (0,1,2,\ldots m))$. $W(r_i, t_j)$ is recorded as $W_{(i,j)} (i = (0,1,2,\ldots n); j = (0,1,2,\ldots m))$. Approximately, replace $\frac{\partial W}{\partial t}, \frac{\partial W}{\partial r}$ with the forward difference quotient and $\frac{\partial^2 W}{\partial r^2}$ with the second central difference quotient, namely:

$$\frac{\partial W}{\partial t} = \frac{W_{(i,j+1)} - W_{(i,j)}}{\Delta t} \tag{6}$$

$$\frac{\partial W}{\partial r} = \frac{W_{(i+1,j)} - W_{(i,j)}}{\Delta h} \tag{7}$$

$$\frac{\partial^2 W}{\partial r^2} = \frac{W_{(i+1,j)} - 2W_{(i,j)} + W_{(i-1,j)}}{\Delta h^2} \tag{8}$$

Substituting (6)~(8) into Equation (3), the difference equation of Equation (3) is:

$$\frac{W_{(i+1,j)} - 2W_{(i,j)} + W_{(i-1,j)}}{\Delta h^2} + \frac{1}{r_i}\frac{W_{(i+1,j)} - W_{(i,j)}}{\Delta h} = \frac{1}{C_f}\frac{W_{(i,j+1)} - W_{(i,j)}}{\Delta t} \tag{9}$$

The difference format for boundary conditions and initial conditions is:

$$W(0, j) = W_s, W(n, j) = W_0, W(i, 0) = W_0 \tag{10}$$

According to the above experimental results, the shale water absorption rate equation is modified, and the water content of the shale formation around the borehole at different times and positions can be obtained using a numerical simulation method.

The formation collapse pressure can be calculated by the following formula [19]:

$$P_b = \frac{1}{2}(3\sigma_H - \sigma_h)(1 - \sin \varphi_0) + \alpha P_p \sin \varphi_0 - C_0 \cos \varphi_0 \tag{11}$$

where $P_b$ is collapse pressure (MPa); $\sigma_H$ is the maximum horizontal stress (MPa); $\sigma_h$ is the minimum horizontal formation stress (MPa); $P_p$ is the formation pore pressure (MPa); $C_0$ is the cohesion (MPa); $\varphi_0$ is the angle of internal friction (°); and $\alpha$ is the effective stress coefficient.

In order to facilitate the design of the drilling fluid density window, the collapse pressure is converted into the form of equivalent density, that is, the drilling fluid density required to just balance the formation collapse pressure, $g/cm^3$.

According to the test results of the rock mechanical parameters soaked in different fluids, the shale deteriorates after drilling fluid soaking due to the existence of cleavage and microfractures [20]. As the conductivity of the structural plane is much higher than that of raw rock, the penetration of drilling fluid along the structural plane will directly cause the weakening of the structural plane strength [21]. As shown in Figure 8, the stress distribution of the vertical well section, the deflection section and the horizontal section of the wellbore in a bedded shale formation is established. The mechanical parameters of the surrounding rock of the wellbore are obtained from the above triaxial compressive strength test results. The change in the cohesion force and the internal friction angle after the core soaking experiment are used to characterize the weakening of the shale's strength, and the force-chemical coupling finite element numerical simulation model of borehole collapse pressure is obtained. It can be clearly seen that stress concentration occurs at the intersection of the well wall and the bedding plane. Under the action of the liquid column pressure in the wellbore, the drilling fluid can easily enter the bedding and increase the collapse pressure around the wellbore.

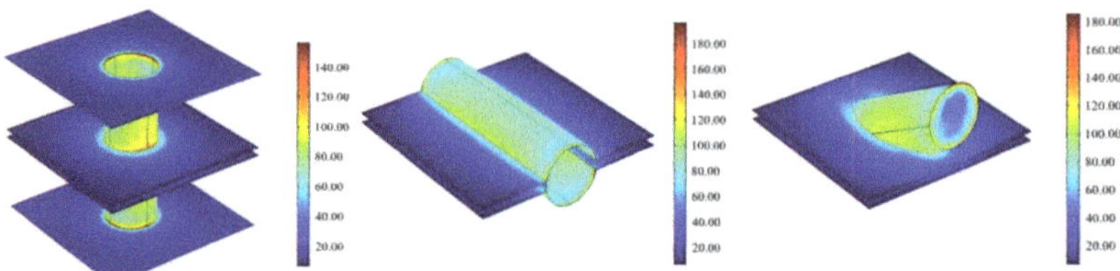

**Figure 8.** The von Mises stress around the borehole wall for different well types and water contents.

In the drilling process of shale formations, the rock strength decreases, and its bearing capacity weakens, and the increased stress leads to an enhanced trend of rock collapse around the well, which is quantitatively reflected in the increase in collapse pressure, leading to wellbore instability [22]. Therefore, finite element numerical simulation is employed to calculate the collapse pressure equivalent density of the shale around the well at different water content times, which can more intuitively analyze the influence of water absorption on the deterioration degree of the shale and the borehole wall stability. Changing the action time of the drilling fluid in the wellbore and the shale bedding plane under a different well inclination and azimuth in Figure 8 until the liquid content of the surrounding rock reaches 5%, 10% and 15%, the derived values were utilized to calculate the maximum collapse pressure and the equivalent density of the collapse pressure of different schemes.

Figure 9 shows the distribution chart of the equivalent density of collapse pressure when the liquid content around the well reaches 5%, 10% and 15% under different well inclination and azimuth conditions. When the shale liquid content around the well is 5%, the variation range of the equivalent density of the collapse pressure under different wellbore trajectories is 1.02~1.53 g/cm$^3$. When the liquid content of the shale reservoir increases, the shale strength changes, and the equivalent density of the borehole collapse pressure decreases rapidly. The collapse density increases from 0.15 to 0.22 g/cm$^3$ and from 0.31 to 0.45 g/cm$^3$ when the liquid content is 10% and 15%, respectively. At a certain time, the water absorption of the shale decreases with the increasing distance from the borehole wall. A hydration zone is formed in the shale formation around the borehole; however, its liquid content is similar to the original liquid content after a certain distance. At a certain distance, the longer the time, the more water absorption of the shale occurs, but at a certain time, saturation and stability will occur.

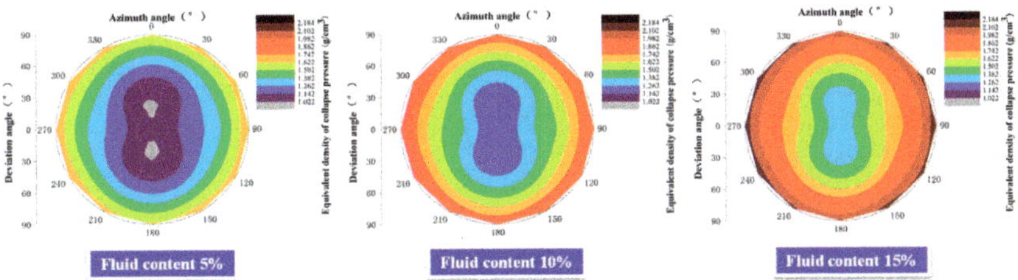


**Figure 9.** Collapse pressure distribution under different well types and water contents.

After the ground layer is drilled, driven by the pressure difference and chemical potential difference between the drilling fluid and the formation pore fluid, the drilling fluid filtrate enters the formation, causing a change in the mechanical properties of the shale [23]. The loss of filtrate from the drilling fluid increases the water content of the formation around the borehole, which leads to a series of changes in the mechanical properties of the formation. For example, the elastic modulus sharply decreases with increasing formation water content, and Poisson's ratio increases with increasing formation water content, whereas the cohesion and internal friction angle of the formation strength decrease with increasing formation water content [24]. These changes are particularly pronounced in near-wellbore formations. The fluid loss caused a softening zone around the borehole wall, which greatly affected the stability of the borehole wall. Under a certain pressure and temperature, the interaction between the shale and the drilling fluid produces a hydration zone around the borehole [25]. The size of the hydration zone depends on the formation and drilling fluid characteristics, drilling fluid column pressure, interaction time between the drilling fluid and formation, interaction time between formation and the drilling fluid temperature, etc. Because of the formation hydration zone around the borehole, the rock around the borehole becomes a complex rock mass medium with a variable water content, modulus and strength that vary with radius and time. As a result, the strength of the shale in certain areas around the well is reduced, and a plastic zone appears or further expands to cause wellhole collapse.

## 4. Field Case Analysis

A horizontal well in the Nanpu block of the Jidong Oilfield was unstable during drilling, which caused serious downhole accidents, such as well collapse and stuck drilling. The shale in this area has obvious bedding, so if the conventional model is selected to analyze the borehole wall collapse problem, the result is very different from the actual situation. Therefore, the calculation model of the equivalent density of collapse pressure with different water contents is used to correct the results, and a reasonable safe density window of drilling fluid is obtained.

The well depth of the well to be designed is 4996 m, and the length of the horizontal section is 1200 m. Adjacent wells were drilled with a maximum drilling fluid density of 1.45 g/cm$^3$. Due to the low density of the drilling fluid, five wellbore collapse and sticking accidents occurred. It took 32 days to deal with complex downhole accidents, and 700.77 m$^3$ of drilling fluid was lost. An analysis of the reasons for the incidents showed that shale microfissures in the Dongying Formation and the Shahejie Formation developed. Most shale formations are strong enough to withstand borehole stress for a certain period as the bit cuts through them. Over time, however, due to the hydration of the shale, even though the water uptake may be small, regardless of the type of drilling mud used, the strength of the shale will significantly decrease, creating a softening zone around the well and causing delayed rock failure. The collapse pressure is corrected according to the well trajectory and drilling construction period, and the safe drilling fluid density window is obtained as shown in Figure 10.

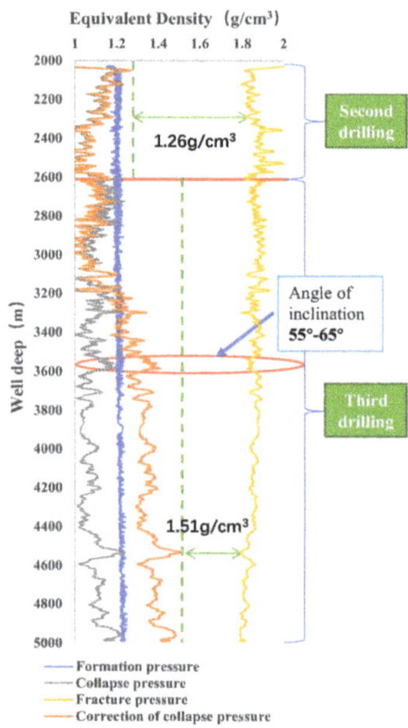


**Figure 10.** Corrected safe drilling fluid density window.

The well was drilled to the Shahejie Formation with water-based drilling fluid in the second drilling. The depth of the well is 2615 m. The estimated construction period is 12 days. The liquid density is 1.26 $g/cm^3$. In the third drilling, oil-based drilling fluid was used to drill to a well depth of 4996 m, and the construction period was 34 days. After calculating the rock deterioration, the maximum collapse pressure appeared in the well inclination angle of 55°~65° and the horizontal reservoir. The maximum collapse pressure of this well section is 1.51 $g/cm^3$, the minimum rupture pressure is 1.79 $g/cm^3$, and the density window is narrow. The recommended drilling fluid density is 1.51 $g/cm^3$. Drilling under the density of the drilling fluid, the well wall did not collapse, and the drilling was successfully completed.

It can be seen that the weakening strength of the shale bedding surface (increased water content) has a significant effect on the wellbore stability of horizontal wells. For any bedding surface occurrence, the collapse pressure increases with increasing water content, and the collapse pressure caused by the weakening of the bedding surface increases by approximately 0.32 $g/cm^3$ compared with that without weakening. The weakening of shale planes is the main controlling factor of shaft wall collapse, and its influence cannot be disregarded. The drilling fluid intrudes into the stratum along the bedding under the action of osmosis, which reduces the cohesion and the internal friction angle of the bedding plane and increases the risk of shear slip of the borehole wall rock along the bedding, thus aggravating the risk of borehole wall collapse and instability. For drilling sites, if the influence of the shale strength deterioration in the drilling fluid on collapse pressure is not taken into account, the drilling fluid density may then be low, causing the collapse of the borehole wall at an early stage of drilling.

## 5. Conclusions

(1) The shale in the Nanpu Depression is mainly argillaceous with dense and hard structures. Scanning electron microscopy showed that the pores were not developed, and the micropores were sporadically distributed, but the microcracks and micropores were relatively developed.

(2) Considering the collapse and rupture of the wellbore, the collapse belongs to bulk shear when the well inclination is $0°{\sim}30°$, and the rupture pressure and collapse pressure are both low. When the well inclination is $30°{\sim}60°$, the drilling fluid density window is the narrowest. Precise pressure control should be ensured. When the inclination angle is $60°{\sim}90°$, the collapse belongs to bedding splitting, and both the collapse pressure and the rupture pressure increase.

(3) Through the oil–water immersion strength deterioration experiment, the effect of shale oil-based drilling fluid in the Nanpu block is better than that of water-based drilling fluid.

(4) An analysis revealed that after the drilling fluid contacts the shale formation, the formation near the borehole wall quickly absorbs water, and its water content quickly reaches the saturation value, which substantially changes the mechanical parameters of the rock, which is manifested as a sharp decrease in rock strength. The formation near the borehole wall becomes a softening zone, which hinders borehole wall instability.

**Author Contributions:** Conceptualization, X.X. and K.Z.; methodology, C.C.; software, C.W.; validation, Y.Z., J.P. and W.S.; formal analysis, C.W.; investigation, X.X. and C.C.; resources, C.C.; data curation, S.L.; writing—original draft preparation, C.W.; writing—review and editing, C.W.; visualization, C.W.; supervision, X.X.; project administration, S.L.; funding acquisition, S.L. All authors have read and agreed to the published version of the manuscript.

**Funding:** This research was funded by [Natural Science Foundation of China], grant number [51874098].

**Data Availability Statement:** Not applicable.

**Conflicts of Interest:** The authors declare no conflict of interest.

# References

1. Zhao, W.; Liu, Y.; Wang, T.; Ranjith, P.G.; Zhang, Y. Stability Analysis of Wellbore for Multiple Weakness Planes in Shale Formations. *Geomech. Geophys. Geo-Energy Geo-Resour.* **2021**, *7*, 44. [CrossRef]
2. Liu, X.; Zeng, W.; Liang, L.; Lei, M. Wellbore Stability Analysis for Horizontal Wells in Shale Formations. *J. Nat. Gas Sci. Eng.* **2016**, *31*, 1–8. [CrossRef]
3. Li, Y.; Fu, Y.; Tang, G.; She, C.; Guo, J.; Zhang, J. Effect of Weak Bedding Planes on Wellbore Stability for Shale Gas Wells. In Proceedings of the IADC/SPE Asia Pacific Drilling Technology Conference and Exhibition, Tianjin, China, 9–11 July 2012.
4. Heng, S.; Guo, Y.; Yang, C.; Daemen, J.J.K.; Li, Z. Experimental and Theoretical Study of the Anisotropic Properties of Shale. *Int. J. Rock Mech. Min. Sci.* **2015**, *74*, 58–68. [CrossRef]
5. Al-Bazali, T. The Impact of Water Content and Ionic Diffusion on the Uniaxial Compressive Strength of Shale. *Egypt. J. Pet.* **2013**, *22*, 249–260. [CrossRef]
6. Zhang, Q.; Fan, X.; Chen, P.; Ma, T.; Zeng, F. Geomechanical Behaviors of Shale after Water Absorption Considering the Combined Effect of Anisotropy and Hydration. *Eng. Geol.* **2020**, *269*, 105547. [CrossRef]
7. Wang, Y.; Liu, X.; Liang, L.; Xiong, J. Experimental Study on the Damage of Organic-Rich Shale during Water-Shale Interaction. *J. Nat. Gas Sci. Eng.* **2020**, *74*, 103103. [CrossRef]
8. Abdideh, M.; Mafakher, A. Wellbore Stability Analysis in Oil and Gas Drilling by Mechanical, Chemical and Thermal Coupling (Case Study in the South of Iran). *Geotech. Geol. Eng.* **2021**, *39*, 3115–3131. [CrossRef]
9. Zou, D.; Han, Z.; Huang, H. Sensitivity Analysis of Time Dependent Wellbore Instability and Application. In Proceedings of the International Field Exploration and Development Conference 2019, Chengdu, China, 16–18 September 2019; Springer: Singapore, 2020; pp. 534–550.
10. Changhao, W.; Ling, Z.; Shibin, L.; Huizhi, Z.; Kai, L.; Xiaoming, W.; Chunhua, W. Time-Sensitive Characteristics of Bedding Shale Deterioration under the Action of Drilling Fluid. *Lithosphere* **2022**, *2022*, 3019090. [CrossRef]

11. Ding, L.; Lv, J.; Wang, Z.; Liu, B. Borehole Stability Analysis: Considering the Upper Limit of Shear Failure Criteria to Determine the Safe Borehole Pressure Window. *J. Pet. Sci. Eng.* **2022**, *212*, 110219. [CrossRef]
12. Peng, S.; Shevchenko, P.; Periwal, P.; Reed, R.M. Water-Oil Displacement in Shale: New Insights from a Comparative Study Integrating Imbibition Tests and Multiscale Imaging. *SPE J.* **2021**, *26*, 3285–3299. [CrossRef]
13. Zhang, S.W.; Shou, K.J.; Xian, X.F.; Zhou, J.P.; Liu, G.J. Fractal Characteristics and Acoustic Emission of Anisotropic Shale in Brazilian Tests. *Tunn. Undergr. Space Technol.* **2018**, *71*, 298–308. [CrossRef]
14. Saleh, T.A. Advanced Trends of Shale Inhibitors for Enhanced Properties of Water-Based Drilling Fluid. *Upstream Oil Gas Technol.* **2022**, *8*, 100069. [CrossRef]
15. Jiang, Z.; Yu, S.; Deng, H.; Deng, J.; Zhou, K. Investigation on Microstructure and Damage of Sandstone Under Cyclic Dynamic Impact. *IEEE Access* **2019**, *7*, 145–158. [CrossRef]
16. He, S.; Liang, L.; Zeng, Y.; Ding, Y.; Lin, Y.; Liu, X. The Influence of Water-Based Drilling Fluid on Mechanical Property of Shale and the Wellbore Stability. *Petroleum* **2016**, *2*, 61–66. [CrossRef]
17. Zhang, Q.; Yao, B.; Fan, X.; Li, Y.; Li, M.; Zeng, F.; Zhao, P. A Modified Hoek-Brown Failure Criterion for Unsaturated Intact Shale Considering the Effects of Anisotropy and Hydration. *Eng. Fract. Mech.* **2021**, *241*, 107369. [CrossRef]
18. Yew, C.H.; Chenevert, M.E.; Wang, C.L.; Osisanya, S.O. Wellbore Stress Distribution Produced by Moisture Adsorption. *SPE Drill. Eng.* **1990**, *5*, 311–316. [CrossRef]
19. Moos, D.; Peska, P.; Finkbeiner, T.; Zoback, M. Comprehensive Wellbore Stability Analysis Utilizing Quantitative Risk Assessment. *J. Pet. Sci. Eng.* **2003**, *38*, 97–109. [CrossRef]
20. Huang, T.; Cao, L.; Cai, J.; Xu, P. Experimental Investigation on Rock Structure and Chemical Properties of Hard Brittle Shale under Different Drilling Fluids. *J. Pet. Sci. Eng.* **2019**, *181*, 106185. [CrossRef]
21. Dokhani, V.; Yu, M.; Bloys, B. A Wellbore Stability Model for Shale Formations: Accounting for Strength Anisotropy and Fluid Induced Instability. *J. Nat. Gas Sci. Eng.* **2016**, *32*, 174–184. [CrossRef]
22. Ma, T.; Chen, P. Influence of Shale Bedding Plane on Wellbore Stability for Horizontal Wells. *J. Southwest Pet. Univ. (Sci. Technol. Ed.)* **2014**, *36*, 97–104. [CrossRef]
23. Yan, C.; Deng, J.; Yu, B. Wellbore Stability in Oil and Gas Drilling with Chemical-Mechanical Coupling. *Sci. World J.* **2013**, *2013*, e720271. [CrossRef] [PubMed]
24. Zhang, F.; Xie, S.Y.; Hu, D.W.; Shao, J.F.; Gatmiri, B. Effect of Water Content and Structural Anisotropy on Mechanical Property of Claystone. *Appl. Clay Sci.* **2012**, *69*, 79–86. [CrossRef]
25. Liu, J.; Yang, Z.; Sun, J.; Dai, Z.; You, Q. Experimental Investigation on Hydration Mechanism of Sichuan Shale (China). *J. Pet. Sci. Eng.* **2021**, *201*, 108421. [CrossRef]

*Article*

# Effect of Wire Design (Profile) on Sand Retention Parameters of Wire-Wrapped Screens for Conventional Production: Prepack Sand Retention Testing Results

Dmitry Tananykhin [1,*], Maxim Grigorev [1], Elena Simonova [2], Maxim Korolev [3], Ilya Stecyuk [4] and Linar Farrakhov [4]

1 Development and Exploitation of Oil and Gas Fields Department, Saint Petersburg Mining University, 199106 Saint Petersburg, Russia
2 LLC «Alfa Horizon», 190031 Saint Petersburg, Russia
3 Higher School of Oil Department, Yugra State Technical University, 628011 Khanty-Mansiysk, Russia
4 LLC «RN-Purneftegaz», 629830 Gubkinskiy, Russia
* Correspondence: tananykhin_ds@pers.spmi.ru

**Abstract:** There are many technologies to implement sand control in sand-prone wells, drilled in either weakly or nonconsolidated sandstones. Technologies that are used to prevent sanding can be divided into the following groups: screens (wire-wrapped screens, slotted liners, premium screens, and mesh screens), gravel packs, chemical consolidation, and technological ways (oriented perforation and bottomhole pressure limitation) of sanding prevention. Each particular technology in these groups has their own design and construction features. Today, slotted liners are the most well-studied technology in terms of design, however, this type of sand control screen is not always accessible, and some companies tend towards using wire-wrapped screens over slotted liners. This paper aims to study the design criteria of wire-wrapped screens and provides new data regarding the way in which wire design affects the sanding process. Wires with triangular (wedge), trapezoidal, and drop-shaped profiles were tested using prepack sand retention test methodology to measure the possible impact of wire profile on sand retention capabilities and other parameters of the sand control screen. It was concluded that a trapezoidal profile of wire has shown the best result both in terms of sand production (small amount of suspended particles in the effluent) and in particle size distribution in the effluent, that is, they are the smallest compared to other wire profiles. As for retained permeability, in the current series of experiments, high sand retention did not affect retained permeability, although it can be speculated that this is mostly due to the relatively high particle size distribution of the reservoir.

**Keywords:** sand production; sanding; sand control; wire-wrapped screens; prepack sand retention test

**Citation:** Tananykhin, D.; Grigorev, M.; Simonova, E.; Korolev, M.; Stecyuk, I.; Farrakhov, L. Effect of Wire Design (Profile) on Sand Retention Parameters of Wire-Wrapped Screens for Conventional Production: Prepack Sand Retention Testing Results. *Energies* **2023**, *16*, 2438. https://doi.org/10.3390/en16052438

Academic Editors: Xingguang Xu, Kun Xie and Yang Yang

Received: 22 November 2022
Revised: 21 February 2023
Accepted: 28 February 2023
Published: 3 March 2023

## 1. Introduction

Sand control technologies are widely used to prevent sand production. Thus, reducing costs and time loss due to well workovers in many areas worldwide. The application of sand control technologies is dictated by the necessity to protect both subsurface and surface equipment from erosion provided by sand grains. A hypothetical "perfect" screen would completely eliminate sand influx into the well (or reduce it to a negligible scale) and allow free passage to the fluid. In reality, screens are designed with the intent to let some sand grains pass through screen slots in order to prevent plugging of the slots, which will result in an overall decrease in production rates. Screens also cause flow redistributions (and velocity increase [1]) due to possible impairment of the slots [2], which results in fluid moving toward the slot opening in the near-screen area. This leads to additional pressure drop in the "bottomhole zone–wellbore" system.

In general, the performance assessment of any screen must be evaluated by the following parameters:

1.  Open to flow area (OFA);
2.  Sand production decrease due to screen implementation;
3.  Retained permeability of the screen (or "bottomhole zone–screen" system);
4.  Particle size distribution of the filtered particles.

There are also criteria to follow in terms of the mechanical integrity of the screen [3]:

1.  Screen installation (torque and other strains);
2.  Screen operation (resistance to erosion and corrosion).

Some authors also note that a screen should provide stability of the bottomhole formation zone [4] and this is essentially achieved by retaining sand grains with a size >50 microns, which hold most of the overburden loadout [5].

Fattahpour [6] argues that instead of using OFA as a parameter in screen performance assessment, aperture size should be used due to the possible misinterpretation of OFA. The screen can have a very high opening, but a limited number of slots (or slots per column (SPC)), and it will have the same OFA as a screen with narrow openings and a large amount of these openings. However, the difference in the amount of sand that each of these screens can "produce" is obvious.

Some researchers argue that an effective screen should not produce more than 0.12 lbs/ft$^2$ (580 grams/m$^2$) [2]; or 0.15 lbs/ft$^2$ (732 grams/m$^2$) [7] of sand. However, this raises a concern. Are these thresholds about equipment protection (by restricting solids production we eliminate erosion) or about preventing sand plugs from forming? Equipment used by different companies is different (in terms of resistivity to corrosion or overall strength) and that also depends on the manufacturer's capabilities and other features of surface and subsurface equipment [8,9]. As for sand plugs forming, this process depends on the rate of liquid production. Thus, extremely productive wells can also have high solids production. However, this is rarely a problem due to the high velocity of fluids in the production string, which prevents sand plugs from forming.

One of the parameters of the screen that affects 3/4 of the listed parameters is slot (wire) geometry. Bennion [10] provided an exquisite explanation of the slot geometry effect for slotted liners (SL), however, there is no study regarding the effect of wire profile for wire-wrapped screens (WWS).

In addition to the properties of the sand control technology implemented, the process of sanding is basically affected by the following properties and parameters of the fluid and reservoir:

1.  Fluid viscosity—the higher the viscosity, the higher is the amount of produced sand [11,12];
2.  Gas content [5,6,13–15];
3.  Particle size distribution of the reservoir and shape parameters of particles [16,17];
4.  Clay content [4,18];
5.  Stress distribution in the bottomhole formation zone [19–21];
6.  Pore pressure [22,23];
7.  Bottomhole pressure, production rate, and velocity of fluids in the reservoir [1,5,24–26];
8.  Flow regime [13,21,27];
9.  Reservoir depletion [15,28,29];
10.  Ramp-up rate [15,16,30];
11.  Water cut [31,32];
12.  Properties of the reservoir—the amount of reservoir cement, rock strength, etc. [33–36].

Essentially, there are two most widely used and adopted screens, namely wire-wrapped screens and slotted liners. Both have their pros and cons, whilst slotted liners tend to have a higher strength, low cost, and ease of production [37–39]. Wire-wrapped screens have a higher OFA (which means they are less prone to plugging and create less pressure drop) [6,35,40] and provide more options in terms of applicability.

On the other hand, OFA is a criterion that does not properly describe sand control screens in terms of overall efficiency due to the nature of OFA itself. It combines both sand control (through screen aperture size) and flow efficiency (through slots per column or opening width) parameters, which act against one another in most cases. In general, higher OFA means higher sand production [12,41], but lower additional pressure drop due to flow convergence [5,25] and lower plugging tendency [4,35].

Slotted liners are widely used in SAGD-wells in Canada, where flowrates in the producer ("bottom well") are not that high. Thus, a high open to flow area does not necessarily means that this will affect amount of produced sand. Reservoirs comprised from weakly consolidated sandstones, tend to have very high permeabilities [42–44]. Thus, wells can have flowrates as high as 500 tons/day (approx. 3500 bbl/day). In this case, OFA is a crucial parameter, the importance of which cannot be underestimated. The only thing that is left to change in terms of sand retention is the geometry of the screen itself.

Mahmoudi [3] has shown that under the effect of high flowrates, perforation tunnels in a base pipe deform from a round to an oval shape in corrosive-active environments. Thus, the base pipe can be considered a weak component. This is mostly the case for slotted liners since the screen itself is a base pipe with openings of a particular form. Furthermore, erosion can also cause a change in metal profile creating spots of increased aperture size, that allow high sand production.

Bennion [10] has shown that the geometry of the slotted liner slots affects the efficiency of the slotted liner in terms of sand control (solids production) and pressure drop. Rolled/seamed top slots (Figure 1) tend to have the lowest additional pressure drop and the lowest sand production. The mechanism of that is not actually explained, although the author argues that plugging happens on top of the slot, rather than in the slot itself in most cases.

**Figure 1.** Slotted liner slot configuration [by authors].

Different slot configurations and patterns were studied by Xie [37]. It is noted that for strength-related reasons (withstanding strain deformations and torque) OFA of slotted liners is designed to be much lower than it should be for effective production [45]. Although calculations for OFA should be different for different types of slots, for straight slots it does not matter which diameter you choose, however, for keystone or rolled/seamed slots, it is better to use the outer diameter.

There is a great difference in the basis of slotted liners and wire-wrapped screens. While a slotted liner must withstand all the stresses and loads and still maintain high enough production capabilities, this role in WWS is played by the base pipe, thus, creating a lot of space for wire designing. For example, wires can be made out of stainless steel or any other alloy that can ensure long-term operation under corrosive-active conditions [39]. This also helps to prevent plugging either by corrosion byproducts or by fine particles, that attach to these byproducts [46,47].

The main concern in the field of slot geometry for slotted liners is whether the enlargement of the opening towards the inner part of the screen results in a better performance in terms of sand retaining and the plugging tendency of the slot [48]. This paper is aimed at studying the same effect, but for wire-wrapped screens (Figure 2).

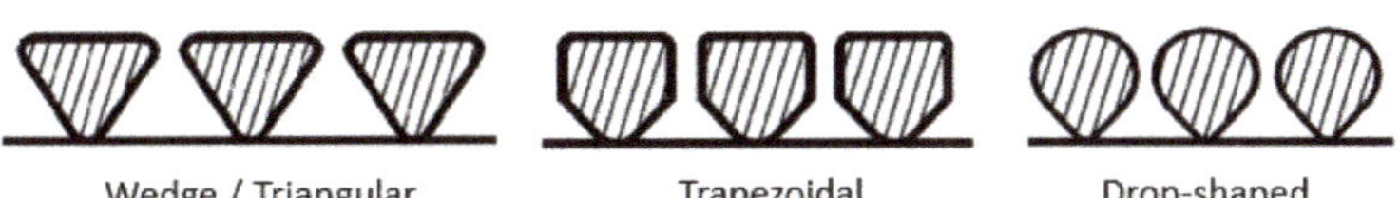


**Figure 2.** Different wire profiles [by authors].

## 2. Materials and Methods

Conventional production does not require accounting for such difficulties in sand control testing as high temperatures, thermal expansion, and some other quirks of SAGD production that are widely studied by different authors [49].

Laboratory testing for this publication was made in the form of a Prepack sand retention test (Prepack SRT). It features conditions when a large portion of the bottomhole formation zone collapses into the wellbore, which is highly likely due to stress distributions and the impossibility of the formation to withstand stresses exerted by reservoir drawdown in weakly consolidated or nonconsolidated reservoirs [50].

A series of tests were run to measure the efficiency of different wire profiles for implementing them in wire-wrapped screens. It is important to note that there is not a single work examining this effect. Images of different wire-wrapped screens with different wire profiles can be seen in Figure 3.

**Figure 3.** Wire-wrapped screens with different wire profiles [by authors] (wires are welded to elements of longitudinal supporting ribs).

These are screens with an aperture size of 150 microns. The same screens, though with an aperture of 200 microns, were also tested.

The methodology of sand preparation, experiment procedure, and other features of Prepack SRT have been covered in many papers. The authors have used a methodology that was previously implemented [42,50]. It is important to note that in this series of tests, sand, which was extracted from the bottomhole formation zone, was used to prepare the core rock models. Core rock models were also saturated with reservoir water before an experiment. The particle size distribution (PSD) of the reservoir rock can be seen in Figure 4.

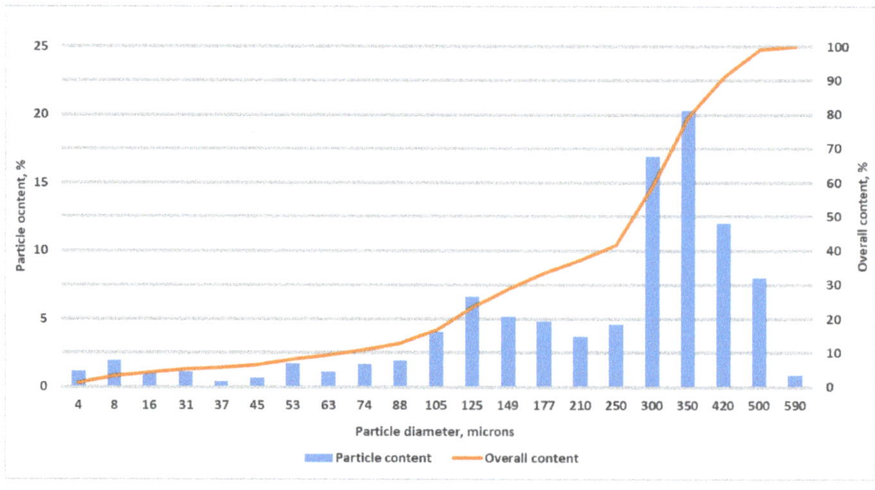


**Figure 4.** Formation particle size distribution.

Note that reservoir PSD is governed by sand particles with high particle diameters. Applying the criterion proposed by Fermaniuk, Coberly, Suman, Markestad, and Gillespie for wire-wrapped screens gives a summary of screen aperture sizes for particular PSD mentioned in this paper (Table 1):

**Table 1.** WWS aperture sizes by different authors.

| Author | Criterion | Screen Aperture Size, Microns |
| --- | --- | --- |
| Fermaniuk | 2·D30 < aperture < 3.5·D50 | 300 < aperture < 900 |
| Coberly | 2·D10 | 140 |
| Suman | D10 | 70 |
| Markestad | D10 < aperture < 2·D10 | 70 < aperture < 140 |
| Gillespie | Aperture < 2·D50 | Aperture < 500 |

Screen aperture sizes of 150 and 200 microns follow Coberly's and Gillespie's criteria.

Researchers note that recalculation of overburden pressure (rock pressure) from the reservoir to laboratory conditions is a challenge that should be carefully examined. Anderson [51] was applying the same rock pressure as in the reservoir. In this series, the authors decided to apply rock pressure that will make it possible for the reservoir rock models to have the same permeability as in reservoir conditions. Thus, during preliminary tests, rock pressure in the coreholder was set to 450 psi (3 MPa approx.), which is approximately 3.5 times smaller than the actual overburden (rock) pressure acting in the reservoir.

In this study, reservoir fluids were simulated with suitable fluids. Oil was modeled by mixing polymethylsiloxane oils with viscosities of 5 and 50 mPa·s to reach a viscosity of 9 mPa·s, which corresponds to oil viscosity in reservoir conditions. Reservoir water was prepared using distilled water and corresponding salts to match the content of different salts in reservoir conditions.

In current field conditions, lots of wells operate with water cuts equal to 70% or higher. Thus, models of oil and reservoir water were injected into the model at a ratio of 30% "oil" and 70% "water".

A schematic representation of the installation, that was used to conduct experiments, is shown in Figure 5.

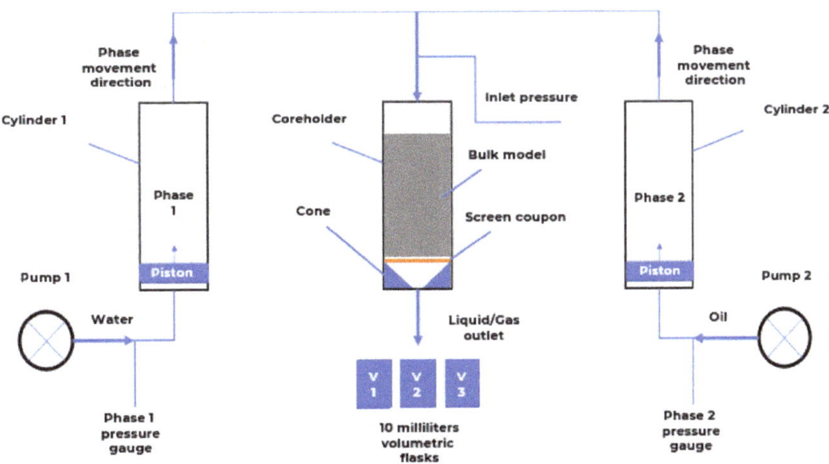


**Figure 5.** Scheme of the installation for Prepack SRT implementation [by authors].

Associated gas content is relatively small with values from 10–15 cubic meters per ton of oil. Thus, it was decided to not account for it and use only 2 pumps (for "oil" and for "water"). The installation used in these experiments allows the use of core rock models, which were saturated before an experiment. However, the size (volume) of the core rock model is relatively small compared to models of the reservoir in the linear or radial sand control evaluation test, however, the small scale allows for the perfect simulation of reservoir properties on a constant basis.

## 3. Results

During the experiments, three key parameters of the sand retention media were measured: suspended particles content (SPC); particle size distribution (PSD) of the particles, that passed through the screen; and retained permeability of the screen.

Results of the data analysis regarding SPC and PSD of particles in the effluent can be seen in Figures 6 and 7.

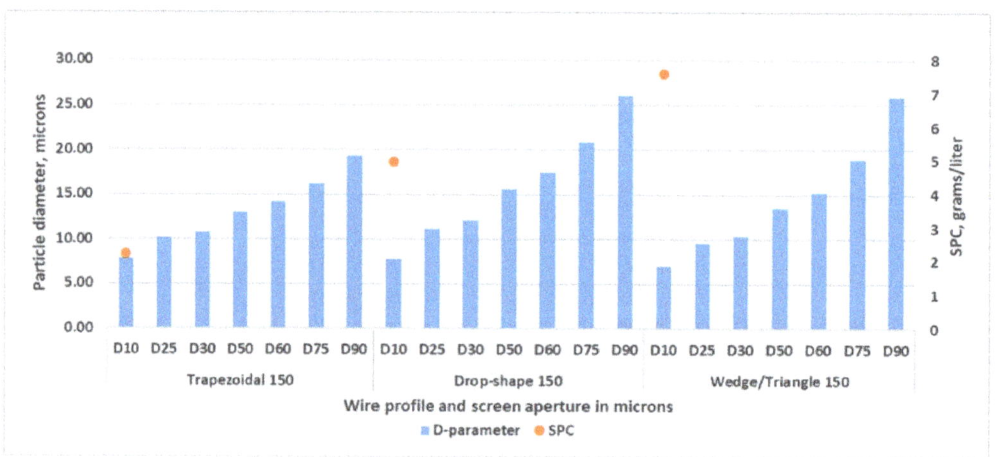

**Figure 6.** Experimental results for screens with an aperture size of 150 microns.

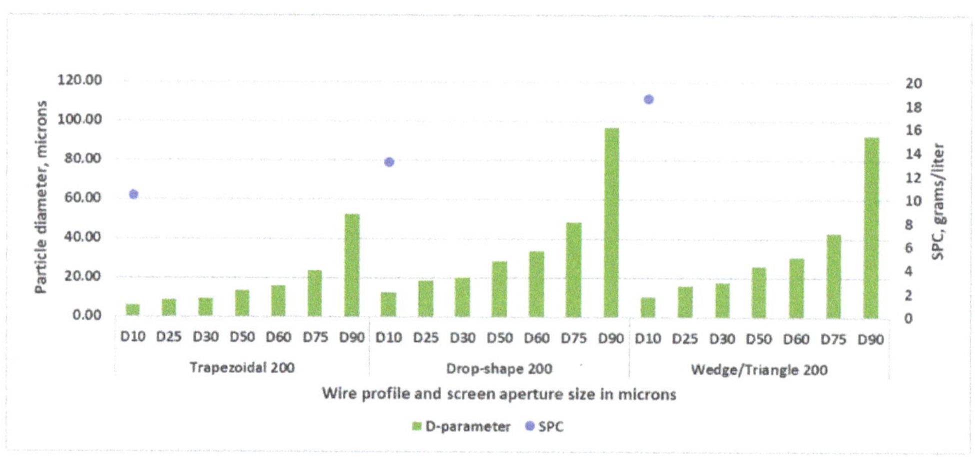

**Figure 7.** Experimental results for screens with an aperture size of 200 microns.

It is worth noting that even though the screen's aperture increased by 50 microns (which is 1/3 of the original 150 microns screen value), the resulting diameter of washed out particles increased by 2–3 times for drop-shape and wedge/triangle wire profiles. This is mostly the case for a larger part of the D-parameter (D75–D90) of particles, though.

A trapezoidal wire profile has shown outstanding performance both in terms of suspended particle content (SPC) and in terms of the washed-out particles' diameter. This might be crucial for wells that work in a periodic regime (for example, 1 h of work and 1 h of accumulation) since such a regime does not provide a sand screen with suitable conditions to form stable sand arches (or bridges), which reduce sand production naturally. For wells that work in a stable regime, sand production stabilizes over time with a certain suspended particle content [2,32,52].

Note that an increase in screen aperture to 50 microns resulted in an almost five times increase in SPC for the trapezoidal screen, almost three times for the drop-shape screen, and in 2.5 times for the screen with a wedge-shaped profile (which is a "conventional" wire-wrapped screen).

This had to affect retained permeability, since larger particles must have formed a bridge on the surface of the screen, thus, decreasing its permeability. The retained permeability of the screens is shown in Figures 8 and 9.

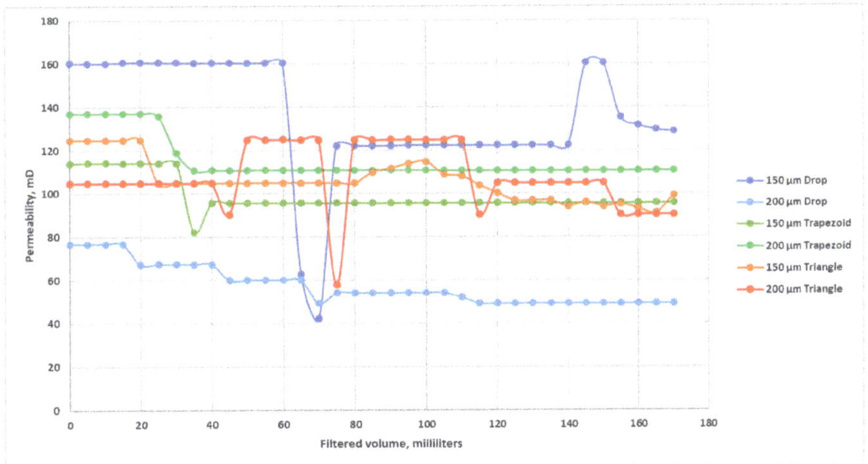

**Figure 8.** Retained permeability of screens in millidarcies.

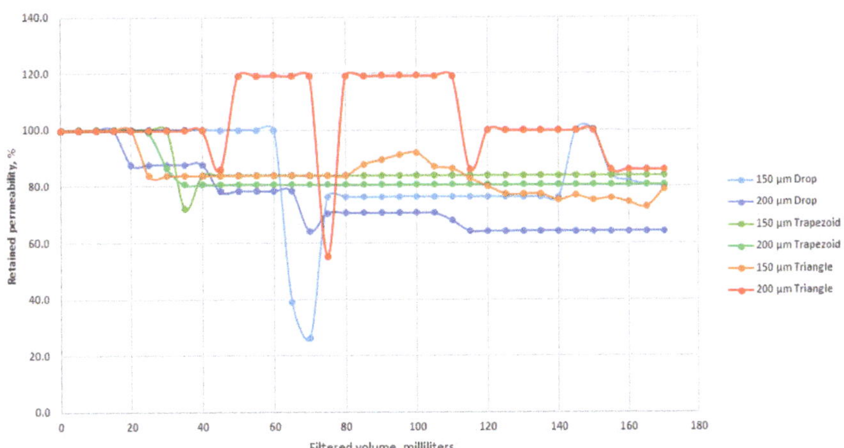

**Figure 9.** Retained permeability of screens in percent.

Figure 8 represents the pressure difference during the tests for each screen. Since core rock models have varying permeability, it is better to compare screens with different wire profiles in relative changes in permeability, as it is presented in Figure 9. The first point on the graph (for zero filtered volume) represents a situation where pressure reached the required value (for each core model it was different, ranging from ~345,000 pascals (50 psi) to ~518,000 pascals (75 psi)). Thus, zero filtered volume is not actually zero.

Although large fluctuations in permeability can be described via the working principle of the pressure transducer (due to a mixture of water and oil, the transducer sensor might translate the wrong data), in general, screens with a lower aperture size (150 microns (μm)) show the lowest retained permeability. In the 150 μm group, the trapezoidal screen shows the highest permeability. In the 200 μm group, it shows the worst permeability. The drawdown pressure (and permeability, as follows) can be also explained via fines migration. A large portion of particles have a high diameter, thus, pore channels also have a relatively large diameter, which allows particles of a smaller size to completely block them, which would explain the decrease in permeability.

## 4. Discussion

There is no referent article to study the effect of the profile of wire in a wire-wrapped screen and how it can possibly affect the results of a sand retention test. It is possible to find an explanation for the phenomenon in other topics. Fattahpour [38] argues that the difference in results between linear and radial sand control evaluation (LSCE/RSCE) tests may be due to the fact that bridging (or the forming of sand arches) occurs more easily on curved surfaces than on the flat surface. Trapezoidal and triangular wires have the same form at the entrance point—they are both flat. This means the prolongation of the wires is what makes the difference in the tests.

The prolongation of the wires allows the particle to have more chances to be "stuck" in the screen media if a particle of sand goes through the bridge (or there is no bridge formed yet on a particular point on the screen's surface). Figures 10–12 represent different scenarios for the wire-wrapped screen operation. The contact line (screen with particles) is the smallest for the wedge/triangular screen; it is essentially a point. Once a particle passes through that point, it can no longer be captured by the screen.

**Figure 10.** Wedge/Triangular wire profile and its interaction with particles.

**Figure 11.** Trapezoidal wire profile and its interaction with particles.

**Figure 12.** Drop-shaped wire profile and its interaction with particles.

The contact line of the trapezoidal profile is the largest. Thus, particles can be stuck easily two or more times during their path through the screen. The drop-shaped profile has a mediocre size of the contact line. Thus, it should be positioned between the previous two wire profiles in terms of produced sand and retained permeability [53].

The difference for produced sand can be observed in Figures 6 and 7, and one can say that it is fair, however, the retained permeability is not that simple due to the complex nature of the sanding process and the inability of scientists to duplicate the results on a constant basis due to sand mixture during core rock model preparations.

Another item of concern is a particle's "roundness", which can be measured via the aspect ratio and sphericity ratio of the particle [17]. Roundness should have a higher effect on screens with higher outer diameters of the screen opening, compared to inner diameters. This mostly applies to a drop-shaped profile of the wire. This is mostly due to the high open to flow area of the drop-shaped screen on the outer area. Thus, if it will be properly blocked by a particle of an irregular shape, it will have higher permeability compared to other wire profiles (and screens made of them) in general.

The height of the opening also increases the overall strength of the wire-wrapped screen (due to the higher content of metal, compared to other screens), which makes it more reliable in terms of installation failures and helps to ensure the screen's efficiency, saving even in emergency situations.

In this study, core rock models were made of the same rock. Thus, it was impossible to study the effect of particle shape parameters on the screen efficiency of different wire profiles. Due to the differences in inner and outer diameters of the wire-wrapped screens of all profiles, the results would differ between them. From a logical point of view, a drop-shaped wire profile of wire-wrapped screens would be the best choice for particles of irregular shape since it has the same actual screen aperture (in the "middle-part"), though it has a higher OFA due to increase in aperture towards the outer part of the screen.

A large topic of discussion is the erosion of wire-wrapped screens. Since erosion depends on many variables, such as the velocity of the fluid [54–56], fines content [57], and degree of plugging (through localized velocity increase) [58]. Among them stands the impact angle. According to [59], impact angle is defined as the angle between the targeted surface and the direction of particle impact velocity. In a study reported by [60], it was shown that the maximum erosion rate is reached when the impact angle is close to the surface normal (i.e., 90°). There are also publications [55] that suggest that the highest erosion rate is reached when the impact angle is close to 45°. The difference can be explained by different methodologies in the experiments since an angle of 90° is reported for liquid flow, while the 45° angle was achieved via sand–air mixture blasting. Zhang [61] widely discussed the effect of turbulence on the erosion of different sand control screens. The authors concluded that turbulent flow attributes to the constant change in impact angle without any connection to the angle between the perforation tunnel or pore-space exit opposite to the screen surface. They also noted that different impact angle causes different types of wear (erosion). The low impact angle causes microcuttings of the screen surface, whereas high angles and speeds cause the particle to apply cracks or extrusive lips on the screen surface. Given that the drop-shaped wire profile has a curved surface, it should be less prone to erosion from liquid flows and more prone to erosion from gas-induced erosion (high speeds and turbulent flows). A trapezoidal wire profile is still prone to liquid-flow induced erosion, just as the triangular (basic) wire-wrapped screen is. Extra plugging caused by the drop-shaped wire-wrapped screen (hence the high probability of localized erosion) can be balanced out with the screen curvature.

Another thing of concern is the form (or shape) of the PSD curve. In this study, the PSD curve was steadily growing (in the overall content of particles) towards the higher size of particles. However, there are also PSD curves that have a bell-shaped profile or even a reverse bell shape, which would also affect the results of the screen tests.

As far as wire production is concerned, it is a rather flexible process in which the shape of the wire can be changed depending on the results of laboratory experiments and any other necessary conditions. The wire is produced by cold rolling, and the only component of the process that needs to be changed is the shape of the die. The economics of wire production depends on the supplier's price for the cold rolling of a wire.

## 5. Conclusions

The development of sand control screens has led to a new field of research in their design and geometry. The main aspect of these screens is slot geometry (opening profile). Slotted liners are well studied in that matter, however, they are not always applicable. This paper studies the effect of wire profiles on sand control capabilities and other parameters of sand screens during prepack sand retention testing. A trapezoidal profile shows outperforming results compared to other screens in both screen groups. Compared to other screens, it has the lowest amount of suspended particles in the effluent, and these particles are the smallest in size.

A screen with a trapezoidal wire profile with an aperture size of 150 microns had 2.5 times less suspended particles content (SPC) than a screen with a drop-shape wire profile and almost four times less SPC than a "conventional" wedge-shaped wire wrapped screen. In comparing the results of screens with different wire profiles with 200 microns aperture sizes, the trapezoidal wire profile also showed the best results, even though in this group, the difference in results was not that notable—1.4 times less than the drop-shaped screen and 1.8 times less than a "conventional" wedge-shaped wire wrapped screen.

Retained permeability of the "core rock model–screen" system for a screen with a trapezoidal wire profile (for both apertures) was approximately 80% of the initial, and it is just a bit smaller than the best results shown by the "conventional" wire wrapped screen with a triangular (wedge) wire profile, which was approximately 85%.

**Author Contributions:** Conceptualization, E.S. and D.T.; methodology, M.G. and M.K.; validation, I.S. and L.F.; investigation, M.G.; resources, D.T. and E.S.; data curation, M.G.; writing—original draft preparation, M.G.; writing—review and editing, D.T., E.S., M.K., I.S. and L.F.; visualization, M.G.; supervision, D.T.; project administration, E.S. and M.G.; All authors have read and agreed to the published version of the manuscript.

**Funding:** This research received no external funding.

**Data Availability Statement:** Not applicable.

**Acknowledgments:** Authors would like to acknowledge LLC «Alfa Horizon» for providing wire-wrapped screen coupons.

**Conflicts of Interest:** The authors declare no conflict of interest.

# References

1. Cameron, J.; Zaki, K.; Jones, C.; Lazo, A. Enhanced Flux Management for Sand Control Completions. In Proceedings of the SPE Annual Technical Conference and Exhibition, Dallas, TX, USA, 24 September 2018; p. 16.
2. Chanpura, R.A.; Hodge, R.M.; Andrews, J.S.; Toffanin, E.P.; Moen, T.; Parlar, M. A Review of Screen Selection for Standalone Applications and a New Methodology. *SPE Drill. Complet.* **2011**, *26*, 84–95. [CrossRef]
3. Mahmoudi, M.; Roostaei, M.; Fattahpour, V.; Sutton, C.; Fermaniuk, B.; Zhu, D.; Jung, H.; Li, J.; Sun, C.; Gong, L.; et al. SPE-191553-MS Standalone Sand Control Failure: Review of Slotted Liner, Wire Wrap Screen, and Premium Mesh Screen Failure Mechanism. In Proceedings of the SPE Annual Technical Conference and Exhibition, SPE Annual Technical Conference and Exhibition, Dallas, TX, USA, 24 September 2018; pp. 1–26.

4.   Guo, Y.; Nouri, A.; Nejadi, S. Effect of Slot Width and Density on Slotted Liner Performance in SAGD Operations. *Energies* **2020**, *13*, 268. [CrossRef]

5.   Romanova, U.G.; Gillespie, G.; Sladic, J.; Weatherford, T.M.; Solvoll, T.A.; Andrews, J.S. A Comparative Study of Wire Wrapped Screens vs. Slotted Liners for Steam Assisted Gravity Drainage Operations. In Proceedings of the World Heavy Oil Congress, New Orleans, LA, USA, 5 March 2014; p. 24.

6.   Fattahpour, V.; Mahmoudi, M.; Wang, C.; Kotb, O.; Roostaei, M.; Nouri, A.; Fermaniuk, B.; Sauve, A.; Sutton, C. Comparative Study on the Performance of Different Stand-Alone Sand Control Screens in Thermal Wells. In Proceedings of the SPE International Conference and Exhibition on Formation Damage Control, Lafayette, LA, USA, 7–9 February 2018; Volume 2.

7.   Adams, P.R.; Davis, E.R.; Hodge, R.M.; Burton, R.C.; Ledlow, L.; Procyk, A.D.; Crissman, S.C. Current State of the Premium Screen Industry: Buyer Beware, Methodical Testing and Qualification Shows You Don't Always Get What You Paid For. *SPE Drill. Complet.* **2009**, *24*, 362–372. [CrossRef]

8.   Dvoynikov, M.; Sidorov, D.; Kambulov, E.; Rose, F.; Ahiyarov, R. Salt Deposits and Brine Blowout: Development of a Cross-Linking Composition for Blocking Formations and Methodology for Its Testing. *Energies* **2022**, *15*, 7415. [CrossRef]

9.   Rogachev, M.; Aleksandrov, A. Justification of a Comprehensive Technology for Preventing the Formation of Asphalt-Resin-Paraffin Deposits during the Production of Highlyparaffinic Oil by Electric Submersible Pumps from Multiformation Deposits. *J. Min. Inst.* **2021**, *250*, 596–605. [CrossRef]

10.   Bennion, D.B. Protocols for Slotted Liner Design for Optimum SAGD Operation. *J. Can. Pet.* **2009**, *48*, 21–46. [CrossRef]

11.   Dong, C.; Zhang, Q.; Gao, K.; Yang, K.; Feng, X.; Zhou, C. Screen Sand Retaining Precision Optimization Experiment and a New Empirical Design Model. *Pet. Explor. Dev.* **2016**, *43*, 1082–1088. [CrossRef]

12.   Devere-Bennett, N. Using Prepack Sand-Retention Tests (SRT's) to Narrow Down Liner/Screen Sizing in SAGD Wells. In Proceedings of the Thermal Well Integrity and Design Symposium, SPE-178443-MS, Baniff, AB, Canada, 24 November 2015.

13.   Ahad, N.A.; Jami, M.; Tyson, S. A Review of Experimental Studies on Sand Screen Selection for Unconsolidated Sandstone Reservoirs. *J. Pet. Explor. Prod. Technol.* **2020**, *10*, 1675–1688. [CrossRef]

14.   Andrews, J.S.; Jøranson, H.; Raaen, A.M. OTC 19130 Oriented Perforating as a Sand Prevention Measure-Case Studies from a Decade of Field Experience Validating the Method Offshore Norway. In Proceedings of the Offshore Technology Conference, Houston, TX, USA, 5 May 2008; p. 12.

15.   Isehunwa, S.; Farotade, A. Sand Failure Mechanism and Sanding Parameters in Niger Delta Oil Reservoirs. *J. Eng. Sci. Technol.* **2010**, *2*, 777–782.

16.   Eshiet, K.I.-I.; Sheng, Y. Investigating Sand Production Phenomena: An Appraisal of Past and Emerging Laboratory Experiments and Analytical Models. *Geotechnics* **2021**, *1*, 492–533. [CrossRef]

17.   Cespedes, A.E.M.; Roostaei, M.; Uzcátegui, A.A.; Gomez, D.M.; Mora, E.; Alpire, J.; Torres, J.; Fattahpour, V. SPE-199062-MS Sand Control Optimization for Rubiales Field: Trade-Off Between Sand, Flow Performance and Mechanical Integrity. In Proceedings of the SPE Latin American and Caribbean Petroleum Engineering Conference, Virtual, 27 July 2020; p. 31.

18.   Gillespie, G.; Deem, K.; Malbrel, C. SPE 64398 Screen Selection for Sand Control Based on Laboratory Tests. In Proceedings of the SPE Asia Pacific Oil and Gas Conference and Exhibition, Brisbane, Australia, 16–18 October 2000.

19.   Salahi, A.; Dehghan, A.N.; Sheikhzakariaee, S.J.; Davarpanah, A. Sand Production Control Mechanisms during Oil Well Production and Construction. *Pet. Res.* **2021**, *6*, 361–367. [CrossRef]

20.   Salehi, M.B.; Moghadam, A.M.; Marandi, S.Z. Polyacrylamide Hydrogel Application in Sand Control with Compressive Strength Testing. *Pet. Sci.* **2019**, *16*, 94–104. [CrossRef]

21.   Ikporo, B.; Sylvester, O. Effect of Sand Invasion on Oil Well Production: A Case Study of Garon Field in the Niger Delta. *Int. J. Eng. Sci.* **2015**, *4*, 64–72.

22.   Kuncoro, B.; Ulumuddin, B.; Palar, S. *Sand Control for Unconsolidated Reservoirs*; IATMI: Jakarta, Indonesia, 3 October 2001; p. 7.

23.   Bouteca, M.J.; Sarda, J.-P.; Vincke, O. Constitutive Law for Permeability Evolution of Sandstones During Depletion. In Proceedings of theSPE International Symposium on Formation Damage Control, Lafayette, LA, USA, 23 February 2000; p. 8.

24.   Al-Awad, M.N.J.; El-Sayed, A.-A.H.; El-Din, S.; Desouky, M.; Soliman, M.A.; Ibrahim, A.A. Factors Affecting Sand Production from Unconsolidated Sandstone Saudi Oil and Gas Reservoir. *J. King Saud Univ.* **1998**, *11*, 163–181. [CrossRef]

25.   Montero, J.D.; Chissonde, S.; Kotb, O.; Wang, C.; Roostaei, M.; Nouri, A. SPE-189773-MS A Critical Review of Sand Control Evaluation Testing for SAGD Applications. In Proceedings of the SPE Canada Heavy Oil Technical Conference, Calgary, AB, Canada, 13 March 2018; p. 21.

26.   Nouri, A.; Belhaj, H.; Rafiqul Islam, M. Comprehensive Transient Modeling of Sand Production in Horizontal Wellbores. *SPE J.* **2007**, *12*, 468–474. [CrossRef]

27.   Sultanbekov, R.; Islamov, S.; Mardashov, D.; Beloglazov, I.; Hemmingsen, T. Research of the Influence of Marine Residual Fuel Composition on Sedimentation Due to Incompatibility. *J. Mar. Sci. Eng.* **2021**, *9*, 1067. [CrossRef]

28.   Garolera, D.; Carol, I.; Papanastasiou, P. Micromechanical Analysis of Sand Production. *Int. J. Numer. Methods Eng.* **2018**, *43*, 1207–1229. [CrossRef]

29.   McPhee, C.; Farrow, C.; McCurdy, P. Challenging Convention in Sand Control: Southern North Sea Examples. *SPE Prod. Oper.* **2007**, *22*, 223–230. [CrossRef]

30.   Mardashov, D. Development of Blocking Compositions with a Bridging Agent for Oil Well Killing in Conditions of Abnormally Low Formation Pressure and Carbonate Reservoir Rocks. *J. Min. Inst.* **2021**, *251*, 617–626. [CrossRef]

31. Luo, W.; Xu, S.; Torabi, F. Laboratory Study of Sand Production in Unconsolidated Reservoir. In Proceedings of the SPE Annual Technical Conference and Exhibition, San-Antonio, TX, USA, 8 October 2012; p. 14.

32. Bianco, L.C.B.; Halleck, P.M. Mechanisms of Arch Instability and Sand Production in Two-Phase Saturated Poorly Sandstones. In Proceedings of the SPE European Formation Damage Conference, SPE 68932, The Hague, The Netherlands, 21 May 2001; p. 10.

33. Fattahpour, V.; Moosavi, M.; Mehranpour, M. An Experimental Investigation on the Effect of Grain Size on Oil-Well Sand Production. *Pet. Sci.* **2012**, *9*, 343–353. [CrossRef]

34. Hisham, B.M.; Leong, V.H.; Lestariono, Y. Sand Production: A Smart Control Framework for Risk Mitigation. *Petroleum* **2020**, *6*, 1–13.

35. Khamehchi, E.; Ameri, O.; Alizadeh, A. Choosing an Optimum Sand Control Method. *Egypt. J. Pet.* **2015**, *24*, 193–202. [CrossRef]

36. Dvoynikov, M.; Buslaev, G.; Kunshin, A.; Sidorov, D.; Kraslawski, A.; Budovskaya, M. New Concepts of Hydrogen Production and Storage in Arctic Region. *Resources* **2021**, *10*, 3. [CrossRef]

37. Xie, J. Slotted Liner Design Optimization for Sand Control in SAGD Wells. In Proceedings of the SPE Thermal Well Integrity and Design Symposium, Banff, AB, Canada, 23–25 November 2015; pp. 23–25. [CrossRef]

38. Fattahpour, V.; Roostaei, M.; Hosseini, S.A.; Soroush, M.; Berner, K.; Mahmoudi, M.; Al-Hadhrami, A.; Ghalambor, A. Experiments with Stand-Alone Sand-Screen Specimens for Thermal Projects. *SPE Drill. Complet.* **2021**, *36*, 188–207. [CrossRef]

39. Van Vliet, J.; Hughes, M.J.; Limited, V.C. Comparison of Direct-Wrap and Slip-On Wire Wrap Screens for Thermal Introduction to Wire Wrap Screen. In Proceedings of the SPE Thermal Well Integrity and Design Symposium, Banff, AB, Canada, 23–25 November 2015; pp. 1–14. [CrossRef]

40. Matanovic, D.; Cikes, M.; Moslavac, B. *Sand Control in Well Construction and Operation*; Springer Environmental Science and Engineering; Springer: Berlin/Heidelberg, Germany, 2012.

41. Mahmoudi, M.; Roostaei, M. SPE-179036-MS Sand Screen Design and Optimization for Horizontal Wells Using Reservoir Grain Size Distribution Mapping. In Proceedings of the SPE International Conference and Exhibition on Formation Damage Control, Lafayette, LA, USA, 24–26 February 2016.

42. Tananykhin, D.; Grigorev, M.; Korolev, M.; Solovyev, T.; Mikhailov, N.; Nesterov, M. Experimental Evaluation of the Multiphase Flow Effect on Sand Production Process: Prepack Sand Retention Testing Results. *Energies* **2022**, *15*, 4657. [CrossRef]

43. Palyanitsina, A.; Safiullina, E.; Byazrov, R.; Podoprigora, D.; Alekseenko, A. Environmentally Safe Technology to Increase Efficiency of High-Viscosity Oil Production for the Objects with Advanced Water Cut. *Energies* **2022**, *15*, 753. [CrossRef]

44. Petrakov, D.G.; Penkov, G.M.; Zolotukhin, A.B. Experimental Study on the Effect of Rock Pressure on Sandstone Permeability. *J. Min. Inst.* **2022**, *254*, 244–251. [CrossRef]

45. Xie, J.; Fan, C.; Wagg, B.; Zahacy, T.A.; Matthews, C.M.; Fang, Y. Wire Wrapped Screens in SAGD Wells. *World Oil* **2008**, *229*, 115–120.

46. Romanova, U.; Ma, T. SPE 165111 An Investigation of the Plugging Mechanisms in a Slotted Liner from the Steam Assisted Gravity Operations. In Proceedings of the SPE European Formation Damage Conference & Exhibition, Noordwijk, The Netherlands, 5 June 2013; p. 8.

47. David, J.; Pallares, M.; Wang, C.; Haftani, M.; Pang, Y. SPE-193375-MS Experimental Assessment of Wire-Wrapped Screens Performance in SAGD Production Wells. In Proceedings of the SPE Thermal Well Integrity and Design Symposium, Banff, AB, Canada, 27–29 November 2018. [CrossRef]

48. Korotenko, V.A.; Grachev, S.I.; Kushakova, N.P.; Mulyavin, S.F. Assessment of the Influence of Water Saturation and Capillary Pressure Gradients on Size Formation of Two-Phase Filtration Zone in Compressed Low-Permeable Reservoir. *J. Min. Inst.* **2020**, *245*, 569–581. [CrossRef]

49. Fattahpour, V.; Azadbakht, S.; Mahmoudi, M.; Guo, Y.; Nouri, A.; Leitch, M. Effect of near Wellbore Effective Stress on the Performance of Slotted Liner Completions in SAGD Operations. In Proceedings of the SPE Thermal Well Integrity and Design Symposium, Banff, AB, Canada, 28 November–1 December 2016; pp. 142–153. [CrossRef]

50. Tananykhin, D.; Korolev, M.; Stecyuk, I.; Grigorev, M. An Investigation into Current Sand Control Methodologies Taking into Account Geomechanical, Field and Laboratory Data Analysis. *Resources* **2021**, *10*, 125. [CrossRef]

51. Anderson, M. SPE-185967-MS SAGD Sand Control: Large Scale Testing Results. In Proceedings of the SPE Canada Heavy Oil Technical Conference, Calgary, AB, Canada, 15 February 2017; pp. 15–16.

52. Chernykh, V.I.; Martyushev, D.A.; Ponomareva, I.N. Study of the Formation of a Well Borehole Zone When Opening Carbonate Reservoirs Taking Into Account Their Mineral Composition. *Bull. Tomsk Polytech. Univ. Geo Assets Eng.* **2022**, *333*, 129–139. [CrossRef]

53. Galkin, V.I.; Martyushev, D.A.; Ponomareva, I.N.; Chernykh, I.A. Developing Features of the Near-Bottomhole Zones in Productive Formations at Fields with High Gas Saturation of Formation Oil. *J. Min. Inst.* **2021**, *249*, 386–392. [CrossRef]

54. Greene, R.; Moen, T. The Impact of Annular Flows on High Rate Completions. In Proceedings of the SPE International Symposium and Exhibition on Formation Damage Control, Lafayette, LA, USA, 15–17 February 2012; Volume 2, pp. 886–897. [CrossRef]

55. Kanesan, D.; Mohyaldinn, M.E.; Ismail, N.I.; Chandran, D.; Liang, C.J. An Experimental Study on the Erosion of Stainless Steel Wire Mesh Sand Screen Using Sand Blasting Technique. *J. Nat. Gas Sci. Eng.* **2019**, *65*, 267–274. [CrossRef]

56. Cameron, J.; Jones, C. Development, Verification and Application of a Screen Erosion Model. In Proceedings of the European Formation Damage Conference, Scheveningen, The Netherlands, 30 May–1 June 2007. [CrossRef]

57. Abduljabbar, A.; Mohyaldinn, M.; Younis, O.; Alghurabi, A. A Numerical CFD Investigation of Sand Screen Erosion in Gas Wells: Effect of Fine Content and Particle Size Distribution. *J. Nat. Gas Sci. Eng.* **2021**, *95*, 104228. [CrossRef]
58. Hamid, S.; Ali, S.A. Causes of Sand Control Screen Failures and Their Remedies. In Proceedings of the SPE European Formation Damage Conference, The Hague, Netherlands, 2–3 June 1997; pp. 445–458. [CrossRef]
59. Abduljabbar, A.; Mohyaldinn, M.E.; Younis, O.; Alghurabi, A.; Alakbari, F.S. Erosion of Sand Screens by Solid Particles: A Review of Experimental Investigations. *J. Pet. Explor. Prod. Technol.* **2022**, *12*, 2329–2345. [CrossRef]
60. Liu, Y.; Zhang, J.; Ma, J.; Li, X. Erosion Wear Behavior of Slotted Screen Liner for Sand Control. *Tribology* **2009**, *29*, 283–287.
61. Zhang, R.; Hao, S.; Zhang, C.; Meng, W.; Zhang, G.; Liu, Z.; Gao, D.; Tang, Y. Analysis and Simulation of Erosion of Sand Control Screens in Deep Water Gas Well and Its Practical Application. *J. Pet. Sci. Eng.* **2020**, *189*, 106997. [CrossRef]

*Article*

# Multi-Well Pressure Interference and Gas Channeling Control in W Shale Gas Reservoir Based on Numerical Simulation

Jianliang Xu [1], Yingjie Xu [2,3], Yong Wang [1] and Yong Tang [2,*]

1   CNPC Chuanqing Drilling Engineering Co., Ltd., Chengdu 610051, China
2   State Key Laboratory of Oil and Gas Reservoir Geology and Exploitation, Southwest Petroleum University, Chengdu 610500, China
3   Shale Gas Research Institute, PetroChina Southwest Oil & Gas Field Company, Chengdu 610051, China
*   Correspondence: tangyong2004@126.com

**Abstract:** Well interference has drawn great attention in the development of shale gas reservoirs. In the W shale gas reservoir, well interference increased from 27% to 63% between 2016 and 2019, but the gas production recovery of parent wells was only about 40% between 2018 and 2019. Therefore, the mechanism and influencing factor of well interference degree were analyzed in this study. A numerical model of the W shale gas reservoir was developed for history matching, and the mechanisms of well interference and production recovery were analyzed. Sensitivity analysis about the effect of different parameters on well interference was carried out. Furthermore, the feasibility and effectiveness of gas injection pressure boosting to prevent interference were demonstrated. The results show that the main causes of inter-well interference are: the reservoir energy of the parent well before hydraulic fractures of the child well, well spacing, the fracture connection, etc. The fracture could open under high pressure causing fracturing fluid to flow in, while fracture closure happens under low pressure and the influence on the two-phase seepage in the fracture becomes more serious. The combination of liquid phase retention and fracture closure comprehensively affects the gas phase flow capacity in fractures. Gas injection pressure boosting can effectively prevent fracturing fluids flowing through connected fractures. Before the child well hydraulic fracturing, gas injection and pressurization in the parent well could reduce the stress difference and decrease the degree of well interference. The field case indicates that gas channeling could be effectively prevented through parent well gas injection pressurization.

**Keywords:** shale gas reservoir; well interference; gas injection pressurization; numerical simulation; multi-fractured horizontal well

**Citation:** Xu, J.; Xu, Y.; Wang, Y.; Tang, Y. Multi-Well Pressure Interference and Gas Channeling Control in W Shale Gas Reservoir Based on Numerical Simulation. *Energies* **2023**, *16*, 261. https://doi.org/10.3390/en16010261

Academic Editors: Xingguang Xu, Kun Xie, Yang Yang and Shu Tao

Received: 2 November 2022
Revised: 2 December 2022
Accepted: 10 December 2022
Published: 26 December 2022

## 1. Introduction

Different from conventional oil and gas reservoirs, shale gas reservoirs are characterized by low porosity and permeability, so it is difficult for shale reservoirs to form industrial oil flows without artificial fracture networks. However, with the large-scale hydraulic fracturing and the reduction in well distance, the possibility of inter-well fracture connection and interference becomes larger. The fracture fluid of the child well flows into the parent well, which leads to the water rate rapidly increasing and gas production sharply decreasing in the parent well [1,2]. In severe cases, it could cause well control problems such as casing damage, and even the scrapping of the parent well [1]. The mechanism of inter-well interference is a prerequisite for proposing well-interference preventative technologies. Previous studies clarified the causes and mechanisms of well interference, including stress changes during fracturing [3,4] and the effect on gas phase flow when there is fracturing fluid residue in the fractures. Fracture connectivity between the parent well and child well can be divided into short-term connectivity and long-term connectivity [5]. Short-term fracture connectivity means that the fracture connectivity and fracture fluid flow disappear when the hydraulic fracturing process is completed. Long-term fracture connectivity means that the fractures between the two wells remain connected and fluid exchange is maintained for a long time after the fracturing process is completed [6].

Generally, the inter-well interference phenomenon disappears with the closure of the inter-well connected fractures. However, because of the disturbance phenomenon, even if the production of the disturbed wells recovers after the fracture closure, it is difficult to recover to the original level [7]. Kang et al. identified the main control factors of well production by single-factor and multi-factor analyses, with the EUR set as the production capacity index [8]. Chang et al. explored the final economic analysis for different scenarios of natural fracture (NF) distributions and well spacing using the non-intrusive EDFM (embedded discrete fracture model) method [9,10]. Meanwhile, Huang et al. used the orthogonal experimental design method to optimize the best combination of well spacing and fracture spacing for different well types [11]. He et al. [12] evaluated different factors on pressure interference by gray correlation analysis with the production recovery rate as the evaluation index. Morales et al. [13] illustrated that the excessive extension of fractures of the child well to the parent well is one of the main reasons for well interference. However, the fracture extension involved is a complex problem in which reservoir energy, minimum in situ stress, maximum in situ stress, inter-well spacing [14], and pore pressure within complex fractures are the main factors affecting the shape of the fracture extension [15]. Wang et al. [16] analyzed the variation in fracture process data by optimizing the number of fracture sections and clusters and concluded that the stimulation volume and pumping rate of fracturing fluid are not the main factors controlling well interference. The alignment of fractured sections between the parent and child wells increases the chance of fracture interference, and the non-homogeneity and distance [17] between the two wells has to be considered as well. The possibility of well interference could increase as well distance becomes smaller and the production time becomes longer [18]. In addition, the study of the main control factors of inter-well interference should not be limited to engineering factors, geological factors should be taken into consideration as well.

The prevention of inter-well interference and the recovery of production after interference is attracting great attention in current research, and it has been suggested that different interference prevention methods should be adopted for different development statuses [19,20]. When severe well interference occurs, periodic well shut-in is needed to help restore the reservoir pressure and output capacity [21–23]. When the reservoir energy is low, the degree of well interference can be reduced by increasing well spacing [24,25], reducing the amount of fracturing fluid and gas injection to boost parent wells. However, Jacobs [26] argued that water injection could not prevent inter-well interference since it is not sufficient to create a pressure shield around the production wells, and he was positive about the $CO_2$ or $CH_4$ injection into parent wells for preventing inter-well interference. In addition, researchers have established models for evaluating the effectiveness of different fluid injection strategies (parent well pressurization) on well interference prevention [27,28]. Gala et al. [29] modeled the interaction between a parent well and child well during pressurization with different fracturing fluids, and it was found that the interference could not be eliminated completely, but the impact degree of well interference on production could be minimized. Swanson et al. [30] proposed a chemical treatment method based on a mixture of solvent and surfactant, which effectively reduced the well interference in the Woodford Block of the Anadarko Basin. Except for gas injection, Paryani et al. [31] proposed an integrated workflow combining geology, geophysics, and asymmetric fracturing models together. This workflow is capable of adjusting the fracture design at each stage based on in situ geological and geomechanical variability and estimating the impact of the fracture design on the final fracture geometry. This validated strain model enables the identification of fracturing stages that may cause interference. Additionally, a new fracturing fluid design would be a wise choice for the shale gas industry to prevent inter-well interference by adding a high concentration of reducer to the slip water or adding a high load of sand filler to reduce the amount of water in the fracturing fluid and keep the fracture open but close to the extension distance so that the quality of fracture modification can be maintained while avoiding inter-well interference [32]. For example, Johnson et al. [33] showed that a far-field steering agent consisting of 325-purpose silica could reduce the communication between the parent and child wells by 80%. Qin et al. [34] developed an integrated approach combining pressure and rate transient analysis for well interference diagnosis considering complex fracture networks of a parent well and child well in unconventional reservoirs. Han et al. [35] proposed the pressure derivative curve to diagnose the connection form of wells for the prevention of inter-well interference. In summary, the current research on shale gas inter-well pressure interference is not clear enough, and there is still a lack of feasible and effective technologies to prevent gas channeling.

## 2. History Matching of Wells in W Shale Gas Reservoir

The numerical model with a multi-fractured horizontal well in a shale gas reservoir is established using the numerical simulation software CMG (model dimension: 574.56 m × 2250 m × 36.5 m). The grid dimensions are 19 × 30.24 m and 45 × 50 m in the x and y directions. There are five layers in the z direction, with the size of 20 m, 6 m, 4 m, 5 m, and 1.5 m, respectively. Figure 1 displays the 3D grid division and plane grid division of this numerical model. The initial reservoir pressure is 45.6 MPa and the initial reservoir temperature is 105 °C.

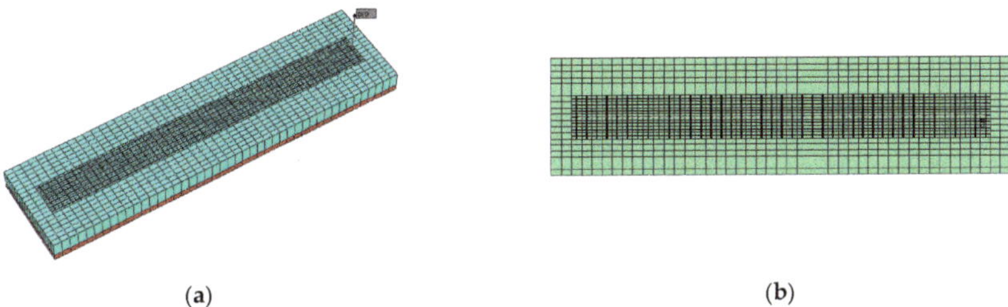

(a)                                               (b)

**Figure 1.** Grid distribution of numerical model. (**a**) 3D view of numerical model. (**b**) Top view of numerical model.

The multi-fractured horizontal wells in W Block were first injected with fracturing fluids at high pressure and then shut in. After two months of shut-in, the production wells were turned on and produced water and gas at the same time. The model focuses on history matching of the gas production and wellhead pressure of Well #W2-A and #W2-B in the W2 block. For Well #W2-A, the fitting time was from January 2018 to January 2020 and the results are shown in Figure 2. It can be seen that the history matching accuracy is quite high.

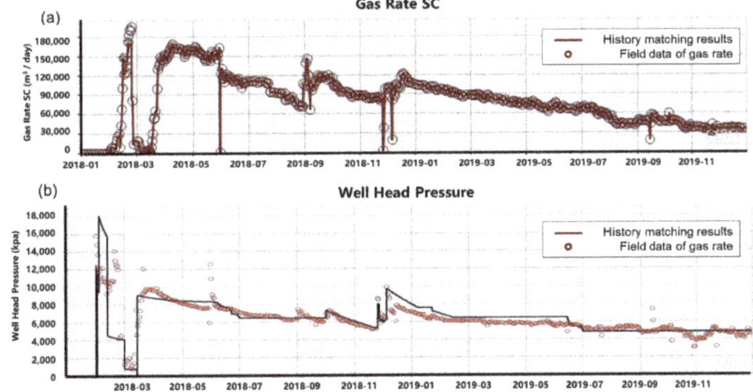


**Figure 2.** History matching results of (**a**) gas production and (**b**) wellhead pressure of Well #W2-A.

The production data of Well #W2-B were also used for historical matching, and the fitting time was from January 2018 to January 2020. The results of the gas rate and wellhead pressure of Well #W2-B are shown in Figure 3, and it can be seen that the history matching accuracy is quite high.

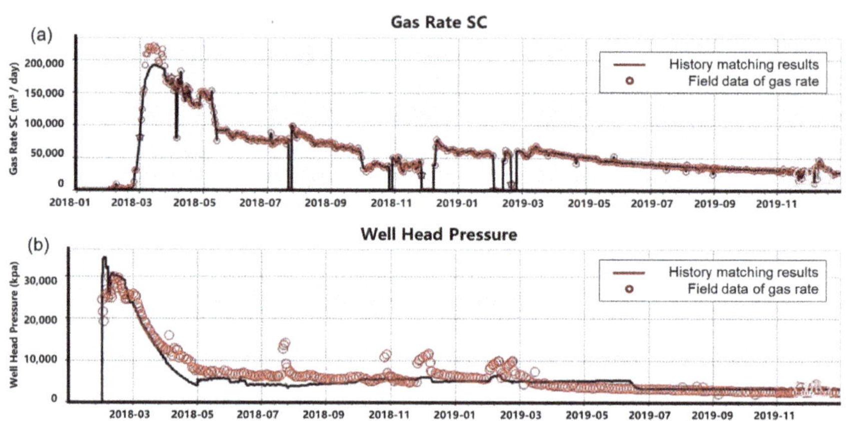


**Figure 3.** History matching results of (**a**) gas production and (**b**) wellhead pressure of Well #W2-B.

## 3. Multi-Well Pressure Interference Analysis in W Shale Gas Reservoir

### *3.1. Numerical Modeling*

The numerical model with a multi-fractured horizontal well in a shale gas reservoir is established using the numerical simulation software CMG (model dimension: 907.2 m × 2250 m × 36.5 m). The grid dimensions are 30 × 30.24 m and 45 × 50 m in the x and y directions. There are 5 layers in the z direction, with the size of 20 m, 6 m, 4 m, 5 m, and 1.5 m, respectively. Figure 4 displays the 3D grid mesh division of this numerical model. The basic parameters of W Block are shown in Table 1.

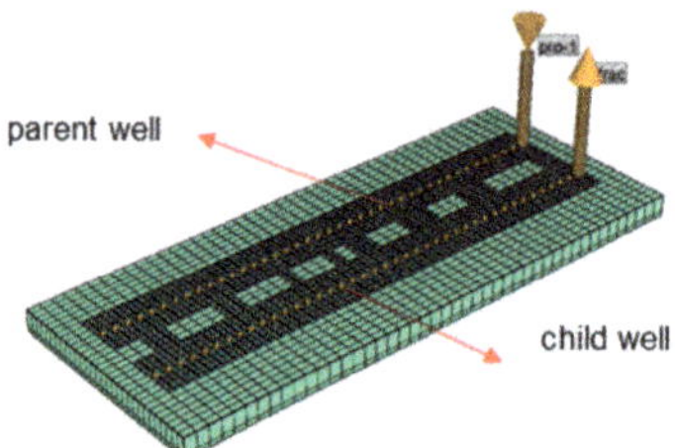


**Figure 4.** D grid mesh division of numerical model of parent well and child well.

**Table 1.** Basic information of numerical model of W Block.

| Reservoir Parameters | Value | Reservoir Parameters | Value |
| --- | --- | --- | --- |
| Reservoir thickness, m | 7 | Matrix permeability, mD | 0.0001 |
| Fracture half-length, m | 105 | Fracture permeability, mD | 0.025 |
| Fracture width, m | 0.0005 | Gas adsorption capacity, m³/t | 1.29 |
| Fracture permeability, mD | 50 | Langmuir pressure, MPa | 2.4 |
| Hydraulic fracturing stage | 40 | Reservoir temperature, °C | 105 |
| Matrix porosity, % | 6.2 | Reservoir initial pressure, MPa | 45.6 |
| Fracture porosity, % | 0.1 | | |

The shale gas adsorption parameters were taken from the adsorption experimental data and the adsorption curve is shown in Figure 5.

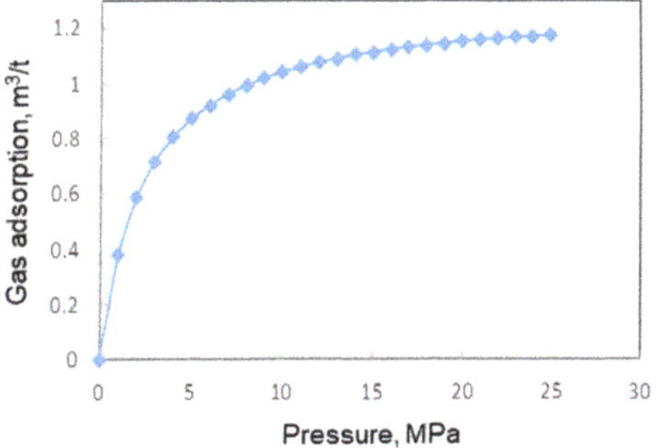


**Figure 5.** Shale gas adsorption curve of W Block based on experiments.

To analyze the effect of different parameters on well interference, the following scenarios are set for further discussion. The summary of the model setup of different scenarios is tabulated in Table 2.

**Table 2.** The summary of the model setup of different scenarios.

| | Well interference caused by hydraulic fracturing of neighboring wells |
|---|---|
| Scenario 1 | The parent well depletes after hydraulic fracturing |
| Scenario 2 | After one year's depletion of parent well, the child well starts hydraulic fracturing |
| | Well interference caused by frac-hits |
| Scenario 3 | The fracture connectivity reaches 40% |
| Scenario 4 | The fracture connectivity reaches 80% |
| | Interference degree under difference well distances |
| Scenario 5 | The well distance equals 300 m |
| Scenario 6 | The well distance equals 510 m |
| | Effect of hydraulic fracturing timing on production |
| Scenario 7 | After one year's depletion of parent well, the child well starts hydraulic fracturing |
| Scenario 8 | The depletion time of the parent well before child well fracturing increases to two years |

### 3.2. Well Interference Caused by Hydraulic Fracturing of Neighboring Wells

The gas and water production data of Scenarios 1 and 2 are compared to analyze the well interference caused by hydraulic fracturing. Scenario 1: The parent well depletes after hydraulic fracturing. Scenario 2: After one year's depletion of the parent well, the child well starts hydraulic fracturing. In Scenarios 1 and 2, the injection rate of fracturing fluids was 500 m$^3$/d and the injection duration was one month. The bottom-hole pressure of the production well was 15 MPa. The gas and water production rates for Scenarios 1 and 2 are shown in Figure 6a,b, respectively. Due to the hydraulic fracturing of the child well, the gas rate decreased rapidly and the water rate increased dramatically from 2 to 40 m$^3$/d of the parent well. This is because the hydraulic fracture of the child well caused the fracture connection between two wells, and the fracturing fluids flowed from the child well to the parent well, which seriously influenced the production of the parent well. Later on, the connecting fractures gradually closed, and the daily water rate gradually decreased, while the gas well slowly recovered. However, since the interconnected fractures were not fully-closed and the fracturing fluids were retained, the relative permeability of the gas phase was affected, resulting in an irreversible decrease in gas production.

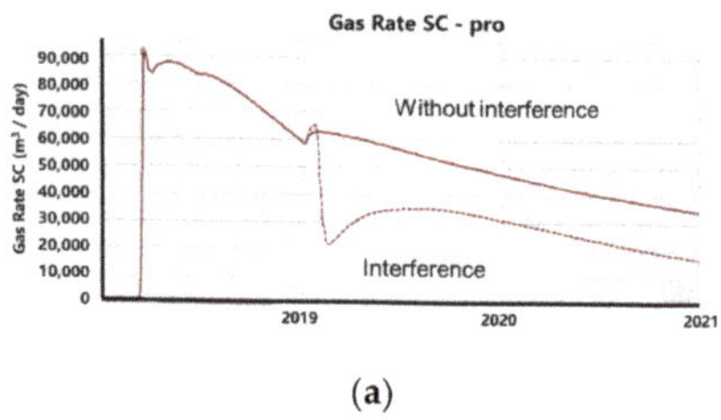


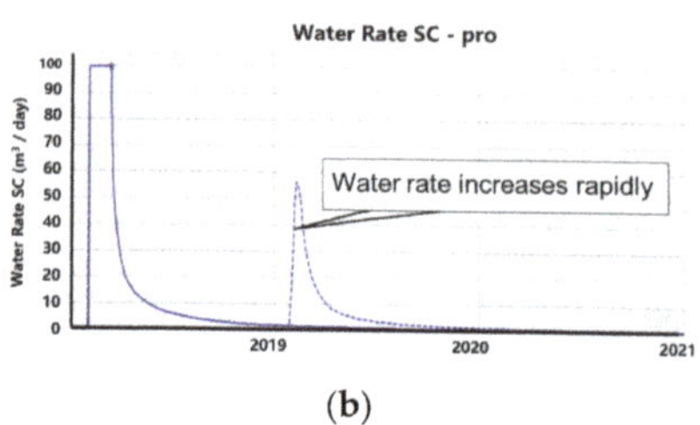


(**a**)  (**b**)

**Figure 6.** Comparison of gas rate and water rate of Scenarios 1 and 2. (**a**) Gas rate. (**b**) Water rate.

### 3.3. Well Interference Caused by Frac-Hits

The effect of fracture connectivity on well interference is analyzed comparing Scenarios 3 and 4. In both cases, the child well starts hydraulic fracturing after one year's depletion of the parent well. The fracture connectivity reaches 40% in Scenario 3 and 80% in Scenario 4. The simulation models of Scenarios 3 and 4 are shown in Figure 7a,b, and the simulation results of these two scenarios are shown in Figure 8a,b. For wells with higher fracture connectivity degree, it would be easier for the fracturing fluids to escape to the parent well from the child well (Figure 9). Moreover, the gas rate decrease rapidly and the water rate increases sharply due to the well interferences. Additionally, the recovery of the parent well gas production would be less compared with the scenario with less fracture connectivity.

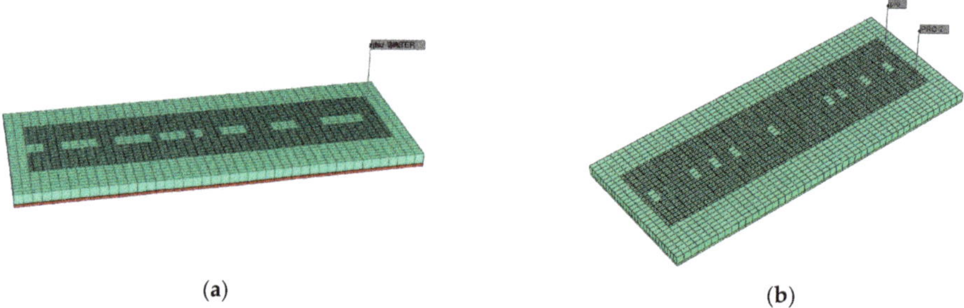

(**a**)  (**b**)

**Figure 7.** D grid mesh division of numerical models of Scenarios 3 and 4. (**a**) Fracture connectivity of 40%. (**b**) Fracture connectivity of 80%.

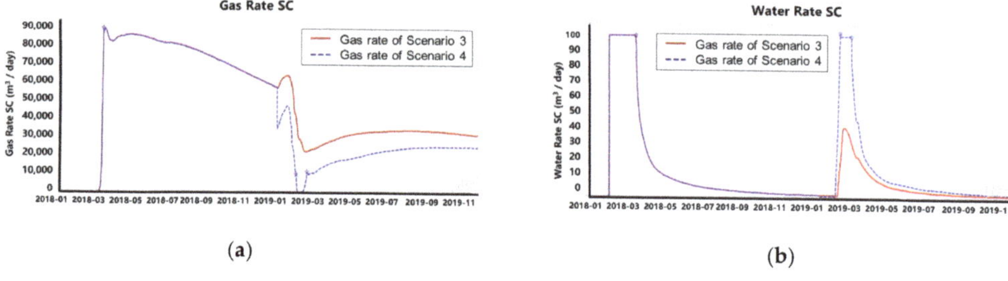


(**a**)  (**b**)

**Figure 8.** Comparison of gas rate and water rate of different scenarios. (**a**) Gas rate. (**b**) Water rate.

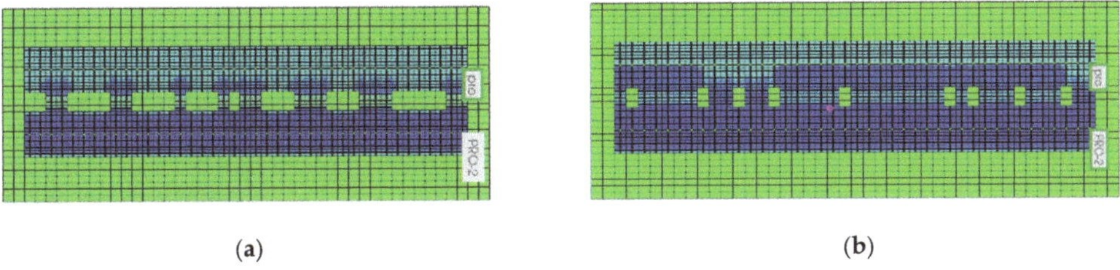

(**a**)           (**b**)

**Figure 9.** Water saturation distribution of multi-well system. (**a**) Fracture connectivity of 40%. (**b**) Fracture connectivity of 80%.

### 3.4. Interference Degree under Different Well Distances

The effect of well spacing on well interference is discussed by comparing Scenarios 5 and 6 (Figure 10a,b). In Scenario 5, the well distance equals 300 m, while in Scenario 6, it increase to 510 m. The gas and water production are shown in Figure 11a,b. When the well distance equals 300 m, the water rate of the parent well increases sharply with the child well hydraulic fracturing, while the gas production decreases due to the well interference. When the well spacing equals 500 m, the inter-well fracture connectivity becomes smaller, and the hydraulic fracturing of the child well has less effect on the parent well.

(**a**)           (**b**)

**Figure 10.** 3D grid mesh division of numerical models of Scenarios 5 and 6. (**a**) Well spacing is 300 m. (**b**) Well spacing is 510 m.

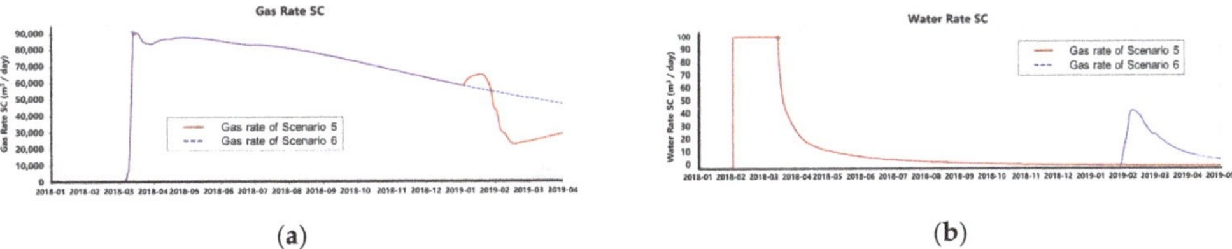

(**a**)           (**b**)

**Figure 11.** Comparison of gas rate and water rates of different scenarios. (**a**) Gas rate. (**b**) Water rate.

### 3.5. Effect of Hydraulic Fracturing Timing on Production

The effect of fracturing timing on production is analyzed in Scenarios 7 and 8. In Scenario 7, after one year's depletion of the parent well, the child well starts hydraulic fracturing. In Scenario 8, the depletion time of the parent well before the child well fracturing increases to two years. The gas rates of Scenarios 7 and 8 are shown in Figure 12a,b, separately. After the child well fractured, the gas rates of both Scenarios 7 and 8 decreased and then started to recover. The recovery degree of the gas rate reached 60% in Scenario 7, while it was only 40% in Scenario 8. When the fracturing timing of the child well happens later, the reservoir pressure near the parent well decreases due to the energy depletion. In that case, the differential pressure between the child well and the parent well becomes larger, and the well interference is more severe since more fracturing fluids are retained in the connected fractures (Figure 13b). When the reservoir energy is high enough (Figure 13a), the recovery degree of the parent well is higher after well interference.

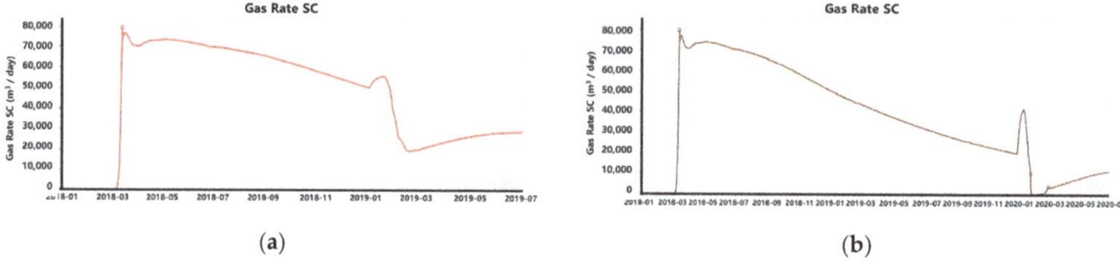

**Figure 12.** Gas rates of (**a**) Scenario 7 and (**b**) Scenario 8.

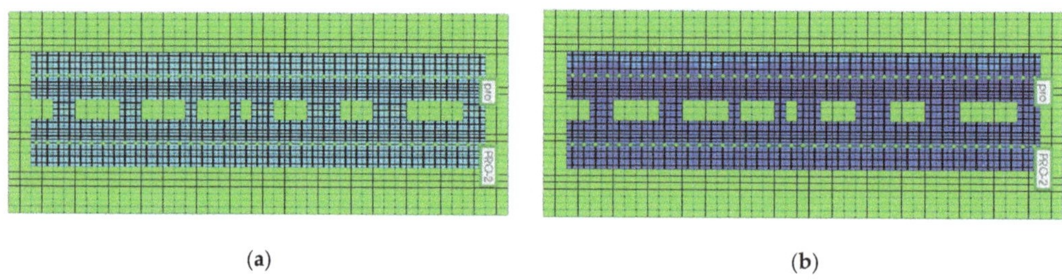

**Figure 13.** Water saturation distribution after production recovery. (**a**) The reservoir energy level is high. (**b**) The reservoir energy level is low.

## 4. Channeling Control by Gas Injection Pressure Boosting of Newly Fractured Wells

Water starts to be produced from the parent well after 20 days of hydraulic fracturing of the child well. The gas rate of the parent well before fracturing was 57,700 m$^3$/d. However, it sharply reduced to 0 after well interference. Therefore, the interference degree turned out to be 100%. After 5 months of recovery, the gas production rate was restored to 37,500 m$^3$/d, and the recovery degree turned out to be 65%, as shown in Figures 14 and 15. The fracture could open under high pressure and fracturing fluid flows in, while fracture closure happens under low pressure and the influence on the two-phase seepage in the fracture becomes more serious. In summary, the combination of liquid phase retention and fracture closure comprehensively affects gas phase flow capacity in fractures.

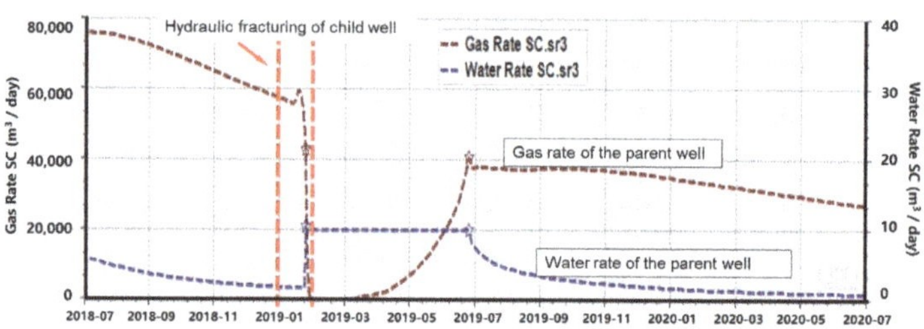

**Figure 14.** Gas and water rates of the parent well before and after hydraulic fracturing of child well.

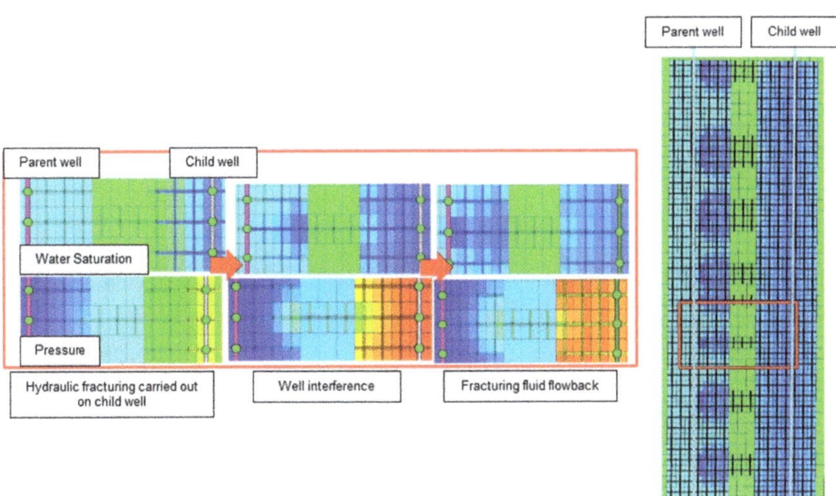


**Figure 15.** Water saturation and pressure distribution after well interference.

In order to theoretically verify the effectiveness of the gas injection pressure boosting technology, a comparison between with and without gas injection was conducted. Figure 16 shows that after gas injection and pressurization of the parent well, the pressure difference between the wells reduces and the fracture propagation is restrained. Therefore, the risk of well interference is reduced.

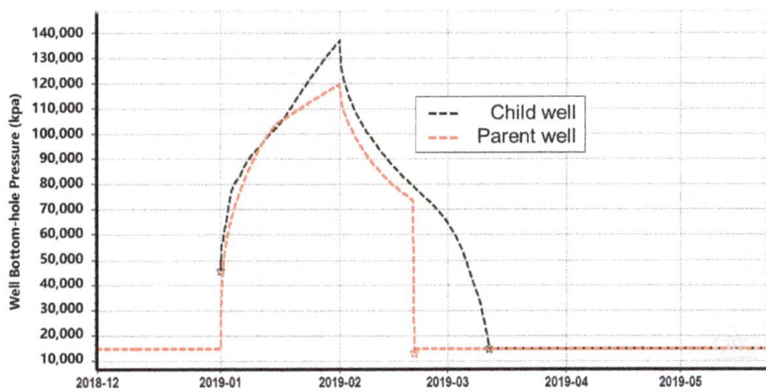


**Figure 16.** Well bottom-hole pressure of parent well and child well.

The comparison of water saturation and pressure distribution between the cases with and without gas injection and pressurization are shown in Figures 17 and 18. Compared with the case without gas injection and pressurization, the water saturation was lower and pressure was higher near the parent well after gas injection and pressurization, indicating that gas injection and pressurization is effective in gas channeling.

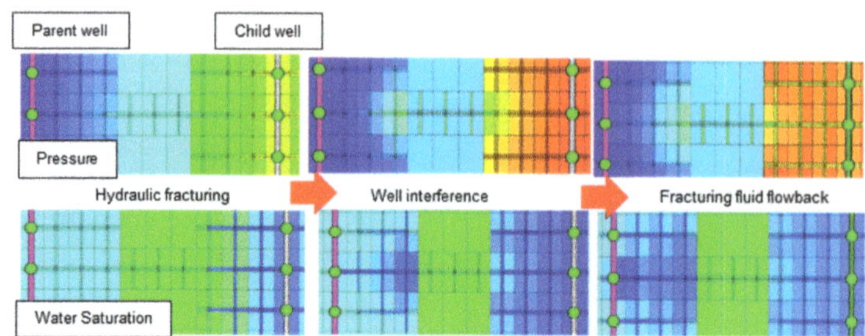



**Figure 17.** Water saturation and pressure distribution without gas injection pressure boosting of newly fractured wells.

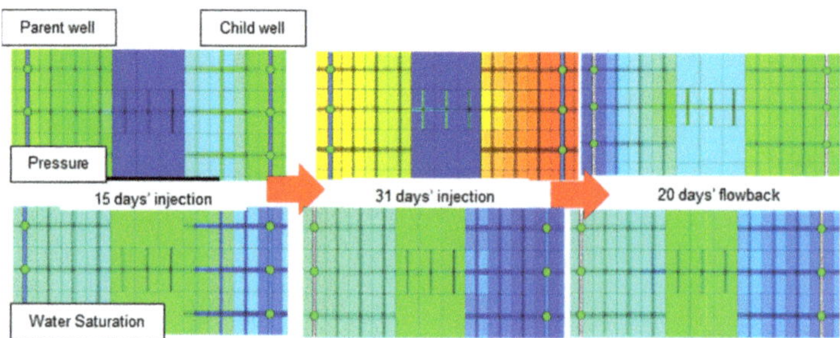



**Figure 18.** Water saturation and pressure distribution with gas injection pressure boosting of newly fractured wells.

In order to optimize the effect of gas injection and pressure boosting, the effect of injection rate on gas production during gas injection and pressure boosting was analyzed. The gas rate, water rate, and cumulative gas production during gas injection are shown in Figures 19–21. With the increase in gas injection rate, the water rate of the parent well decreases, and the channeling control results become better. Gas injection and pressure boosting can effectively prevent fracturing fluid flows through connecting fractures.

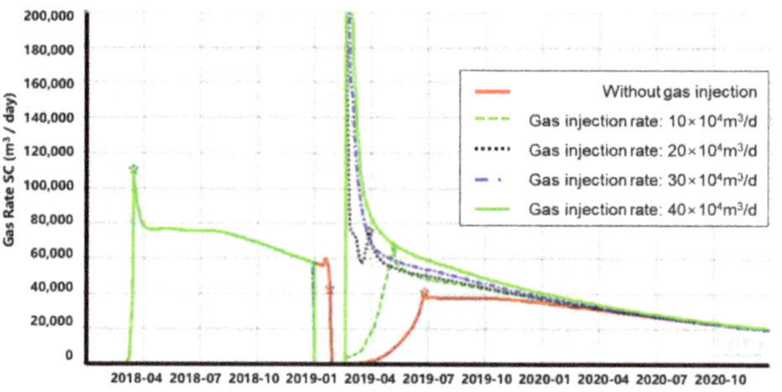



**Figure 19.** Effect of injection rate of pressurization on gas rate.

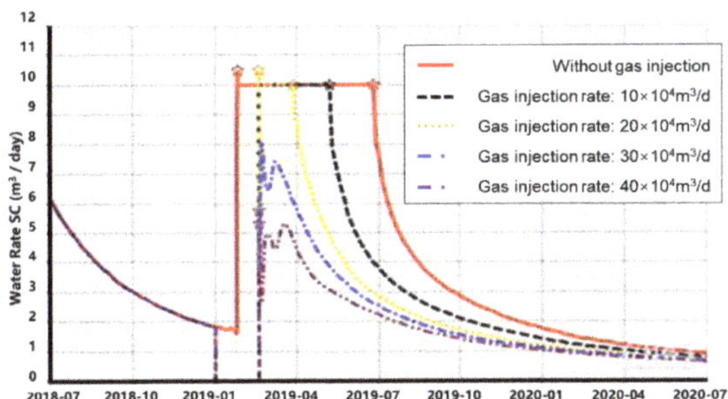

**Figure 20.** Effect of injection rate of pressurization on water rate.

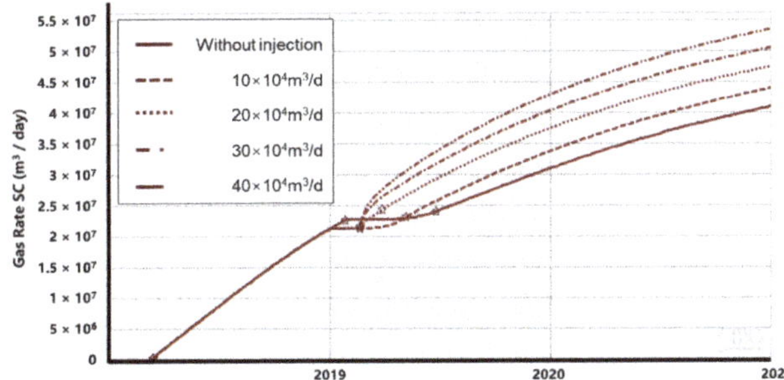

**Figure 21.** Effect of injection rate of pressurization on cumulative gas production.

## 5. Conclusions

The well interference caused by child well hydraulic fracturing could lead to a rapid decrease in gas production and an increase in water production in the parent well, which would lead to lower cumulative gas production and gas recovery. As the fracture connectivity degree increases, the well interference between the parent well and child well becomes stronger. In addition, smaller well spacing, longer production history of the parent well before child well fracturing, and lower reservoir energy could cause more severe well interference.

Gas injection pressure boosting can effectively prevent fracturing fluids flowing through connected fractures. Before the child well hydraulic fracturing, gas injection and pressurization in the parent well could reduce the stress difference and decrease the degree of well interference.

**Author Contributions:** J.X.: study design and established model. Y.X.: data analysis. Y.W.: sensitivity analysis and paper writing. Y.T.: supervision and study design. All authors have read and agreed to the published version of the manuscript.

**Funding:** This research received no external funding.

**Institutional Review Board Statement:** Not applicable.

**Acknowledgments:** We would like to acknowledge Computer Modeling Group Ltd. for providing the CMG software.

**Conflicts of Interest:** The authors declare no conflict of interest.

## References

1. Jacobs, T. Oil and Gas Producers Find Frac Hits in Shale Wells a Major Challenge. *J. Pet. Technol.* **2017**, *69*, 29–34. [CrossRef]
2. Kurtoglu, B.; Salman, A. How to utilize hydraulic fracture interference to improve unconventional development. In Proceedings of the Abu Dhabi International Petroleum Exhibition and Conference, Abu Dhabi, United Arab Emirates, 9–12 November 2015.

3. Rungamornrat, J.; Wheeler, M.F.; Mear, M.E. A Numerical Technique for Simulating Nonplanar Evolution of Hydraulic Fractures. In Proceedings of the SPE Annual Technical Conference and Exhibition, Dallas, TX, USA, 9–12 October 2005.

4. Fiallos, M.X.; Yu, W.; Ganjdanesh, R.; Kerr, E.; Sepehrnoori, K.; Miao, J.; Ambrose, R. Modeling Interwell Interference Due to Complex Fracture Hits in Eagle Ford Using EDFM. In Proceedings of the International Petroleum Technology Conference, Beijing, China, 26–28 March 2019.

5. Jia, C.; Huang, Z.; Sepehrnoori, K.; Yao, J. Modification of two-scale continuum model and numerical studies for carbonate matrix acidizing. *J. Pet. Sci. Eng.* **2020**, *197*, 107972. [CrossRef]

6. Daneshy, A.; King, G. Horizontal Well Frac-Driven Interactions: Types, Consequences, and Damage Mitigation. *J. Pet. Technol.* **2019**, *71*, 45–47. [CrossRef]

7. Kang, L.; Guo, W.; Zhang, X.; Liu, Y.; Shao, Z. Differentiation and Prediction of Shale Gas Production in Horizontal Wells: A Case Study of the Weiyuan Shale Gas Field, China. *Energies* **2022**, *15*, 6161. [CrossRef]

8. Shi, W.; Jiang, Z.; Gao, M.; Liu, Y.; Tao, L.; Bai, J.; Zhu, Q.; Cheng, J. Analytical model for transient pressure analysis in a horizontal well intercepting with multiple faults in karst carbonate reservoirs. *J. Pet. Sci. Eng.* **2023**, *220*, 111183. [CrossRef]

9. Chang, C.; Li, Y.; Li, X.; Liu, C.; Torres, M.F.; Yu, W. Effect of Complex Natural Fractures on Economic Well Spacing Optimization in Shale Gas Reservoir with Gas-Water Two-Phase Flow. *Energies* **2020**, *13*, 2853. [CrossRef]

10. Chang, C.; Liu, C.; Li, Y.; Li, X.; Yu, W.; Miao, J.; Sepehrnoori, K. A Novel Optimization Workflow Coupling Statistics-Based Methods to Determine Optimal Well Spacing and Economics in Shale Gas Reservoir with Complex Natural Fractures. *Energies* **2020**, *13*, 3965. [CrossRef]

11. Huang, T.; Liao, X.; Huang, Z.; Song, F.; Wang, R. Numerical Simulation of Well Type Optimization in Tridimensional Development of Multi-Layer Shale Gas Reservoir. *Energies* **2022**, *15*, 6529. [CrossRef]

12. He, L.; Yuan, C.; Gong, W. Influencing factors and preventing measures of intra-well frac hit in shale gas. *Reserv. Eval. Dev.* **2020**, *10*, 63–69.

13. Morales, A.; Zhang, K.; Gakhar, K.; Porcu, M.M.; Lee, D.; Shan, D.; Malpani, R.; Pope, T.; Sobernheim, D.; Acock, A. Advanced modeling of interwell fracturing interference: An Eagle Ford Shale oil study-refracturing. In Proceedings of the SPE Hydraulic Fracturing Technology Conference, The Woodlands, TX, USA, 9–11 February 2016.

14. Khodabakhshnejad, A. Impact of frac hits on production performance-a case study in Marcellus Shale. In Proceedings of the SPE Western Regional Meeting, San Jose, CA, USA, 23–26 April 2019.

15. Pankaj, P.; Shukla, P.; Kavousi, P.; Carr, T. Determining optimal well spacing in the Marcellus shale: A case study using an integrated workflow. In Proceedings of the SPE Argentina Exploration and Production of Unconventional Resources Symposium, Neuquen, Argentina, 14–16 August 2018.

16. Wang, L. Frac Hit in Complex Tight Oil Reservoir in Ordos Basin: The Challenges, the Root Causes and the Cure. In Proceedings of the SPE Annual Technical Conference and Exhibition, Virtual, 5–7 October 2020.

17. Ajani, A.; Kelkar, M. Interference study in shale plays. In Proceedings of the SPE Hydraulic Fracturing Technology Conference, The Woodlands, TX, USA, 6–8 February 2012.

18. Magneres, R.D.; Castello, M.; Bertoldi, F.; Griffin, M. Vaca Muerta frac-hit occurrence and magnitude forecast methodology. In Proceedings of the Latin America Unconventional Resources Technology Conference, Buenos Aires, Argentina, 16–18 November 2020.

19. Vincent, M. Restimulation of unconventional reservoirs: When are refracs beneficial? *J. Can. Pet. Technol.* **2011**, *50*, 36–52. [CrossRef]

20. Barree, R.D.; Miskimins, J.L.; Svatek, K.J. Reservoir and Completion Considerations for the Refracturing of Horizontal Wells. *SPE Prod. Oper.* **2017**, *33*, 1–11.

21. Gao, D.; Liu, Y.; Wang, D.; Han, G. Numerical Analysis of Transient Pressure Behaviors with Shale Gas MFHWs Interference. *Energies* **2019**, *12*, 262. [CrossRef]

22. He, Y.; Barree, A. Shale Gas Production Evaluation Framework based on Data-driven Models. *Pet. Sci.* **2022**. [CrossRef]

23. Manchanda, R.; Hwang, J.; Bhardwaj, P.; Sharma, M.M.; Maguire, M.; Greenwald, J. Strategies for improving the performance of child wells in the permian basin. In Proceedings of the Unconventional Resources Technology Conference, Houston, TX, USA, 23–25 July 2018.

24. Jia, C.; Sepehrnoori, K.; Huang, Z.; Yao, J. Modeling and Analysis of Carbonate Matrix Acidizing Using a New Two-Scale Continuum Model. *SPE J.* **2021**, *26*, 2570–2599. [CrossRef]

25. Wei, Y.; Han, G. Optimization of shale gas well pattern and spacing. *Nat. Gas Ind.* **2022**, *38*, 129–137.

26. Jacobs, T. In the Battle Against Frac Hits, Shale Producers Go to New Extremes. *J. Pet. Technol.* **2018**, *70*, 35–38. [CrossRef]

27. Bommer, P.A.; Bayne, M.A. Active Well Defense in the Bakken: Case Study of a Ten-Well Frac Defense Project, McKenzie County, ND. In Proceedings of the SPE Hydraulic Fracturing Technology Conference and Exhibition, The Woodlands, TX, USA, 23 January 2018.

28. Zheng, S.; Manchanda, R.; Gala, D.; Sharma, M. Preloading Depleted Parent Wells To Avoid Fracture Hits: Some Important Design Considerations. *SPE Drill. Complet.* **2020**, *36*, 170–187. [CrossRef]

29. Gala, D.P.; Manchanda, R.; Sharma, M.M. Modeling of fluid injection in depleted parent wells to minimize damage due to frac-hits. In Proceedings of the SPE/AAPG/SEG Unconventional Resources Technology Conference, Houston, TX, USA, 23–25 July 2018.

30. Swanson, C.; Hill, W.A.; Nilson, G.; Griman, C.; Hill, R.; Sullivan, P.; Aften, C.; Jimenez, J.C.; Pietrangeli, G.; Shedd, D.C.; et al. Post-frac-hit mitigation and well remediation of Woodford horizontal wells with solvent/surfactant chemistry blend. In Proceedings of the Unconventional Resources Technology Conference, Houston, TX, USA, 23–25 July 2018.
31. Paryani, M.; Smaoui, R.; Poludasu, S.; Attia, B.; Umholtz, N.; Ahmed, I.; Ouenes, A. Adaptive fracturing to avoid frac hits and interference: A Wolfcamp shale case study. In Proceedings of the SPE Unconventional Resources Conference, Calgary, AB, Canada, 15–16 February 2017.
32. Johnson, D.C.; Yeager, B.B.; Roberts, C.D.; Fowler, B.W. Offset fracture events made simple: An operator's collaborative approach to observe parent child interactions, measure frac hit severity and test mitigation strategies. In Proceedings of the SPE Hydraulic Fracturing Technology Conference and Exhibition, The Woodlands, TX, USA, 4 February 2020.
33. Shi, W.; Cheng, J.; Liu, Y.; Gao, M.; Tao, L.; Bai, J.; Zhu, Q. Pressure transient analysis of horizontal wells in multibranched fault-karst carbonate reservoirs: Model and application in SHB oilfield. *J. Pet. Sci. Eng.* **2023**, *220*, 111167. [CrossRef]
34. Qin, J.-Z.; Zhong, Q.-H.; Tang, Y.; Yu, W.; Sepehrnoori, K. Well interference evaluation considering complex fracture networks through pressure and rate transient analysis in unconventional reservoirs. *Pet. Sci.* **2022**. [CrossRef]
35. Han, G.; Liu, Y.; Liu, W.; Gao, D. Investigation on Interference Test for Wells Connected by a Large Fracture. *Appl. Sci.* **2019**, *9*, 206. [CrossRef]

*energies*

*Article*

# Study on Shear Velocity Profile Inversion Using an Improved High Frequency Constrained Algorithm

Qing Ye [1], Huafeng Sun [2,*], Zhiqiang Jin [1] and Bing Wang [1,*]

1   State Key Laboratory of Petroleum Resources and Prospecting, China University of Petroleum, Beijing 102249, China
2   Cores and Samples Center of Natural Resources, China Geological Survey, Beijing 100083, China
*   Correspondence: sunhf@mail.cgs.gov.cn (H.S.); wangbing@cup.edu.cn (B.W.)

**Abstract:** The formation shear-wave (S-wave)'s velocity information around a borehole is of great importance in evaluating borehole stability, reflecting fluid invasion, and selecting perforation positions. Dipole acoustic logging is an effective method for determining a formation S-wave's velocity radial profile around the borehole. Currently, the formation S-wave's radial-profile inversion methods are mainly based on the impacts of radial velocity changes of formations outside the borehole on the dispersion characteristics of dipole waveforms, without considering the impacts of an acoustic tool on the dispersion curves in the inversion methods. Accordingly, the inversion accuracy is greatly impacted in practical data-processing applications. In this paper, a novel inversion algorithm, which introduces equivalent-tool theory into the shear-velocity radial profile constrained-inversion method, is proposed to obtain the S-wave's slowness radial profile. Based on the equivalent-tool theory, the acoustic tool can be modeled using two parameters, radius and elastic modulus. The tool's impact on the dipole waveform's dispersion is eliminated first by using the equivalent-tool theory. Then, the corrected dispersion curve is used to carry out the constrained inversion processing. The results of this processing on the simulation data and the real logging data show the validity of the proposed algorithm.

**Keywords:** oil and gas exploration; logging tool; velocity radial profile; constrained inversion method; equivalent-tool theory

**Citation:** Ye, Q.; Sun, H.; Jin, Z.; Wang, B. Study on Shear Velocity Profile Inversion Using an Improved High Frequency Constrained Algorithm. *Energies* **2023**, *16*, 59. https://doi.org/10.3390/en16010059

Academic Editors: Xingguang Xu, Kun Xie and Yang Yang

Received: 29 November 2022
Revised: 14 December 2022
Accepted: 19 December 2022
Published: 21 December 2022



## 1. Introduction

The rock properties of formations around a borehole are of great significance in the exploration and development of oil and gas. The velocity information of the formations around the borehole, which is one of the most important rock properties, can be addressed with an acoustic logging method. However, abnormal in-situ stress, mud invasion, and borehole damage would result in radial changes in formation velocity around the borehole, thereby bringing inaccurate interpretation results [1]. Based on a dipole waveform obtained by multipole array acoustic logging data, a formation's shear-velocity radial profile can be inverted for evaluating the borehole stability, estimating in-situ stress, and reflecting mud invasion [2–7]. Consequently, it is crucial to study the inversion method of the formation's S-wave velocity radial profile around the borehole.

When the formation's S-wave velocity changes with radial depth, the measured dipole's S-wave dispersion characteristics will also change accordingly. The existing inversion methods regarding the formation's S-wave velocity radial profile are mainly based on the above relationship. Generally, there are two inversion methods. One is Sinha's method, which combines a perturbation method and the Backus–Gilbert (B-G) theory to invert the formation's S-wave velocity radial profile [8]. This is a compromise method between radial distance and velocity change, resulting in an average radial velocity profile that is closest to the actual situation [9,10]. Subsequently, this inversion method is successfully applied in well completions, stress parameters, and velocity estimations of Transversely Isotropic with a Vertical axis of symmetry (TIV) media [11–13]. For example, some researchers have improved this method by using a parametric velocity-characterization equation, and this method was successfully implemented in many cased wells [14,15]. However, when the signal-to-noise ratio is low, there may be deviations between the inversion results and a real formation model, failing to reach a high resolution and high precision. Furthermore, the convergence of this algorithm is seriously affected by the selecting of the initial values. The other inversion method is the constrained inversion method using a high-frequency band, which obtains the conclusion that the radial shear profile is mainly affected by the formation's shear velocity and with regards to the radial variation formation as a formation described by two parameters [16]. This method assumes that the radial change of the formation's S-wave velocity is mainly controlled by two parameters, which are the velocity change and the thickness change of a transition zone (TZ), and an inversion objective function is constructed based on these two parameters. This method constrains a high-frequency band and increases the contribution of a high-frequency component in the objective function, which can effectively improve the accuracy of inversion the results, especially in the formation around the borehole. This constrained inversion method is then applied in a logging-while-drilling quadrupole waveform processing to obtain the formation radial velocity, and to assess the rock-brittleness and fracability of the near-borehole formation [4,5]. In real data processing, the acoustic tool in the borehole has a great impact on the dispersion characteristics of the dipole waveform. Meanwhile, the existing two inversion methods of the formation's S-wave velocity radial profile do not consider the impacts of the tool on the dispersion, which affects the processing effect of the real data. In this case, an equivalent-tool theory was proposed to eliminate impacts of the tool on the flexural wave dispersion in the dipole acoustic logging, and it has been effectively applied in filed data processing and formation information-extraction [17,18].

In this paper, based on the equivalent-tool theory, an improved high-frequency constrained inversion method is proposed to invert the formation's S-wave velocity radial profile around the borehole. The impacts of the tool on the dipole dispersion characteristics are eliminated by equivalently treating the acoustic tool as an elastic rod characterized by two parameters, which are the radius and the equivalent elastic modulus. The improved high-frequency constrained inversion method can eliminate the impacts of the tool and obtain a high-precision formation S-wave velocity radial profile. By processing and comparing the theoretical simulation data and the measured data, the reliability of the improved inversion method is validated.

## 2. High-Frequency Constrained Inversion Method of Formation Shear-Velocity Radial Profile

### 2.1. Basic Method Principle

Tang and Patterson found that a dispersion curve of a radially continuously changing formation is very similar to that of a single-layer altered formation [16]. According to this characteristic of the dispersion curve, the formation with a radial change of S-wave velocity is regarded as a double-layer medium formation with a thickness around the borehole and a velocity difference from the undisturbed formation. In cases of constantly changing values, multiple theoretical models were established, and then the dispersion curve of each theoretical model is compared with the dispersion curve of the real formation. When the comparison results are closest, it can be considered that the theoretical model in this case is the real formation, and an objective function can be constructed:

$$E(\Delta r, \Delta V) = \sum_{\Omega}[V_m(\omega, \Delta r, \Delta V) - V_d(\omega)]^2 \tag{1}$$

where $V_m$ is a dispersion curve calculated by the theoretical model established using $\Delta r$ and $\Delta V$; $V_d$ is a dispersion curve of a real radially changing formation; $\Omega$ is a frequency range; and $\omega$ is the circular frequency. When the value of the objective function (1) is the smallest, the velocity variation $\Delta V$ and the TZ thickness $\Delta r$ can be used to represent the velocity profile of the real radially changing formation.

The objective function (1) can be used to invert a relatively accurate velocity profile, but there is a large error for the model's calculation of a near-borehole formation. In order to obtain more accurate results, Tang and Patterson delved into this study, and found that a phase velocity of a low-frequency part of the dispersion curve in a radially changing model is close to the shear-velocity of the undisturbed zone [16]. When the S-wave velocity of a homogeneous model is equal to the near-borehole S-wave velocity of an inhomogeneous formation model, the high-frequency parts of the dispersion curves of the two formation models are very close. This shows that the low-frequency part of the flexural wave is influenced by the undisturbed zone, and the high-frequency part is influenced by the formation near the borehole. This is because the bending wave is a dispersive wave, leading to a longer wavelength and a further detection range, whereas a shorter wavelength travels along the shallow formation area. From this, Tang and Patterson proposed the high-frequency constrained inversion method to reduce the near-bore error of the S-wave's slowness profile by constraining the high-frequency band [16]. The objective function (1) is modified to:

$$E(\Delta r, \Delta V) = \sum_{\Omega}[V_m(\omega, \Delta r, \Delta V) - V_d(\omega)]^2 + \lambda \sum_{\Omega'}[V_m(\omega, \Delta r, \Delta V) - V_h(\omega, V_1)]^2 \quad (2)$$

where $V_h$ is the dispersion curve calculated by establishing a uniform formation model; $V_1$ is the wellbore formation shear-velocity; $\Omega'$ is the high-frequency range for constraint, usually (9~10 kHz); and $\lambda$ is used to adjust the relative contribution of the high-frequency term.

The radial change of a formation's S-wave velocity is usually monotonically increasing or monotonically decreasing. According to this phenomenon, the formation's S-wave velocity radial-change model is established [4]:

$$V(r) = V_0 - \left(\frac{\overline{\Delta V}}{1 - e^{-1}}\right) \exp\left(-\frac{r - r_0}{\Delta r}\right), (r \geq r_0) \quad (3)$$

where $V_0$ is the S-wave velocity of the undisturbed formation and $r_0$ is the borehole's radius. The shear-velocity profile can be obtained by substituting the result obtained from Equation (2) into Equation (3).

### 2.2. Theoretical Model Validation

A radial layered formation was modeled by the numerical simulation method, and the high-frequency constrained inversion method was used for the inversion calculation. The relevant parameters of this formation model are shown in Table 1. In the model, the number of sampling points for each receiver was 400, the spacing was 0.1524 m, the time-sampling interval was 36 μs, and the center frequency was 3 kHz. The velocity alteration schematic is shown in Figure 1.

**Table 1.** Parameters of radial variation-formation model.

| Formation Properties | Radial Depth (m) | P-Wave Velocity (m/s) | S-Wave Velocity (m/s) | Density (kg/m$^3$) |
|---|---|---|---|---|
| Borehole fluid | r < 0.1 | 1500 | - | 1000 |
| TZ1 | 0.10 < r < 0.15 | 2400 | 1000 | 2500 |
| TZ2 | 0.15 < r < 0.20 | 2400 | 1050 | 2500 |
| TZ3 | 0.20 < r < 0.25 | 2400 | 1100 | 2500 |
| TZ4 | 0.25 < r < 0.30 | 2400 | 1150 | 2500 |
| Undisturbed Formation | r > 0.30 | 2400 | 1200 | 2500 |

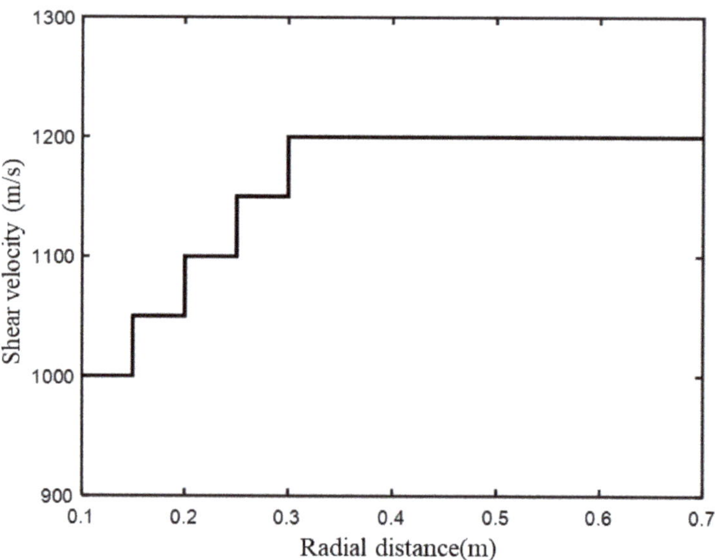

**Figure 1.** Structure diagram of numerical simulation model.

For the model in Table 1, a real-axis-integration method was used to simulate its waveform [19], as shown in Figure 2:

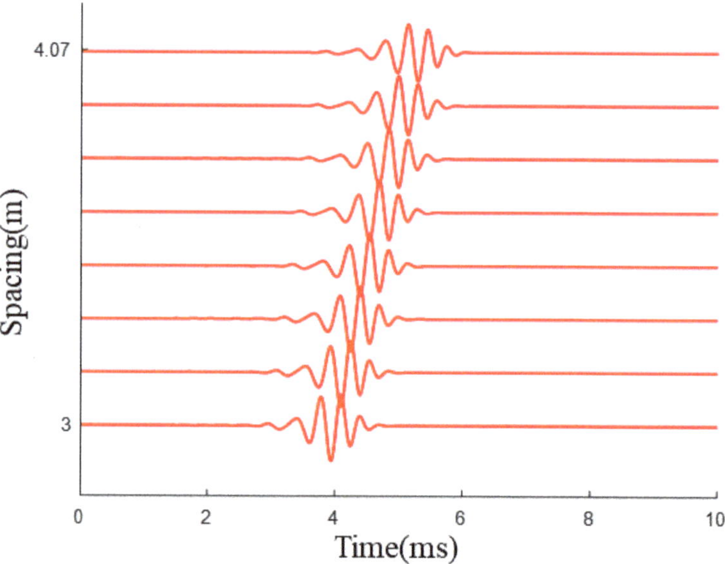

**Figure 2.** Waveforms of the numerical simulation model.

The dispersion curve was calculated using a weighted spectral semblance algorithm [19], and the results are as shown in Figure 3a. For the dispersion curve without a frequency range of 8~9 kHz, an arctangent function fitting should be applied to obtain the high frequency range (shown in Figure 3b).

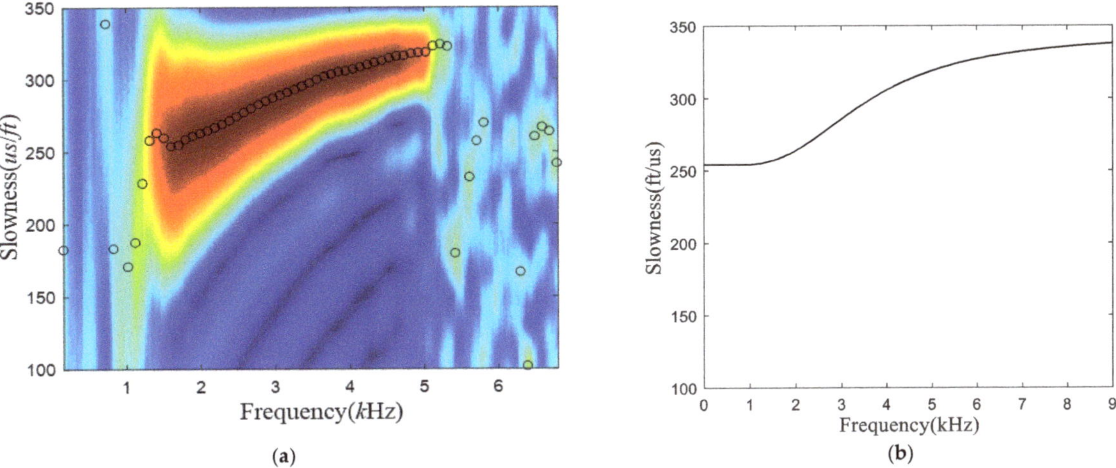

**(a)**

**(b)**

**Figure 3.** (**a**) Dispersion analysis by the weighted spectral semblance method and (**b**) the fitting results.

According to the model parameters (Table 1), the velocity changes of the TZs during the inversion process were set to: 1000 m/s: 2 m/s: 1200 m/s, and the thickness changes of the TZs were set to: 0 m: 0.005 m: 0.3 m. Then, it was necessary to construct $101 \times 61$ theoretical models, calculate the dispersion curve of each theoretical model one by one, and substitute them into Equation (2). The full inversion frequency range was 3~9 kHz, and the high-frequency constraint range was 8~9 kHz. A two-dimensional image of the correlation coefficients of the objective function was obtained, as shown in Figure 4:

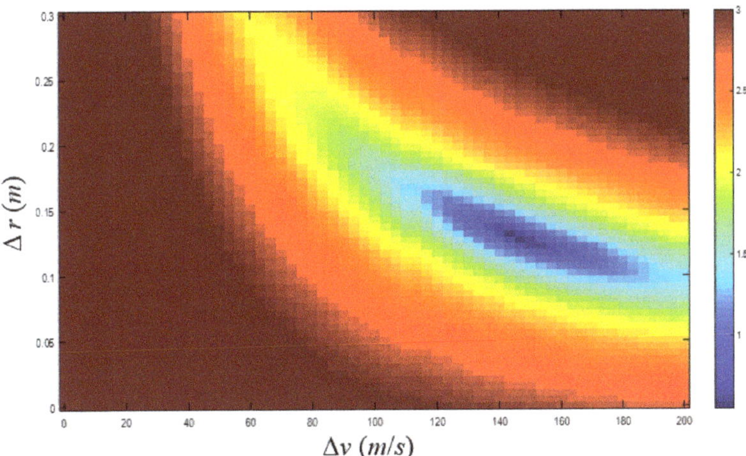

**Figure 4.** Two-dimensional correlation image of objective function in high-frequency constrained inversion.

The ordinate of Figure 4 is the equivalent TZ thickness $\Delta r$, the abscissa is the velocity change $\Delta V$, and the color represents the size of the objective function value. When the objective function value is the smallest, the corresponding $\Delta V$ and $\Delta r$ are 146 m/s and 0.125 m, respectively. By putting these two parameters into Equation (3), the shear-velocity radial profile can be obtained, as shown in Figure 5, and its relative error is shown in Figure 6.

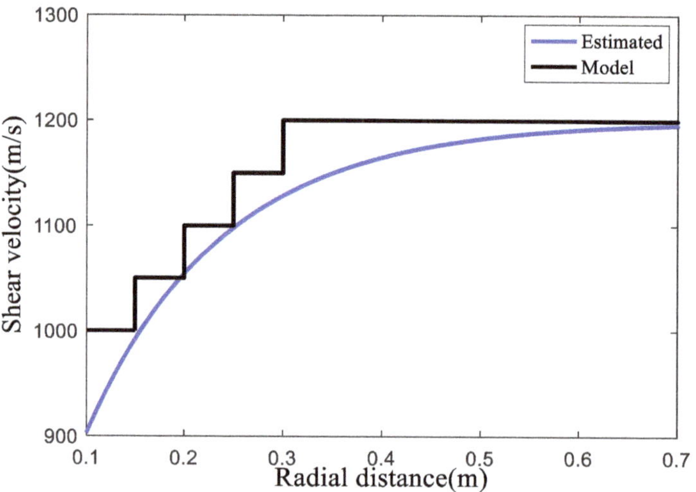

**Figure 5.** Inversion result of high-frequency constrained inversion method.

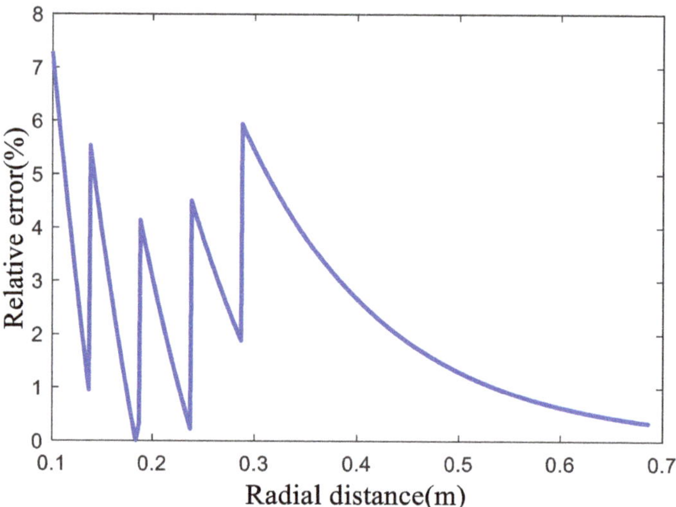

**Figure 6.** Radial distribution of relative error between inversion result and real formation velocity.

From Figure 6, due to the high-frequency constraints, the inversion method had a relatively high accuracy, and the overall error was basically below 4%, and especially in the near-borehole formation a high inversion-accuracy could be obtained.

## 3. Application of Equivalent-Tool Theory in Constrained Inversion

### 3.1. Equivalent-Tool Theory

When using real logging data to invert the formation's S-wave velocity radial profile, the impacts of the acoustic logging tool cannot be ignored. Due to a complex structure, the equivalent-tool theory can be introduced to simulate the impact of the tool [17]. Assuming that the logging tool is a cylinder with a radius of $a$, no matter how complex the structure of the tool is, as long as the wavelength of the bending wave is greater than the radius of the tool, the tool can be equivalent to an elastic rod with an elastic modulus of $M_T$. One parameter can used to simulate the impacts of multiple elastic parameters of the tool. Considering that there is a logging tool in center of the borehole, the displacement ($u$) and pressure ($p$) in the annular fluid between the tool and the borehole caused by the acoustic wave can be expressed as follows:

$$u = \left\{ A_n \left[ \frac{n}{r} I_n(fr) + f I_{n+1}(fr) \right] + B_n \left[ \frac{n}{r} K_n(fr) - f K_{n+1}(fr) \right] \right\} \cos[n(\theta - \phi)] \tag{4}$$

$$p = \left\{ A_n \left[ \rho_f \omega^2 I_n(fr) \right] + B_n \left[ \rho_f \omega^2 K_n(fr) \right] \right\} \cos[n(\theta - \phi)] \tag{5}$$

where $A_n$ is the amplitude coefficients of the incident wave; $B_n$ is the amplitude coefficients of the outgoing wave; n is the sound-source order, $n$ = zero for monopole, $n$ = one for dipole, and $n$ = two for quadrupole; $I_n$ and $K_n$ are the first-kind and second-kind of nth-order variant Bessel functions, respectively; f is the radial wave number of the fluid; k is the axial wavenumber of the mode wave; $V_f$ and $\rho_f$ are the velocity and density of the fluid, respectively; $\omega$ is the circular frequency; and $\theta - \varphi$ is the azimuth of the alignment of the multipoles to a reference azimuth $\varphi$ [19].

The boundary conditions at the fluid–solid interfaces (fluid in the borehole and the logging tool's surface, fluid in the borehole and borehole wall) are fully matched by the acoustic admittance (i.e., $u/p$). At longer wavelengths, the logging tool can be equivalent to the elastic rod with only one modulus. Norris studied the case of the solid elastic rod in the center of the borehole, and used the following acoustic admittance matching for the Stoneley wave between the elastic rod and the borehole wall, and the specific expression is [20]:

$$\left( \frac{u}{p} \right)_{r=a} = -\frac{a}{M_T} \tag{6}$$

Assuming that this result is still applicable to multipole waves, and extending Equation (6) to the frequency domain, it is only necessary to calculate the acoustic admittance of the elastic waves according to Equations (4) and (5) and then match it with the acoustic admittance of the logging tool at the tool surface. This gives the relationship between the unknown coefficients $A_n$ and $B_n$:

$$E_{tool} = \frac{B_n}{A_n} = -\frac{\left( \frac{M_T}{a} \right) \left[ \frac{n}{a} I_n(fa) + f I_{n+1}(fa) \right] + \rho_f \omega^2 I_n(fa)}{\left( \frac{M_T}{a} \right) \left[ \frac{n}{a} K_n(fa) - f K_{n+1}(fa) \right] + \rho_f \omega^2 K_n(fa)} \tag{7}$$

where $E_{tool}$ represents the elastic parameter related to the tool, and it can be added to the dispersion equation for the simulation of the impacts of the tool on the dispersion curve. The specific method is: add the terms related to the equivalent tool to two matrix elements $M_{11}$, $M_{21}$ related to the sound field in the borehole in the dispersion equation, which is:

$$M_{11} : \frac{n}{R} I_n(fR) + f I_{n+1}(fR)$$
$$\rightarrow \left[ \frac{n}{R} I_n(fR) + f I_{n+1}(fR) \right] + E_{tool} \left[ \frac{n}{R} K_n(fR) - f K_{n+1}(fR) \right] \tag{8}$$
$$M_{21} : I_n(fR) \rightarrow I_n(fR) + E_{tool} K_n(fR)$$

where the symbol "→" in the above formula means to convert the expression on the left to the expression on the right. From this, the multipole dispersion equation when the logging tool is located in the center of the borehole is obtained:

$$D\left(k, \omega, V_p, V_s, \rho, V_f, \rho_f, R, M_T, a\right) = 0 \tag{9}$$

where $V_p$, $V_s$, $\rho$ are compressional velocity, shear velocity, and the density of the formation, respectively. The Newton–Raphson method is normally used to solve the dispersion equation [19]; thus, the multipole dispersion curve with the impact of the tool is obtained.

$$V_{ph} = \omega/k \tag{10}$$

In order to validate the feasibility of the equivalent-tool theory, an annular heavy-mud column is used to simplify and simulate the wireline dipole acoustic logging tool, and a formation model with the tool in the borehole is established; its parameters are shown in Table 2.

**Table 2.** Parameters of simulated formation and tool.

| Formation Parameters | | Tool Parameters | |
|---|---|---|---|
| P-wave velocity (m/s) | 3200 | Density (kg/m$^3$) | 7850 |
| S-wave velocity (m/s) | 2300 | Inner radius (m) | 0.01 |
| Density (kg/m$^3$) | 2500 | Outer radius (m) | 0.045 |
| Radius (m) | 0.1 | | |

According to the parameters in Table 2, the real-axis integration method is used to simulate and calculate the corresponding waveforms and, respectively, calculate the reference dispersion curve without considering the impacts of the tool, the reference dispersion curve calculated by the equivalent-tool theory, and the dispersion curve calculated by simulating the waveform with the impacts of the tool. The results are shown in Figure 7.

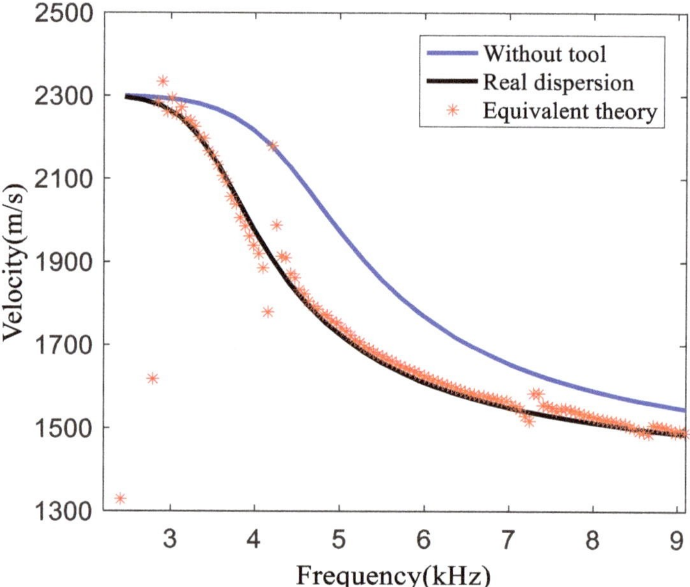


**Figure 7.** Dispersion curve without tool impacts (blue); reference dispersion curve calculated by the equivalent-tool theory (black); dispersion curves of the waveform affected by the tool (red).

In the calculation, the equivalent radius a of the tool is equal to the outer diameter of the heavy-mud column (0.045 m), and the value of the equivalent modulus of the tool $M_T$ is 32 GPa. It can be seen from Figure 7 that in the frequency range of 2–9 kHz, the calculation results using the equivalent-tool theory are consistent with the theoretical curve, which validates the applicability of the equivalent-tool theory. Furthermore, in the frequency band higher than the cutoff frequency, logging tools make the flexural wave dispersion curve shift to low frequencies, which indicates that the impact of the tool must be considered when we use the dipole dispersion curve to carry out any inversion processing.

### 3.2. Application of Equivalent-Tool Theory in Constrained Inversion Method

The parameters of the annular heavy-mud column in Table 2 are used to simulate the formation model with the impacts of the tool. The high-frequency constraint method is used for the inversion calculation. When calculating the dispersion curve of the theoretical model, the equivalent theory is used to calculate the dispersion curve with the impacts of the tool, and a two-dimensional image of the objective function is obtained by substituting it into Equation (2).

Figure 8a,b represents the two-dimensional images of the objective function using the equivalent-tool theory and without using the equivalent-tool theory, respectively. When the value of the objective function is the smallest, the velocity change and equivalent thickness using the equivalent-tool theory are 146 m/s and 0.125 m, respectively, and the velocity change and equivalent thickness without using the equivalent-tool theory are 171 m/s and 0.145 m, respectively. The impacts of the tool make the velocity change and the equivalent thickness larger. By bringing these two results into Equation (3), the obtained dipole formation S-wave velocity radial profile is as follows:

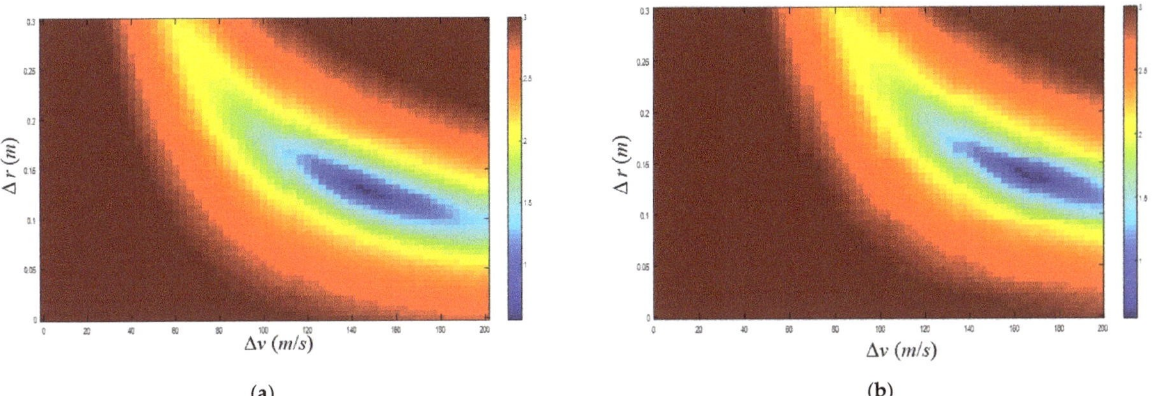

(a)  (b)

**Figure 8.** Two-dimensional image of objective function: (**a**) using equivalent-tool theory and (**b**) without using equivalent-tool theory.

From Figures 9 and 10, it can be seen that the relative error of the inversion results without considering the tool is larger. This is because the tool's existence makes the dispersion curve shift to the low-frequency of the real formation, the velocity change, and the equivalent thickness increase when the objective function value is the smallest, resulting in a generally smaller radial S-wave velocity and a larger relative error. Therefore, the impact of the tool can be eliminated by using the equivalent-tool theory, and the result is more in line with true values.

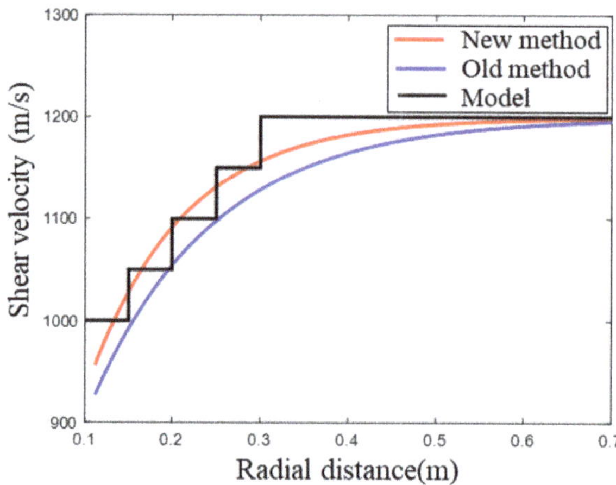

**Figure 9.** Inversion results using equivalent-tool theory (red line) and without using equivalent-tool theory (blue line).

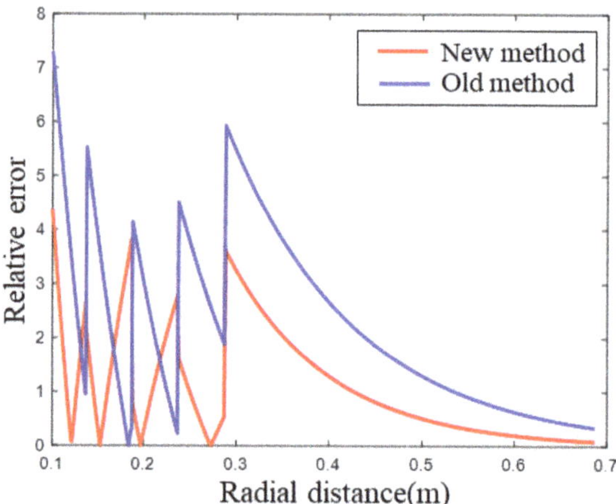

**Figure 10.** Relative error between inversion results and real values using equivalent-tool theory (red line) and without using equivalent-tool theory (blue line).

## 4. Real Data Processing

Based on the previous theoretical analysis, we can identify that the existence of the acoustic tool affects the dipole dispersion characteristic dramatically. The impacts of the tool should be considered carefully when performing the formation S-wave velocity radial profile inversion. For the real data processing, due to unknown tool parameters, the tool would be different for different borehole-logging measurements, and thus constant parameters cannot be used as equivalent to tool impacts. The following procedure is adopted in real data processing:

1. Step 1, select the impermeable and tight interval, extract the dispersion curve of the measured dipole waveform, and compute the theoretical dispersion curve using the known formation velocity, density, and borehole parameters.
2. Step 2, invert the tool parameters $M_T$ and $a$ by matching the real data and the synthetic curve based on the equivalent-tool theory in the impermeable and tight intervals.

3. Step 3, implement the formation's S-wave velocity radial profile inversion algorithm corrected by the equivalent-tool theory in other intervals. The tool equivalent parameters $M_T$ and $a$ in this step use the value from Step 2.

The formation's S-wave velocity radial profile is processed for the real data of an oilfield in Western China, as shown in Figure 11.

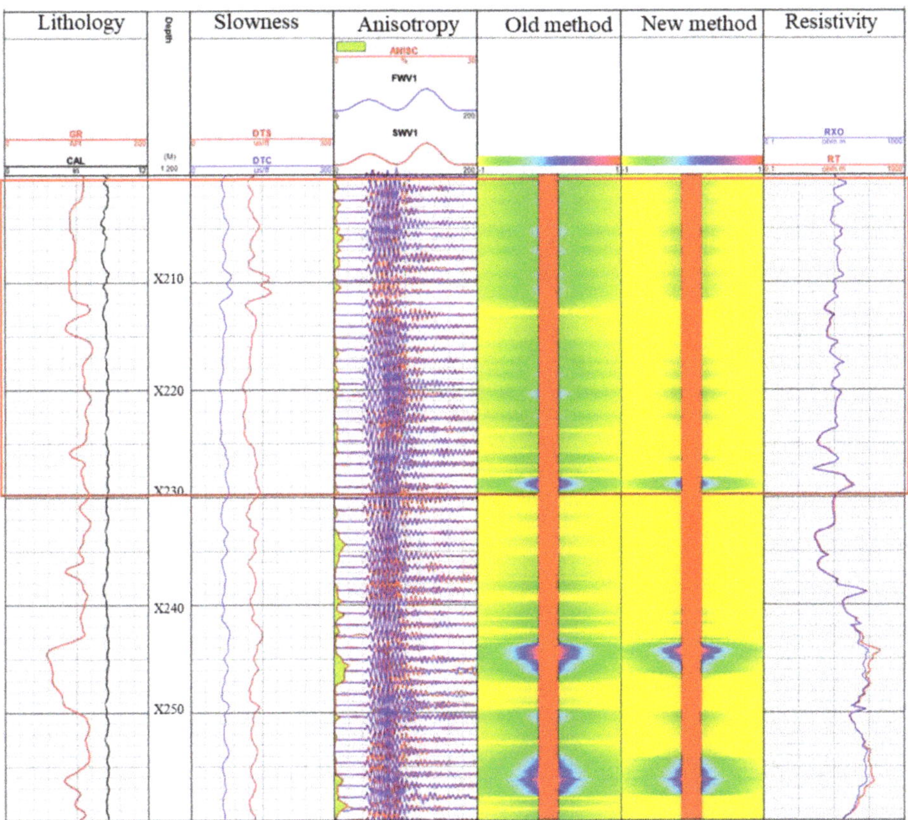

**Figure 11.** Comparison of formation shear-velocity profile radial with several logging results.

The depth of the borehole is X200–X260 m. In Figure 11, Trace 1 is the well diameter curve (black line) and the natural gamma curve (red line), Trace 2 is the time difference between the formation's P and S-waves in this interval, and Trace 3 gives the orthogonal dipole anisotropy analysis and fast and slow S-wave waveforms. Trace 4 is the formation's S-wave velocity radial profile calculated by the high-frequency constrained inversion method. In Trace 4, the shade of color represents the difference with the undisturbed formation velocity. Trace 5 is the formation's S-wave velocity profiles with the improved inversion method. Trace 6 gives the deep and shallow resistivity logging curves. From Figure 11, it can be seen that there are obvious differences between the two velocity profiles in the interval marked by the red frame. In this marked interval, there is no change in the borehole diameter, the anisotropy is not obvious, the GR value has no offset, there is no obvious difference between the deep and shallow resistivity curves, and thus the radial S-wave velocity should not change. Therefore, the result of Trace 5 is more in line with the real formation. At the interval between X245 m and X255 m, the radial S-wave velocity on Trace 5 has obvious changes, while the depth and shallow resistivity curves on Trace 6 are significantly different, the GR value on Trace 1 is small, and the borehole diameter (Trace 1) and anisotropy (Trace 3) have no obvious changes. This may be caused by mud invasion. The radial velocity changes are not obvious in other intervals, but the radial velocity change exists in the whole interval on Trace 4, which is obviously not in line with the actual situation.

### 5. Conclusions

In this paper, the basic theory of the high-frequency constrained inversion method for the formation's S-wave velocity radial profile is studied, and examples of a forward and an inversion simulation are given to illustrate that this method has a large error in the case of considering the impacts of the acoustic logging tool. Then, the equivalent-tool theory is introduced to eliminate the impacts of the tool through theoretical and real data studies. Finally, an improved high-frequency constraint inversion method using the equivalent-tool theory is proposed and used to process the real logging data. The following conclusions are drawn:

1.  The existence of the tool makes the dispersion curve of the dipole acoustic logging move to a low frequency. If the inversion calculation is carried out directly without considering the impacts of the tool, the velocity change and the equivalent thickness obtained by this inversion method will be larger, thereby increasing the error. The equivalent-tool theory can be used to calculate the dispersion curve of the theoretical model by considering the tool, thereby eliminating the impacts of the tool and reducing the inversion error.
2.  The results obtained using the improved high-frequency constrained inversion method show that the radial velocity changes in the entire interval of the formation's S-wave velocity radial profile without using the equivalent-tool theory are not in line with the actual situation, and the formation's S-wave velocity radial profile using the equivalent-tool theory is in good agreement with other curves. This proves the effectiveness of the improved high-frequency constrained formation's S-wave velocity radial profile-inversion method.

**Author Contributions:** Conceptualization, Q.Y. and B.W.; methodology, Q.Y.; software, H.S.; validation, Z.J.; writing—original draft preparation, Q.Y.; writing—review and editing, H.S.; visualization, Z.J.; supervision, B.W. All authors have read and agreed to the published version of the manuscript.

**Funding:** This research was funded by the National Natural Science Foundation of China, grant number 42274141.

**Data Availability Statement:** Not applicable.

**Conflicts of Interest:** The authors declare no conflict of interest.

## References

1.  Baker, L.J. The effect of the invaded zone on full wavetrain acoustic logging. *Geophysics* **1984**, *49*, 686–847. [CrossRef]
2.  Sinha, B.K.; Asvadurov, S. Dispersion and radial depth of investigation of borehole modes. *Geophys. Prospect.* **2004**, *52*, 271–286. [CrossRef]
3.  Sinha, B.K.; Valero, H.; Ikegami, T.; Pabon, J. Borehole flexural waves in formations with radially varying properties. In Proceedings of the IEEE Ultrasonics Symposium, Rotterdam, The Netherlands, 18–21 September 2005; pp. 556–559. [CrossRef]
4.  Su, Y.-D.; Tang, X.-M.; Zhuang, C.-X.; Xu, S.; Zhao, L. Mapping formation shear-velocity variation by inverting logging-while-drilling quadrupole-wave dispersion data. *Geophysics* **2013**, *78*, D491–D498. [CrossRef]
5.  Tang, X.-M.; Xu, S.; Zhuang, C.-X.; Su, Y.-D.; Chen, X.-L. Assessing rock brittleness and fracability from radial variation of elastic wave velocities from borehole acoustic logging. *Geophys. Prospect.* **2016**, *64*, 958–966. [CrossRef]
6.  Li, J.; Innanen, K.A.; Tao, G.; Zhang, K.; Lines, L. Wavefield simulation of 3D borehole dipole radiation. *Geophysics* **2017**, *82*, D155–D169. [CrossRef]
7.  Liu, Y.; Chen, H.; Li, C.; He, X.; Wang, X.; Habibi, D.; Chai, D. Radial profiling of near-borehole formation velocities by a stepwise inversion of acoustic well logging data. *J. Pet. Sci. Eng.* **2021**, *196*, 107648. [CrossRef]
8.  Backus, G.; Gilbert, F. Uniqueness in the inversion of inaccurate gross Earth data. *Philos. Trans. R. Soc. Lond., Math. Phys. Sci.* **1970**, *266*, 123–192. [CrossRef]

9.  Burridge, R.; Sinha, B.K. Inversion for formation shear modulus and radial depth of investigation using borehole flexural waves. In Proceedings of the SEG Technical Program Expanded Abstracts 1996, Denver, CO, USA, 10–15 November 1996; pp. 158–161. [CrossRef]
10. Sinha, B.K.; Burridge, R. Radial Profiling of Formation Shear Velocity from Borehole Flexural Dispersions. In Proceedings of the 2001 IEEE Ultrasonics Symposium, Atlanta, GA, USA, 7–10 October 2001; pp. 391–396. [CrossRef]
11. Sinha, B.K.; Vissapragada, B.; Renlie, L.; Tysse, S. Radial profiling of the three formation shear moduli and its application to well completions. *Geophysics* **2006**, *71*, E65–E77. [CrossRef]
12. Sinha, B.; Bratton, T.; Cryer, J.; Nieting, S.; Ugueto, G.; Bakulin, A.; Hauser, M. Estimation of near-wellbore alteration and formation stress parameters from borehole sonic data. *SPE Reserv. Eval. Eng.* **2008**, *11*, 478–486. [CrossRef]
13. Sunaga, S.; Sinha, B.K.; Endo, T.; Walsh, J. Radial Profiling of Shear Slowness in TIV Formations. In Proceedings of the SEG Technical Program Expanded Abstracts 2009, Houston, TX, USA, 25–30 October 2009; pp. 3481–3485. [CrossRef]
14. Yang, J.; Sinha, B.K.; Habashy, T.M. A theoretical study on formation shear radial profiling in well-bonded cased boreholes using sonic dispersion data based on a parameterized perturbation model. *Geophysics* **2012**, *77*, WA197–WA210. [CrossRef]
15. Wang, B.; Ma, M.-M.; Liu, H.; Liu, Z.-J. Inversion of the shear velocity radial profile in a cased borehole. *Chin. J. Geophys.* **2016**, *59*, 4782–4790. [CrossRef]
16. Tang, X.-M.; Patterson, D.J. Mapping formation radial shear-wave velocity variation by a constrained inversion of borehole flexural-wave dispersion data. *Geophysics* **2010**, *75*, E183–E190. [CrossRef]
17. Su, Y.-D.; Tang, X.-M.; Hei, C.; Zhuang, C.-X. An equivalent-tool theory for acoustic logging and applications. *Appl. Geophys.* **2011**, *8*, 69–78. [CrossRef]
18. Xu, S.; Tang, X.-M.; Su, Y.; Zhuang, C. Determining formation shear wave transverse isotropy jointly from borehole Stoneley- and flexural-wave data. *Chin. J. Geophys.* **2018**, *61*, 5105–5114. [CrossRef]
19. Tang, X.M.; Cheng, A. *Quantitative Borehole Acoustic Methods*; Elsevier: Amsterdam, The Netherlands, 2004; pp. 120–125.
20. Norris, A.N. The speed of a tube wave. *J. Acoust. Soc. Am.* **1990**, *87*, 414. [CrossRef]

*Article*

# Numerical Simulation of Multiarea Seepage in Deep Condensate Gas Reservoirs with Natural Fractures

Lijun Zhang [1], Wengang Bu [2,*], Nan Li [1], Xianhong Tan [1] and Yuwei Liu [2]

[1]  CNOOC Research Institute Co., Ltd., Beijing 100028, China
[2]  School of Civil and Resource Engineering, University of Science and Technology Beijing, Beijing 100083, China
*   Correspondence: binbeiu@163.com

**Abstract:** Research into condensate gas reservoirs in the oil and gas industry has been paid much attention and has great research value. There are also many deep condensate gas reservoirs, which is of great significance for exploitation. In this paper, the seepage performance of deep condensate gas reservoirs with natural fractures was studied. Considering that the composition of condensate gas changes during the production process, the component model was used to describe the condensate gas seepage in the fractured reservoir, modeled using the discrete fracture method, and the finite element method was used to conduct numerical simulation to analyze the seepage dynamic. The results show that the advancing speed of the moving pressure boundary can be reduced by 55% due to the existence of threshold pressure gradient. Due to the high-speed flow effect in the near wellbore area, as well as the high mobility of oil, the condensate oil saturation near the wellbore can be reduced by 42.8%. The existence of discrete natural fractures is conducive to improving the degree of formation utilization and producing condensate oil.

**Keywords:** condensate gas reservoir; condensate oil; discrete fracture

**Citation:** Zhang, L.; Bu, W.; Li, N.; Tan, X.; Liu, Y. Numerical Simulation of Multiarea Seepage in Deep Condensate Gas Reservoirs with Natural Fractures. *Energies* **2023**, *16*, 10. https://doi.org/10.3390/en16010010

Academic Editors: Xingguang Xu, Kun Xie and Yang Yang

Received: 16 November 2022
Revised: 12 December 2022
Accepted: 14 December 2022
Published: 20 December 2022



## 1. Introduction

The development of petroleum industrial technologies and the quick growth in demand for oil and gas resources have meant that unconventional oil and gas resources have gradually become a hot spot in the industry, and the development of deep oil and gas reservoirs is gradually increasing. Of the newly discovered recoverable reserves of global condensate oil in 2020, deep and ultra-deep condensate oil accounts for 95% [1], and 100 billion cubic meters of large, deep condensate gas fields exist in Bohai [2]. However, problems such as their great depth and the complex hydrocarbon fluid phase state bring difficulties to their development.

Deep, fractured condensate gas reservoirs have obvious retrograde condensate characteristics in the development process, forming a multiscale, multiphase flow system with matrix fractures. Syed et.al [3] reported a comprehensive review of the complex nature of unconventional reservoirs. Apart from tight reservoirs, deep natural gas reservoirs and geo-pressurized zones, which usually exist under high pressure and high temperature, are also referred to as unconventional reservoirs. The low, dual porosity distribution and ultra-low permeability of those reservoirs are evident. According to an evaluation of core samples [4], the pore size of deep reservoirs can be tiny and heterogeneous. The matrix of deep formation is tight and has a low permeability, and the permeability difference between it and its fractures is large, which causes different flow characteristics to form [5,6]. For low-permeability reservoirs, many research results have shown the impact of threshold pressure gradient on seepage [7–10]. Many studies focus on oil, water, or oil–water phases [11,12], while practice has proved that there is a threshold pressure gradient in the development of low-permeability gas reservoirs [13–15]. Tian et al. [16] pointed out that the threshold pressure gradient of a tight sandstone condensate gas reservoir is positively related to the liquid saturation, and the threshold pressure gradient is significantly increased when fractures have not developed. Martyushev et.al [17] reported that the properties of fractured reservoirs cannot fully recover after damage caused by the stress state increasing. Therefore, there is a risk of a production decline due to reservoir pressure drops and fracture closures [18]. For the study of fractured reservoirs, the dual medium model, equivalent model, and discrete fracture model are mainly used. Due to its advantages with regard to simulation and accuracy, the discrete fracture model is widely used [19–21]. The discrete fracture method is popular for dealing with the matrix-fracture flow of shale gas. Wang et al. [22] established a discrete fracture model to analyze the impact of hydraulic fractures on shale gas development, taking into account non-Darcy flow, and this model can be used for ultra-low permeability reservoirs. Zhao et al. [23] established a discrete fracture network including micro fractures and hydraulic fractures, coupled with the gas exchange between the matrix and fracture, and analyzed the impact of fracture geometry and distribution on production.

The purpose of this paper is to study the seepage performance of deep condensate gas reservoirs with natural fractures and to establish a mathematical model for this subject. The model can be applied to calculate the formation pressure distribution and the expansion dynamic of the moving pressure boundary at different development stages, and to analyze the changes in condensate oil in different areas of the formation and the impact of natural discrete fractures on the seepage field.

## 2. Model and Methods

### 2.1. Physical Model

The seepage problem present in deep fractured condensate gas reservoirs was considered. The porosity and permeability of the formation were assumed to be homogeneous, and the temperature was constant. The seepage in the formation was unsteady; that is, the formation pressure was a function of time and position. The physical model of formation with natural fractures was built using the discrete fracture method, as shown in Figure 1. The model was divided into matrix area (gray entity) and fracture area (black line).

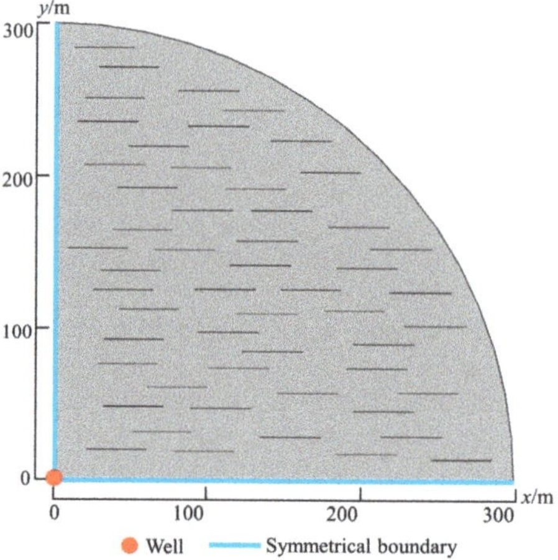


**Figure 1.** Formation model diagram.

### 2.2. Mathematical Description

The seepage field structure of condensate gas reservoirs can be considered as fitting the four-zone flow model [24], as shown in Figure 2. From the boundary to the bottom of the well, the formation is divided into a single-phase gas area, a transition area, an outer two-phase flow area, and an inner two-phase flow area. The formation pressure in the single-phase gas area is higher than the dew pressure, and no condensate is produced. The formation pressure in the transition zone decreases gradually, and there is no flowing condensate oil, only gas-phase flows. Gas and oil two-phase flow exists in both the outer and inner two-phase flow areas, which are divided by Reynolds number.

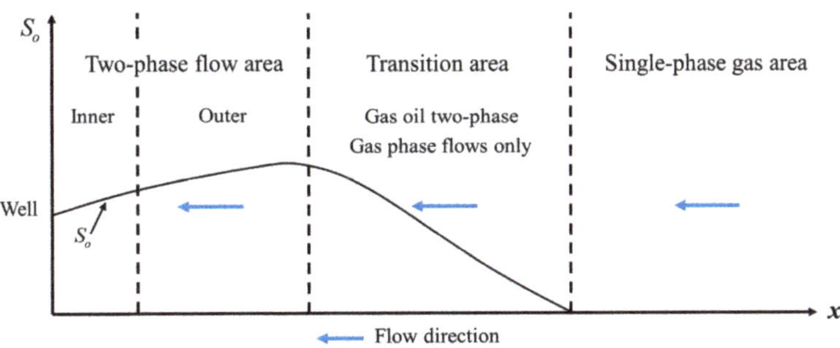


**Figure 2.** Schematic diagram of four-zone flow model.

### 2.2.1. Continuity Equation

The flow of condensate gas belongs to a multicomponent, multiphase flow, and the continuity equation of component $i$ is:

$$\nabla \cdot \left( x_i \rho_o v_o + y_i \rho_g v_g \right) + \frac{\partial}{\partial t} \left[ \phi_j \left( x_i \rho_o S_o + y_i \rho_g S_g \right) \right] - q_i = 0, \tag{1}$$

where $x_i$ and $y_i$ are the mole fractions of component $i$ in the oil phase and gas phase respectively; $t$ is the time; $\phi$ is the porosity; $\rho$ is the fluid density; $S$ is the saturation; $v$ is the velocity vector; $q_i$ is the source and sink item; subscript $j = m$ and $f$, representing matrix and fracture, respectively; and subscripts $o$ and $g$ denote oil and gas phases, respectively.

The dimension of the discrete fracture model can be reduced [25], and the continuity equation of the fracture can be written as:

$$\nabla \cdot \left[ d_f \left( x_i \rho_o v_o + y_i \rho_g v_g \right) \right] + d_f \frac{\partial}{\partial t} \left[ \phi_j \left( x_i \rho_o S_o + y_i \rho_g S_g \right) \right] - d_f q_i = 0, \tag{2}$$

where $d_f$ is the fracture width.

### 2.2.2. Motion Equation

In the single-phase gas area, the transition area, and the outer two-phase flow area, fluid flow can be described by Darcy's law modified by threshold pressure gradient:

$$v_n = -k_{rn} \frac{k_j}{\mu_n} (\nabla p_n - G), \tag{3}$$

where $v$ is the velocity; $k_r$ is the relative permeability; $\mu$ is the viscosity; $p$ is the pressure; $G$ is the threshold pressure gradient; and subscript $n = o$ and $g$.

In the inner two-phase flow area, the influence of the near-well high-speed non-Darcy effect shall be considered [26]. The non-Darcy term was introduced to describe the flow law and is written as:

$$-\left( \nabla p_n - G \right) = \frac{\mu_n}{k_{rn} k_j} v_n + \beta \rho_n v_n^2, \tag{4}$$

where $\beta$ is the inertial coefficient. The inertial coefficient reflects the deviation of seepage from Darcy's law. When oil and gas two-phases coexist in the near-wellbore area, the inertial coefficient can be calculated using [27]:

$$\beta = \frac{3.3 \times 10^{-9}}{\phi^{0.75} k^{1.25} s_g^{4.5}}, \tag{5}$$

The inner and outer two-phase flow areas are divided by Reynolds number, which can be calculated by Kajiahoff Equation (6), and the critical value is 0.2~0.3. The area where Reynolds number is greater than the critical value is the inner two-phase flow area, and the opposite is the outer two-phase flow area.

$$Re = \frac{v\rho\sqrt{k}}{17.5\mu\phi^{1.5}},\tag{6}$$

### 2.2.3. Constraint Equation

The constraint condition of the component is:

$$\sum x_i = \sum y_i = 1,\tag{7}$$

The constraint condition of saturation is:

$$S_o + S_g + S_w = 1,\tag{8}$$

where $S_w$ is the connate water saturation.

The total molar amount of component $i$ is:

$$z_i = Lx_i + Vy_i,\tag{9}$$

where $L$ and $V$ are the mole fractions of the oil phase and gas phase in the system, respectively, as shown in Equation (10).

$$L = \frac{\rho_o S_o}{\rho_o S_o + \rho_g S_g}, V = \frac{\rho_g S_g}{\rho_o S_o + \rho_g S_g},\tag{10}$$

### 2.2.4. Phase Equilibrium Equation

$$f_i^L = f_i^V,\tag{11}$$

where $f_i^L$ and $f_i^V$ represent the liquid fugacity and gas fugacity of component $i$, respectively.

### 2.2.5. State Equation

The state equation of rock is:

$$\phi = \phi_{int}[1 + C_s(p - p_{int})],\tag{12}$$

where $C_s$ is the rock compression coefficient; $\phi_{int}$ is the original porosity; and $p_{int}$ is the original pressure.

The fluid phase density is calculated using:

$$\rho_n = \frac{pM_n}{Z_n RT},\tag{13}$$

where $M$ is the average molecular weight; $Z$ is the deviation factor; $R$ is the ideal gas constant; and $T$ is the temperature.

The PVT properties are described by Peng-Robinson equation [28]:

$$p = \frac{RT}{V_r - b} - \frac{\alpha(T)}{V_r(V_r + b) + b(V_r - b)},\tag{14}$$

$$\alpha(T_c) = 0.45724\frac{(RT_c)^2}{p_c},\tag{15}$$

$$b = 0.0778\frac{RT}{p_c},\tag{16}$$

where $V_r$ is the molar volume; $p_c$ is the contrast pressure; and $T_c$ is the contrast temperature.

2.2.6. Boundary Conditions and Initial Conditions

The closed outer boundary condition is:

$$-\boldsymbol{n}\cdot(\rho\boldsymbol{v}) = 0, \tag{17}$$

where $\boldsymbol{n}$ is the boundary normal vector.

The specified rate condition for the inner boundary is:

$$v|_{\Omega_{in}} = \frac{q_0}{2\pi r_w h}, \tag{18}$$

where $\Omega_{in}$ represents the inner boundary, and $q_0$ is the production rate under formation condition.

The initial conditions can be written as:

$$p|_{t=0} = p_{int}, \tag{19}$$

$$S_o|_{t=0} = 0. \tag{20}$$

## 3. Model Validation

### 3.1. Static Parameters

The modeled typical deep fractured condensate gas reservoir had a depth of 4550.6 m, and a fracture width between 0.1 mm and 0.4 mm, with east-west strike fractures,. The static parameters of the formation are shown in Table 1.

**Table 1.** Formation statics parameters.

| Parameters | Value |
|---|---|
| Initial formation pressure/MPa | 51.3 |
| Formation temperature/°C | 178.4 |
| Matrix porosity | 0.0335 |
| Fracture porosity | 0.0052 |
| Matrix permeability/mD | 0.26 |
| Fracture permeability/mD | 1.79 |

### 3.2. Fluid Properties

The content of $C_1$, $C_2 \sim C_6 + CO_2$, and $C_7+$ in the condensate gas of the gas reservoir accounted for 71.97%, 18.76%, and 9.27%, respectively. The dew pressure was 47.56 MPa at the formation temperature of 178 °C. The P-T phase diagram of formation fluid is shown in Figure 3, in which 0%, 5%, 10%, and 20% represent the phase envelope lines at different condensate oil volumes. The components of condensate gas are complex, and calculations using the complete component will greatly reduce the calculation efficiency. Therefore, $C_2 \sim C_5$ components and $C_6 \sim C_{10}$ components were each combined into pseudo components. The PVT fitting was performed and dew pressure and relative volume fitting results are shown in Figures 4 and 5. The results show that the fitting effect of the pseudo components is good and can be used for calculation.

### 3.3. Validation

The equations were solved using the finite element numerical method. The measured production performance data of a production well within 480 days are matched with the above model, and the result is shown in Figure 6. In practice, the production rate is always changing. Therefore, in the early period of production, the measured production pressure data are unstable, and there are discrepancies between measured data and calculated data. After 390 days, the errors almost disappear, meaning that the model fits well.

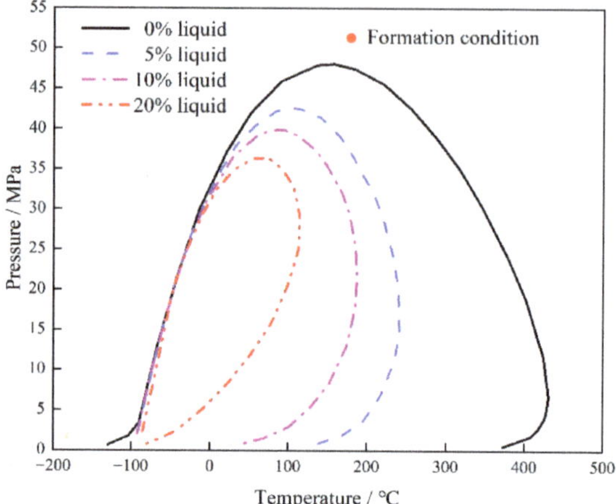

**Figure 3.** P-T phase diagram of formation fluid.

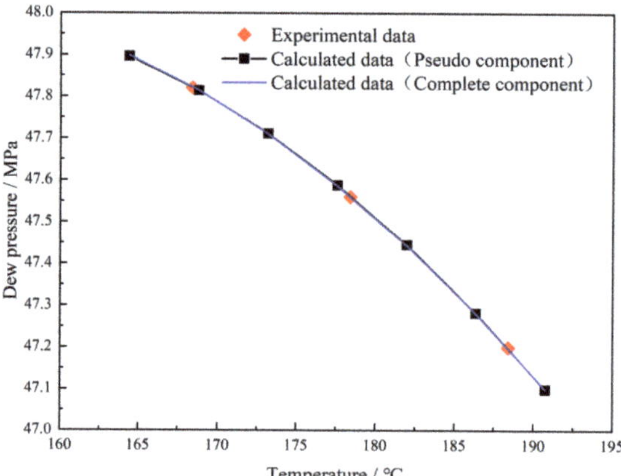

**Figure 4.** Dew pressure fitting diagram.

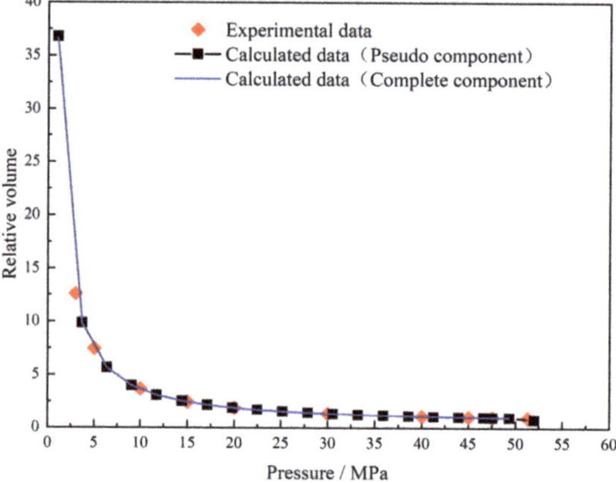

**Figure 5.** Relative volume fitting diagram.

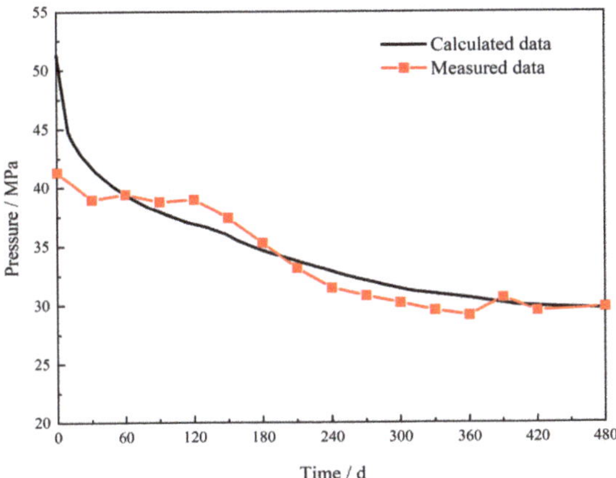

**Figure 6.** Comparison of measured and calculated pressure data.

## 4. Simulation Results and Discussion

In this section, a production simulation was conducted using the specified rate condition with a value of $14.27 \times 10^4$ m$^3$/d from the 480th day to the 1500th day.

### 4.1. Formation Pressure Distribution

Figure 7 shows the radial distribution nephogram of formation pressure at different times. With the development of production, the pressure propagation range increases, meaning that the producing range of the formation gradually expands. As can be seen from the pressure curves shown in Figure 8, the advancing speed of the pressure moving boundary gradually decreases. From 500 days to 750 days, the moving pressure boundary advances 20 m, while from 1250 days to 1500 days, the moving pressure boundary advances only 9 m, showing a relative decrease of 55% compared with the former. The formation energy decreases when the production is conducted, and at the same time, the pressure propagation also needs to overcome the threshold pressure gradient, so the moving boundary moves slowly.

### 4.2. Condensate Oil Distribution

Figure 9 shows the distribution of condensate oil from the wellbore to the far side of formation at different times. Generally speaking, the closer to the wellbore, the lower the bottom hole pressure is, the higher the condensate oil saturation is. However, high saturation always brings high liquidity. With the development of production, the two-phase flow area expands, and the condensate oil is separated and gradually forms continuous phases and flows. In addition, the high-speed gas flow will carry part of condensate oil out of the reservoir, so condensate oil saturation near the wellbore decreases. When the production reaches 1500 days, the condensate oil saturation near the wellbore decreases from 0.07 to 0.04, forming a decrease of 42.8%. The pressure drop spreads to the far side of the formation, thus the transition area also gradually expands. From the 500th day to 1500th day, the transition area boundary expands from 94.4 m to 128.3 m.

### 4.3. Influence of Discrete Natural Fractures

Figure 10 shows the nephogram of formation pressure distribution under different fracture permeability. The discrete natural fractures and matrix are spatially continuous, so the pressure field is continuous. Compared with the initial fracture permeability, the advancing distance of the moving pressure boundary increases from 120 m to 140 m, with a relative increase of 16.7% after the fracture permeability doubled. Figure 11 shows the variation of condensate oil saturation in the presence of fractures with different permeabilities. With an increase in fracture permeability, the condensate oil near the wellbore is further exploited, and the condensate oil saturation decreases. As the moving pressure boundary advances farther, the oil–gas two-phase area is further expanded, and more condensate oil is produced in the middle of the formation (101–135 m).

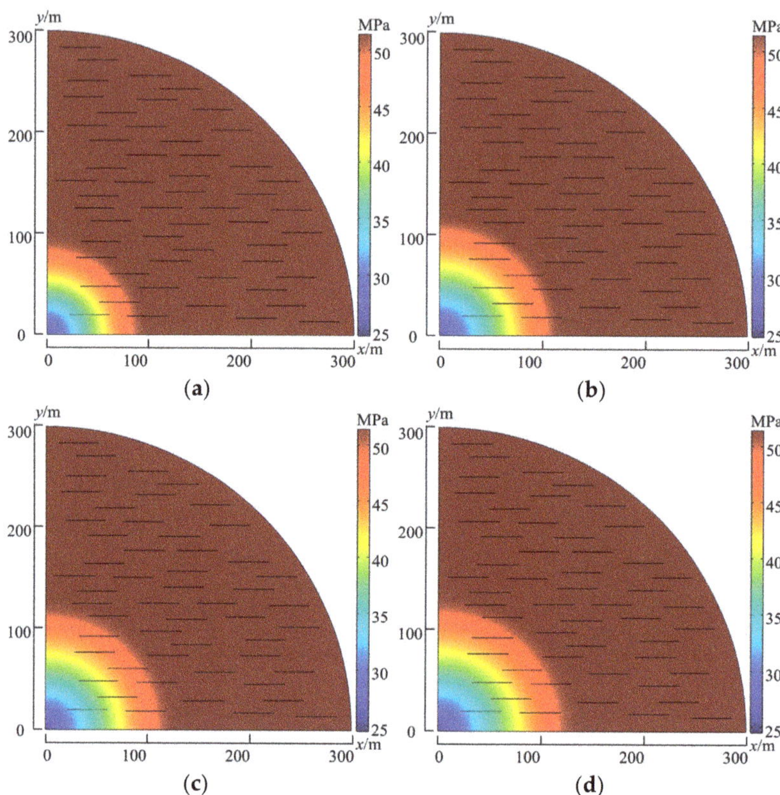

**Figure 7.** Formation pressure distribution at different times. (**a**) 500 d, (**b**) 1000 d, (**c**) 1250 d, (**d**) 1500 d.

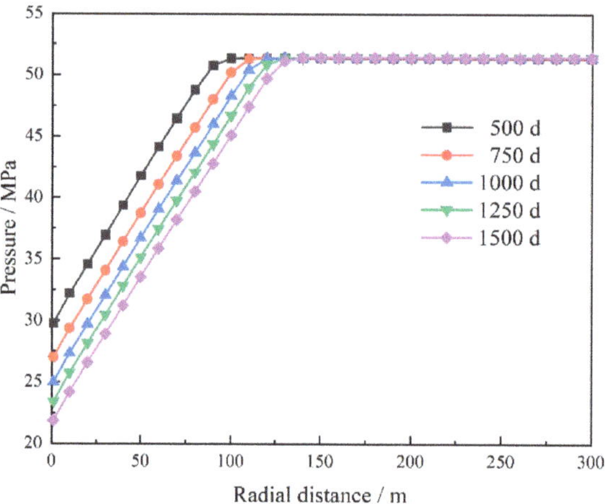

**Figure 8.** Radial pressure distribution at different times.

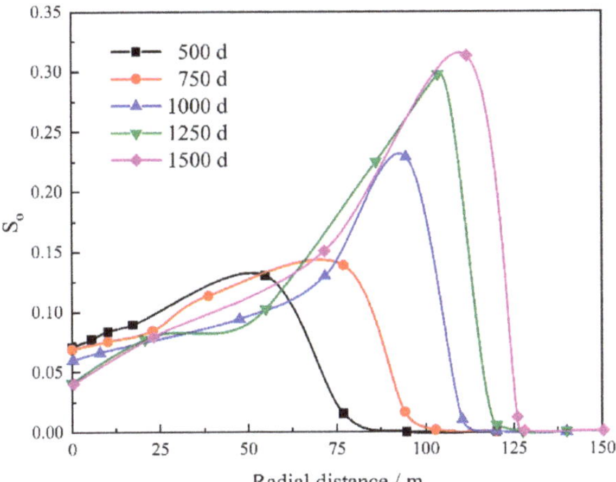

Figure 9. Radial oil saturation distribution at different times.

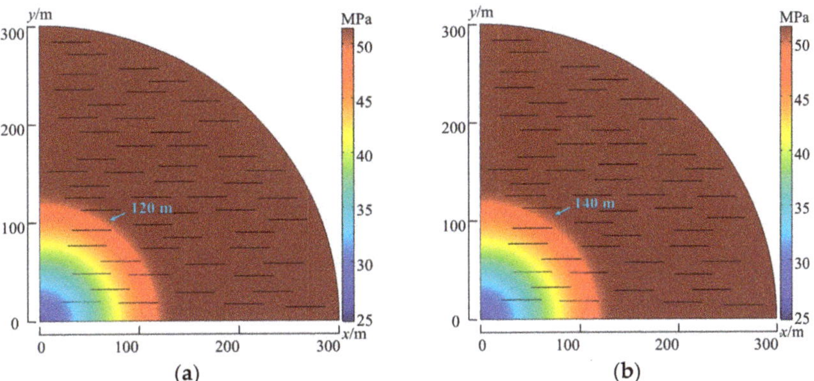

Figure 10. Formation pressure distribution under different fracture permeabilities. (**a**) $k_f$ = 1.79 mD, (**b**) $k_f$ = 3.58 mD.

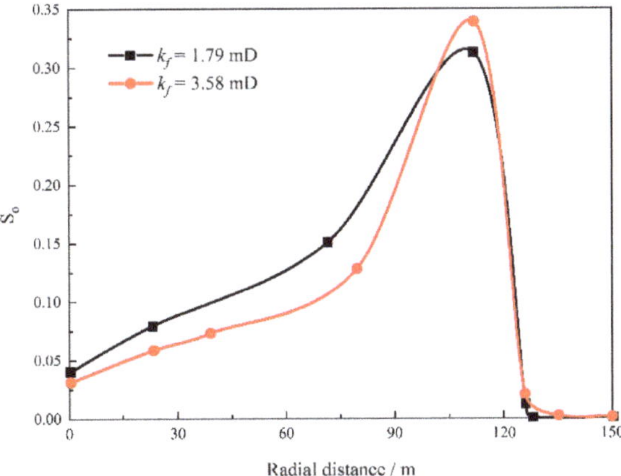

Figure 11. Radial condensate oil saturation under different fracture permeability.

## 5. Conclusions

In this paper, the component equation and discrete fracture method was applied to establish a seepage model of deep condensate gas reservoirs with natural fractures by considering the seepage laws of different flow areas, and the numerical simulation study was carried out to analyze the seepage dynamic. The following conclusions are drawn:

1. The existence of the threshold pressure gradient slows down the propagation speed of pressure to the far side of formation. For the same production interval, the advancing speed of the moving pressure boundary decreases by 55%.
2. The condensate oil saturation is not higher when it is closer to the wellbore. The continuous flow of condensate oil itself and the high-speed gas flow can reduce the condensate oil saturation near the wellbore by up to 42.8%.
3. The existence of natural fractures is conducive to the expansion of the moving pressure boundary and the utilization of formation and condensate oil. The gas–oil two-phase area becomes larger, and the condensate oil saturation in the middle of the formation increases.

**Author Contributions:** Conceptualization, L.Z. and W.B.; Methodology, L.Z., W.B. and Y.L.; Validation, N.L. and X.T.; Writing—original draft, W.B., N.L. and Y.L.; Writing—review and editing, L.Z. and X.T. All authors have read and agreed to the published version of the manuscript.

**Funding:** This research received no external funding.

**Data Availability Statement:** Not applicable.



## References

1. Wang, Z.; Wen, Z.; He, Z.; Song, C.; Liu, X.; Chen, R.; Liu, Z.; Bian, H.; Shi, H. Global condensate oil resource potential and exploration fields. *Acta Pet. Sin.* **2021**, *42*, 1556–1565.
2. Shi, H.; Wang, Q.; Wang, J.; Liu, X.; Feng, C.; Hao, Y.; Pan, W. Discovery and exploration significance of large condensate gas fields in BZ19-6 structure in deep Bozhong sag. *China Pet. Explor.* **2019**, *24*, 36–45.
3. Syed, F.I.; Dahaghi, A.K.; Muther, T. Laboratory to field scale assessment for EOR applicability in tight oil reservoirs. *Pet. Sci.* **2022**, *19*, 2131–2149. [CrossRef]
4. Martyushev, D.A.; Ponomareva, I.N.; Osovetsky, B.M.; Kazymov, K.P.; Tomilina, E.M.; Lebedeva, A.S.; Chukhlov, A.S. Study of the structure and development of oil deposits in carbonate reservoirs using field data and X-ray microtomography. *Georesursy* **2022**, *24*, 114–124.
5. Martyushev, D.A.; Ponomareva, I.N.; Filippov, E.V.; Li, Y. Formation of hydraulic fracturing cracks in complicated carbonate reservoirs with natural fracturing. *Bull. Tomsk Polytech. Univ. Geo Assets Eng.* **2022**, *333*, 85–94.
6. Galkin, S.V.; Martyushev, D.A.; Osovetsky, B.M.; Kazymov, K.P.; Song, H. Evaluation of void space of complicated potentially oil-bearing carbonate formation using X-ray tomography and electron microscopy methods. *Energy Rep.* **2022**, *8*, 6245–6257. [CrossRef]
7. Luo, X.; Wang, X.; Wu, Z.; He, T.; Qiu, X.; Yuan, F.; Tan, C. Study on Stress Sensitivity of Ultra-Low Permeability Sandstone Reservoir Considering Starting Pressure Gradient. *Front. Earth Sci.* **2022**, *10*, 890084. [CrossRef]
8. Li, D.; Zha, W.; Liu, S.; Wang, L.; Lu, D. Pressure transient analysis of low permeability reservoir with pseudo threshold pressure gradient. *J. Pet. Sci. Eng.* **2016**, *147*, 308–316. [CrossRef]
9. Wang, R.; Yang, J.; Wang, M.; Zhao, Y.; Chen, W. Effect of Threshold Pressure Gradients on Control Areas Determination of Production Well in CBM Reservoirs. *Adv. Polym. Technol.* **2019**, *2019*, 3517642. [CrossRef]
10. Ren, S.; Shen, F.; Yang, S.; Zhang, X.; Luo, H.; Feng, C. Analysis and Calculation of Threshold Pressure Gradient Based on Capillary Bundle. *Math. Probl. Eng.* **2021**, *2021*, 5559131. [CrossRef]
11. Shi, X.; Wei, J.; Bo, H.; Zheng, Q.; Yi, F.; Yin, Y.; Chen, Y.; Dong, M.; Zhang, D.; Li, J.; et al. A novel model for oil recovery estimate in heterogeneous low-permeability and tight reservoirs with pseudo threshold pressure gradient. *Energy Rep.* **2021**, *7*, 1416–1423. [CrossRef]
12. Liu, H.; Zhao, T. Theoretical and Experimental Study on Nonlinear Flow Starting Pressure Gradient in Low Permeability Reservoir. In *IOP Conference Series: Earth and Environmental Science*; IOP Publishing: Bristol, UK, 2020; Volume 571, p. 012044.
13. Lu, C.; Qin, X.; Ma, C.; Yu, L.; Geng, L.; Bian, H.; Zhang, K.; Zhou, Y. Investigation of the impact of threshold pressure gradient on gas production from hydrate deposits. *Fuel* **2022**, *319*, 123569. [CrossRef]

14. Bo, N.; Zuping, X.; Xianshan, L.; Zhijun, L.; Zhonghua, C.; Bocai, J.; Xin, Z.; Huan, T.; Xiaolong, C. Production prediction method of horizontal wells in tight gas reservoirs considering threshold pressure gradient and stress sensitivity. *J. Pet. Sci. Eng.* **2020**, *187*, 106750. [CrossRef]
15. Lu, J. Pressure Behavior of a Hydraulic Fractured Well in Tight Gas Formation with Threshold Pressure Gradient. In Proceedings of the SPE Middle East Unconventional Gas Conference and Exhibition, Abu Dhabi, United Arab Emirates, 23–25 March 2012.
16. Tian, W.; Zhu, W.; Zhu, H.; Zhang, X.; An, Z.; Du, S. Condensate gas reservoir threshold pressure gradient for tight sandstone. *J. Cent. South Univ.* **2015**, *46*, 3415–3421.
17. Martyushev, D.A.; Galkin, S.V.; Shelepov, V.V. The influence of the rock stress state on matrix and fracture permeability under conditions of various lithofacial zones of the tournaisian-fammenian oil fields in the Upper Kama Region. *Moscow Univ. Geol. Bull.* **2019**, *74*, 573–581. [CrossRef]
18. Martyushev, D.A.; Yurikov, A. Evaluation of opening of fractures in the Logovskoye carbonate reservoir, Perm Krai, Russia. *Pet. Res.* **2021**, *6*, 137–143. [CrossRef]
19. Xia, Y.; Jin, Y.; Chen, M.; Chen, K.P. An Enriched Approach for Modeling Multiscale Discrete-Fracture/Matrix Interaction for Unconventional-Reservoir Simulations. *SPE J.* **2018**, *24*, 349–374. [CrossRef]
20. Zidane, A.; Firoozabadi, A. An efficient numerical model for multicomponent compressible flow in fractured porous media. *Adv. Water Resour.* **2014**, *74*, 127–147. [CrossRef]
21. Xia, Y.; Jin, Y.; Oswald, J.; Chen, M.; Chen, K. Extended finite element modeling of production from a reservoir embedded with an arbitrary fracture network. *Int. J. Numer. Methods Fluids* **2018**, *86*, 329–345. [CrossRef]
22. Wang, Y.; Shahvali, M. Discrete fracture modeling using Centroidal Voronoi grid for simulation of shale gas plays with coupled nonlinear physics. *Fuel* **2016**, *163*, 65–73. [CrossRef]
23. Zhao, Y.; Lu, G.; Zhang, L.; Wei, Y.; Guo, J.; Chang, C. Numerical simulation of shale gas reservoirs considering discrete fracture network using a coupled multiple transport mechanisms and geomechanics model. *J. Pet. Sci. Eng.* **2020**, *195*, 107588. [CrossRef]
24. Gringarten, A.C.; Al-Lamki, A.; Daungkaew, S.; Mott, R.; Whittle, T.M. Well Test Analysis in Gas-Condensate Reservoirs. In Proceedings of the SPE Annual Technical Conference and Exhibition, Dallas, TX, USA, 1–4 October 2000.
25. Huang, Z.; Yan, X.; Yao, J. A Two-Phase Flow Simulation of Discrete-Fractured Media using Mimetic Finite Difference Method. *Commun. Comput. Phys.* **2014**, *16*, 799–816. [CrossRef]
26. Forchheimer, P. Wasserbewegung durch boden. *Z. Ver. Dtsch. Ing.* **1901**, *45*, 1782–1788.
27. Blom, S.M.P.; Hagoort, J. The Combined Effect of Near-Critical Relative Permeability and Non-Darcy Flow on Well Impairment by Condensate Drop Out. *SPE Reserv. Eval. Eng.* **1998**, *1*, 421–429. [CrossRef]
28. Peng, D.-Y.; Robinson, D.B. A New Two-Constant Equation of State. *Ind. Eng. Chem. Fundam.* **1976**, *15*, 59–64. [CrossRef]

*Article*

# The Control of Sea Level Change over the Development of Favorable Sand Bodies in the Pinghu Formation, Xihu Sag, East China Sea Shelf Basin

Zhong Chen [1], Wei Wei [2,*], Yongchao Lu [1], Jingyu Zhang [1], Shihui Zhang [3] and Si Chen [1]

[1]   School of Earth Resources, China University of Geosciences, Wuhan 430074, China
[2]   School of Earth Science, China University of Geosciences, Wuhan 430074, China
[3]   School of Institute of Geophysics & Geomatics, China University of Geosciences, Wuhan 430074, China
*   Correspondence: weiwei910911@163.com; Tel.: +86-133-498-386-59

**Abstract:** The Pinghu Formation consists primarily of marine-continental transitional deposits. The widely distributed fluvial and tidal transgressive sand bodies comprise the main reservoirs of the Baochu slope zone in the Xihu Sag in the East China Sea Shelf Basin. These sand bodies are deeply buried, laterally discontinuous, and are frequently interrupted by coal-bearing intervals, thereby making it extremely difficult for us to characterize their hydrocarbon potential quantitatively via seismic inversion techniques, such as multi-attribute seismic analysis and post-stack seismic inversion, hindering further hydrocarbon exploration in the Xihu Sag. Here, a prestack seismic inversion approach is applied to the regional seismic data to decipher the spatiotemporal distribution pattern of the sand bodies across the four sequences, i.e., SQ1, SQ2, SQ3, and SQ4, from bottom up, within the Pinghu Formation. In combination with detailed petrology, well log, and seismic facies analysis, the secular evolution of the sedimentary facies distribution pattern during the accumulation of the Pinghu Formation is derived from the sand body prediction results. It is concluded that the sedimentary facies and sand body distribution pattern rely on the interplay between the hydrodynamics of fluvial and tidal driving forces from the continent and open ocean, respectively. Drops in the sea level led to the gradual weakening of tidal driving forces and relative increases in riverine driving forces. The direction of the sand body distribution pattern evolves from NE–SW oriented to NW–SE oriented, and the dominant sand body changes from tidal facies to fluvial facies. In addition, the sea level drop led to the decrease in the water column salinity, redox condition, organic matter composition, and the development of coal seams, all of which directly influenced the quality of reservoir and source rocks. The sand bodies in SQ2 and SQ3 are favorable reservoirs in the Pinghu Formation due to their good reservoir properties and great thickness. The high-quality source rock in SQ1 could provide significant hydrocarbons and get preserved in the sand body within SQ2 and SQ3. This contribution provides an insight into the control of the sea level change over the development of hydrocarbon reservoirs in the petroleum system from marginal-marine environments such as the Xihu Sag.

**Keywords:** reservoir; seismic sedimentology; river delta facies; tidal facies; lacustrine

**Citation:** Chen, Z.; Wei, W.; Lu, Y.; Zhang, J.; Zhang, S.; Chen, S. The Control of Sea Level Change over the Development of Favorable Sand Bodies in the Pinghu Formation, Xihu Sag, East China Sea Shelf Basin. *Energies* **2022**, *15*, 7214. https://doi.org/10.3390/en15197214

Academic Editor: Pål Østebø Andersen

Received: 5 July 2022
Accepted: 27 September 2022
Published: 30 September 2022

**Publisher's Note:** MDPI stays neutral with regard to jurisdictional claims in published maps and institutional affiliations.

## 1. Introduction

The quantitative constraints of sand bodies are critical for assessing the hydrocarbon potential within a petroleum system. Numerous seismic techniques, such as multi-attribute seismic analysis [1–6], post-stack seismic inversion [7–11], and prestack seismic inversion [12–16], have been developed and applied extensively to characterize the distribution of reservoir sand bodies within hydrocarbon-bearing basins and its correlation with external factors, such as tectonic settings and sea level changes. Multi-attribute seismic analysis relies on variations in the reflection amplitude, apparent frequency, continuity, external form, and internal reflection configuration to characterize the sand body formation within a given study interval [17–19]. Furthermore, seismic wave attenuation, known by its high sensitivity to lithology variation in rocks, has a great potential to characterize sand bodies [20,21]. However, the non-uniqueness of seismic attribute interpretations and the lack of well-log constraints often lead to the relatively low accuracy of sand body prediction results. Seismic inversion is currently one of the most important techniques for reservoir prediction [22–25]. Traditional post-stack seismic inversion methods are conducted under the assumption that no distance exists between the source and receiver; the characteristics of the effective reservoir, including its fluid content, are obtained using a single compressional-wave (P-wave) impedance parameter [26,27], making it difficult to identify the effective reservoir and characterize its fluid content via this method. In contrast, prestack inversion methods can incorporate well-log data and provide a more quantitative interpretation of the target reservoir using multiple elastic parameters, including P- and shear-wave (S-wave) impedances and density [6,28,29].

The Xihu Sag is one of the most important petroliferous basins in eastern China; numerous sandstone gas fields have been developed in this area, with the Baochu slope zone representing one of the most petroliferous units [30,31]. The Baochu slope zone covers an area of 9000 km$^2$; it has been investigated extensively using three-dimensional (3D) seismic datasets that have spanned most of the area, as well as 53 well logs and associated geochemical analyses. The Pinghu Formation consists mainly of delta front and tidal channel deposits [32,33]. The widely distributed deltaic and tidal transgressive sand bodies are the main reservoirs in the Baochu slope zone [34]. However, these sand bodies are thin and deeply buried, with poor continuity, and are frequently interrupted by coal-bearing strata. These features make it difficult to accurately predict the distribution pattern of the sand bodies via traditional post-stack seismic inversion methods, hence hindering further petroleum exploration in the Xihu Sag. In additions, in marginal-marine environments, the hydrodynamic activities are associated closely with the sea level change, which may exert significant influence over the sand body distribution within sedimentary basins such as the Xihu Sag [35,36]. Higher resolution sand body prediction is therefore needed for a better understanding of the co-evolution of the sea level change and the tempo-spatial distribution of sand bodies for hydrocarbon-bearing basins such as the Xihu Sag.

The present contribution investigates a block (study area) from the Baochu slope zone in the Xihu Sag via an integrated analysis of well-log, drill-core, and 3D seismic data, while a prestack stochastic seismic inversion technique is applied in order to: (1) quantitatively predict the spatiotemporal sand body and sedimentary facies distributions within the sequence stratigraphic framework of the Pinghu Formation; (2) evaluate the favorable reservoir intervals in the study area; and (3) demonstrate the control of sea level change over the sand body distribution pattern within the Pinghu Formation.

## 2. Geological Setting

The Xihu Sag is situated in the center of the East China Sea Shelf Basin (ECSSB), covering an area of 46,000 km$^2$ and representing one of the most petroliferous provinces in China (Figure 1a) [32,37]. The Xihu Sag is surrounded by several uplifts and mountain belts, with the Diaoyudao uplift to the east and the Hupijiao, Haijiao and Yushan uplifts to the west; the Yangtze and Diaobei sags lie to the north and south, respectively (Figure 1b). The Xihu Sag can be further subdivided into the Baochu slope zone, the Santan deep-depression zone, the Central anticline zone, the Baidi deep-depression zone, and the Tianping fault terrace zone from west to east (Figure 1c). Among these tectonic units, the Baochu slope zone is the most petroliferous unit in the western Xihu Sag [38,39].

The Xihu Sag was deposited within a semi-restricted bay environment, with deltaic–estuarine, tidal flat, and fluvial–lacustrine facies dominating the depositional sequences (Figure 1c) [40]. Multiple tectonic movements occurred in the Xihu Sag, including the Pinghu movement which occurred in the middle of the Eocene, the Yuquan movement at the end of the Eocene, the Huagang movement at the end of the Oligocene, and the Longjing movement at the end of the Miocene. The tectonic evolution history includes a rifting stage, a depression stage, and a regional subsidence stage (Figure 2) [32,41–43]. The

Paleogene to Quaternary sedimentary succession in the Xihu Sag is up to ~10,000 m thick and consists primarily of thick sandstones and mudstones with localized coal seams and limestone deposits [32,40]. The Baoshi and Pinghu formations, deposited primarily during the rifting stage, are dominated by deltaic–estuarine sedimentary facies [44]. The Huagang, Longjing, Yuquan, Liulang, Santan, and Donghai formations, which were deposited during the following depression and regional subsidence stages, developed mainly fluvial–deltaic–lacustrine sedimentary facies (Figure 2) [30,45].

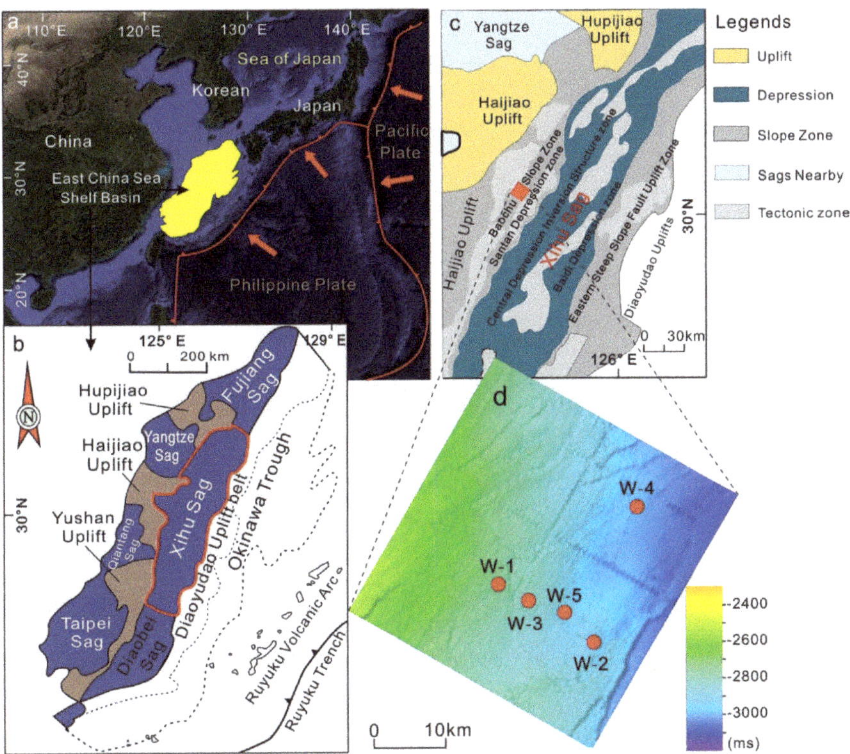


**Figure 1.** (**a**) Location of the study area; (**b**) tectonic units in the East China Sea Shelf Basin; (**c**) tectonic units of the Xihu sag; and (**d**) A block and the location of the study cores. Developed from [37,40].

The study area A block is located in the central Baochu slope zone and covers an area of 163 km$^2$ (Figure 1c). The fault system in A block is dominated by NE–NNE-oriented normal faults [46]. The Eocene Pinghu and Oligocene Huagang formations are the main oil-bearing intervals in the Baochu slope zone [33,47]. Extensive hydrocarbon exploration has been conducted around the Baochu slope, with considerable hydrocarbon potential discovered, especially within A block [46].

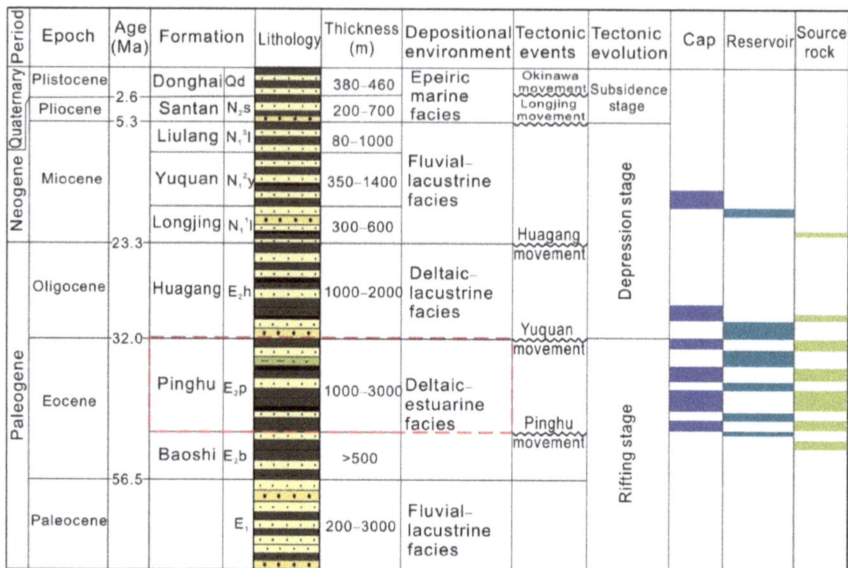


**Figure 2.** Generalized Cenozoic stratigraphy of the Xihu Sag, East China Sea Basin. The dashed red rectangle shows the present study interval (Eocene–Oligocene Pinghu Fm., adapted with permission from [48–52]. 2012, Zhu et al.; 2018, Su et al.; 2022, Quan et al.).

## 3. Sequence Stratigraphy

Identifying the unconformities is critical for sequence stratigraphic analysis [53,54]. In this study, the identification of sequence boundaries was based on geological data from a set of regional 3D seismic profiles (e.g., Figure 3) and five exploration boreholes (Figure 1d), which allowed for the establishment of a sequence stratigraphic framework for the Pinghu Formation. A total of 46 3D seismic profiles from A block were extracted and analyzed using standard seismic processing and sequence analysis theory and methods [55,56]. These seismic data were then complemented by 55 well logs and drill cores from the five boreholes tied to seismic profiles. Unconformities were identified with obvious truncations and surfaces of toplap, onlap, and downlap structures on seismic profiles (Figure 3) [54,56]. Sequence boundaries can be determined by the recognition of incised valleys and lowstand fans. In drill cores, scour surfaces and contacts, characterized by abrupt transitions in lithology, grain size, and porosity/permeability values, can be recognized as possible sequence boundaries. These features often correspond to abrupt changes in seismic reflection characteristics and shifts in well-log profiles.

The wavelet transform is a widely used tool for geological interpretation and stratigraphic sequence division [57–59]. It is a local time–frequency transformation process that can effectively extract hidden information from the signal and conduct multi-scale refinement analysis. The rhythmic characteristics of sediment deposition are formed by the superposition of several sedimentary cycles with different periods. The abrupt change points or regions between the frequency structure sections can then be determined from the processed signals, with these points/regions reflecting changes in the depositional setting of the formation. Here, we selected the Morlet wavelet after a series of tests and analyses, as this wavelet is more applicable for dividing sedimentary cycles. The wavelet coefficient curve of the wavelet transform also yields a corresponding good relationship with the formation lithology. The strong oscillation section of the wavelet coefficient curve corresponds to a sandstone depositional environment, which is expressed as a high-energy deposition section (red) and generally contains coarse-grained sediments, and the gentle oscillation section corresponds to a mudstone depositional environment, which is expressed as a low-energy deposition section (blue) and generally contains fine-grained sediments (Figure 4). Therefore, the vertical changes in the wavelet coefficient curve from gentle to strong oscillations reflect the sediment accumulation mode (progradation, retrogradation, or aggradation), which correspond to changes in the base sea level.

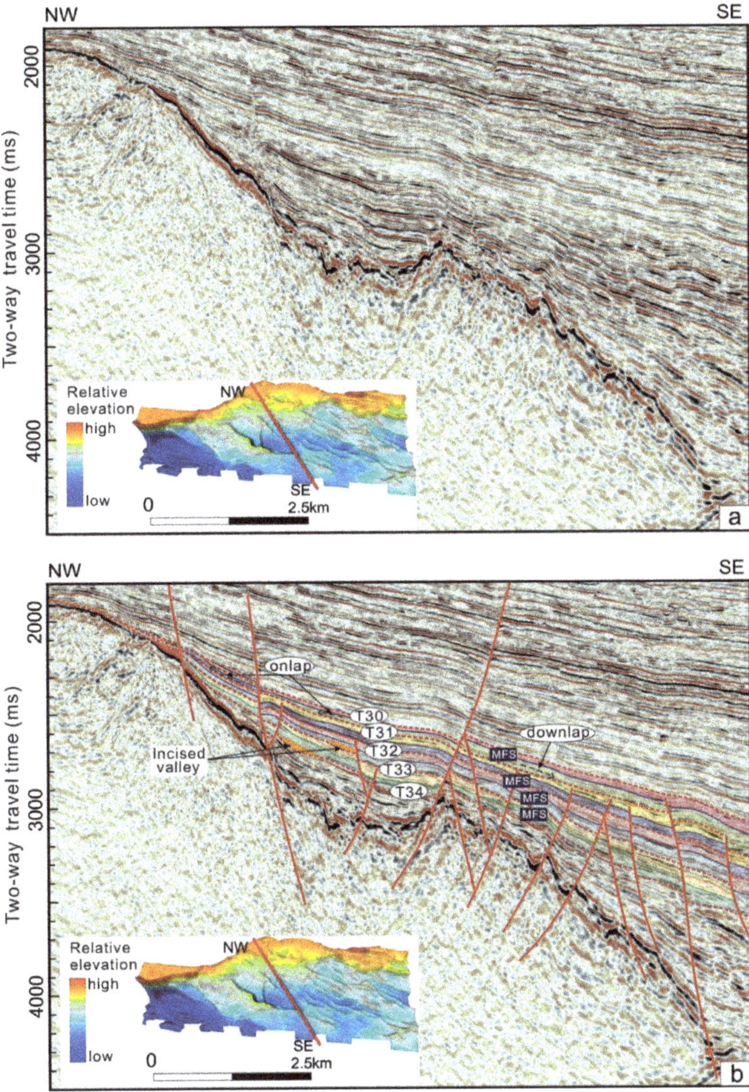


**Figure 3.** (**a**) Uninterpreted NW–SE seismic profile and (**b**) identification of sequence boundaries and division of sequence units in this seismic profile. The inset shows the location of the seismic profile in the Xihu Sag.

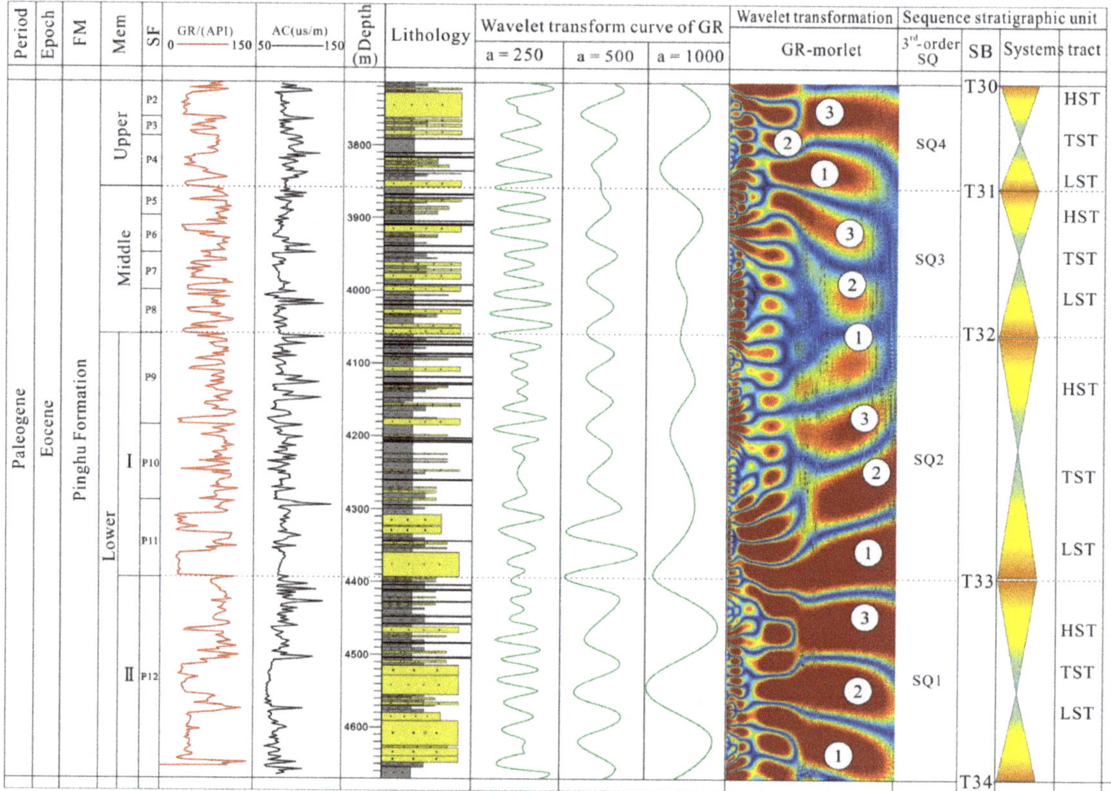

**Figure 4.** W-2 well logging and wavelet analysis of sequence boundaries in the Pinghu Formation of the Xihu Depression. Each third-order sequence exhibits a tripartite wavelet power signature (denoted as circled values 1–3 in the wavelet transform column) that corresponds to the three systems tracts of that sequence.

Stratigraphic successions are commonly assigned to a hierarchy of sequences [17]. Based on the sequence architecture analysis of the Xihu Sag described above, we assigned the full Eocene succession as a single first-order sequence and its constituent formations, i.e., the Pinghu and Baoshi formation, to two second-order sequences (Figure 2). Four third-order sequences were within the Pinghu Formation, SQ1-SQ4 from bottom up. SQ1 and SQ2 represent the lower and upper parts of the lower member of the Pinghu Formation, respectively. SQ3 and SQ4 represent the middle and upper members of the Pinghu Formation, respectively (Figure 2).

The base of the Pinghu Formation represents a second-order sequence boundary (SB1; the T34 surface) and corresponds to the onset of the Pinghu movement; the Baoshi Formation lies below this boundary. The top of the Pinghu Formation is a first-order sequence boundary (SB5; the T30 surface) and corresponds to the Yuquan movement (Figure 2). SB5 (T30) is characterized by an abrupt increase in the gamma ray (GR) values in the well logs and is marked by an abrupt lithological change from sandstone to mudstone in the W-2 drill core (Figure 4). The lower member of the Pinghu Formation is subdivided by SB2 (the T33 surface), which is a third-order sequence boundary. The top of SQ2 is SB3 (the T32 surface), a second-order sequence boundary that characterizes the transition from the fault-rifting stage (lower Pinghu Formation) to the subsidence stage (middle and upper Pinghu Formation). Sequences SQ3 and SQ4 are subdivided by the third-order sequence boundary SB4 (the T31 surface; Figures 3 and 4).

The T34 surface marks the basal boundary of the Pinghu Formation and is a transitional boundary that represents a fault-rifting subsidence unconformity. The upper Baoshi Formation was deposited during the fault-rifting stage, and the lower Pinghu Formation was deposited during the fault-rifting-to-depression transitional stage. The T34 surface is a continuous and strong seismic reflector. Onlap and truncation features are commonly observed, especially along the margins of the basin. The absolute age of this surface is ca. 42.5 Ma based on evidence from the SHRIMP Zircon U-Pb ages of volcanic rocks (Figure 2) [49].

The T33, T32, and T31 surfaces represent the SQ1/SQ2, SQ2/SQ3, and SQ3/SQ4 sequence boundaries, respectively. These boundaries are closely related to the tectonic evolution of the region. The T33 surface possesses a relatively weak lateral continuity across the study area, with the seismic reflection above this surface exhibiting chaotic and wavy characteristics. In seismic profiles, incised valley structures are commonly observed along the margins of the basin (Figure 3). This reflector gradually evolves into a parallel conformable contact in the center of the basin. In well log, this surface coincides with an abrupt lithological change from mudstone to sandstone, and an abrupt decrease in the GR values is observed in the W-2 well log (Figure 4).

The T32 surface is featured by a continuous, large-amplitude reflector, with a weak onlap observed along the margins of the basin. The absolute age of the T32 surface is ca. 36 Ma (Figure 2) [49], with this surface representing the boundary of the lower and middle members of the Pinghu Formation. The T32 surface coincides with an abrupt increase in the amplitude of the GR wavelet transform curve in the well logs (Figure 4), and the lithology changes from mudstone to sandstone across this surface.

The T31 surface possesses obvious onlap and truncation structures, with a continuous, large-amplitude seismic signature in the northern Xihu Sag that weakens to the south. The T31 surface is the interface between the middle and upper members of the Pinghu Formation. Although this surface is not as obvious as the other surfaces due to its relatively weak amplitude and discontinuous seismic signature, onlap and truncation structures can easily be observed in slope zones. This surface is obvious in the drill cores and is identified as a lithological change from coal-bearing mudstone to massive sandstone (Figure 4).

The T30 surface is a first-order tectonic unconformity between the lower Oligocene Huagang Formation and the upper Eocene Pinghu Formation that has an absolute age of 32.4 Ma [49]. Onlap structures are commonly observed along this surface and are indicative of rifting activity. The large scale of these overlap structures is observable across the entire Baochu slope zone (Figure 3) [49].

Systems tracts are recognized within each investigated sequence in the present study via integrated seismic, well-log, and drill core analysis. Each sequence contains a distinct lowstand systems tract (LST), a transgressive systems tract (TST), and a highstand systems tract (HST), with the systems tract analysis conducted using the transgressive surface (TS) and maximum flooding surface (MFS) constraints [56,60]. The well logs and drill cores were instrumental in identifying the small-scale sequence features such as the TS and MFS in the seismic data. The TS and MFS locations were determined on the basis of the distribution of the systems tracts within each sequence, with the energy shifts in the wavelet transform figures also contributing to the TS and MFS identification. For example, three high-energy clusters are observed in each sequence of the Pinghu Formation (Figure 4).

In the lacustrine system, the development of sequence stratigraphy in lacustrine systems from continental–marine transitional areas is dominated by the tectonic evolution of the lake basin, sea level change, and climate change [61–63]. The sequence development of the Pinghu Formation along the Baochu slope zone is a function of tectonics and sea level change. The Pinghu Formation was deposited primarily during the late rifting stage when the tectonic activity evolved gradually from fault rifting to strike-slip faulting (Figure 2). The Xihu Sag then entered the post-rifting depression stage after the Yuquan movement [64,65]. During the Pinghu stage, there was a gradual drop in sea level from SQ1 to SQ4, and the deposits exhibit a finning upward pattern [65]. The SQ1 and SQ2 were deposited during the lower Pinghu stage, when the sea level was relatively high and the Xihu Sag lacustrine system was connected partly to the open ocean. The sediments deposited during this stage are dominated by mudstone and sandstone of tidal, swamp, and estuarine facies [65–67]. The SQ3 and SQ4 were deposited during the middle and upper Pinghu stage, respectively; when the sea level was relatively low, the basin was being uplifted, and the connectivity to the ocean was weakened [65]. The deposits within SQ3 and SQ4 are dominated by fluvial, deltaic, and lacustrine facies [65,67].

It is worth mentioning that the Pinghu formation was often subdivided into three third-order sequences in which the lower, middle, and upper members were assigned as SQ1, SQ2, and SQ3, respectively [59,68]. However, much of the evidence observed in the present study has led to the identification of T33 as the sequence boundary that subdivided the lower member of the Pinghu formation into two third-order sequences, including a strong seismic reflector that is characterized by overlap, an incised valley (Figure 3), strong energy shifts in the wavelet transform diagram (Figure 4), etc.

## 4. Methods

### 4.1. Prestack Seismic Inversion

A stochastic prestack seismic inversion technique was applied to predict the sand body distribution. The stochastic prestack seismic inversion is a combination of the deterministic seismic inversion method and stochastic simulation theory; it is a stochastic prestack seismic inversion based on amplitude vs. angle (AVA) simultaneous inversion and sequential Gauss simulations [69,70]. Based on the sparse pulse inversion and model-based inversion methods, well-seismic calibration and AVA simultaneous inversion were applied to multi-dimensional angle gathers. These procedures produce a data volume of multiple elastic parameters, including Poisson's ratio, P- and S-wave velocities ratios (Vp/Vs), and density. The lithology (sand, mud) of each stratum was recognized and documented firstly; afterwards, the correlation between lithology with all the elastic parameters was analyzed. After these processes, the sequential Gauss simulations were applied in order to obtain the probability distribution of the gathers, which is based on the variational functions and the Kriging method. Random sampling was applied according to the Gauss distribution, and multiple gathers and multiple synthetic seismograms were produced. These results are compared with seismic gathers; the corresponding inversion result from the synthetic seismogram that matches the most is the final realization. With these processes, we can obtain equal probability realizations of the lithology data of the study area.

According to the theory and techniques demonstrated above, the specific workflows are presented in Figure 5. Firstly, denoising was applied to the prestack common reflection point gather data, which were then processed via conventional approaches to extract the super gathers and angle gathers. Different angle gathers, which range in angles from 3 to 36 degrees, were then stacked selectively, thereby producing partially stacked gathers (near-to-middle and far-offset stacks; Figure 5). A low-frequency model constrained by the seismic velocity structure was created based on the prestack seismic data, well logs, and structural interpretations. The partially stacked gathers and low-frequency model were then employed during the prestack elastic wave-impedance inversion process. Multiple elastic parameters, including Poisson's ratio, Vp/Vs, and density, were obtained for the subsequent sand body prediction procedure (Figure 5), with which we can obtain the final lithologic inversion data for the sand body prediction.

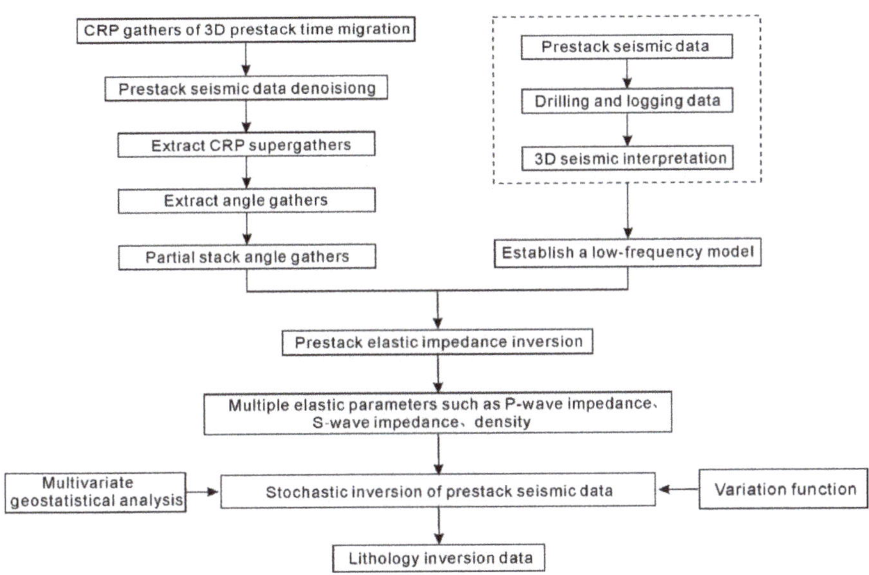

**Figure 5.** Flow chart of the stochastic prestack inversion technique used in the present study.

*4.2. Lithologic Discrimination*

Cross plots of Poisson's ratio, the Vp/Vs ratio, S-wave impedance, and density versus the P-wave impedance and GR data were analyzed to determine the elastic parameters for effective sand body prediction. The cross plots indicate that the Vp/Vs and Poisson's ratios versus the P-wave impedance are most effective in distinguishing the mudstone and sandstone lithologies, with thresholds of Vp/Vs > 1.77 and Poisson's ratio > 0.27 for mudstone, and Vp/Vs < 1.77 and Poisson's ratio < 0.27 for sandstone (Figure 6). The cross plot of Poisson's ratio and the Vp/Vs ratio shows minor overlaps and exhibits strong discrimination between mudstone and sandstone. The overlap of the mudstone and sandstone is ~50% or greater in other cross plots. This also indicates that the sand body distribution characterized by post-stack seismic inversion, which merely uses the P-wave impedance to determine the lithological distribution, features lower accuracy. Therefore, combining the stochastic prestack inversion process (based on Poisson's ratio, P- and S-wave velocity ratios, and density) with transverse and longitudinal variation functions and multivariate geostatistical analyses (based on the Bayesian theory) may help to obtain more reliable lithological data.

The Vp/Vs and Poisson's ratios are therefore used in the seismic statistical inversion to generate the attribute data for lithologies. This article uses Hampson–Russell software for geostatistical inversion and lithofacies prediction. During geostatistical inversion, few modifications were made to the parameters. Well W-1, W-2, W-3, W-4, and W-5 were utilized to constrain the inversion and obtain the lithofacies prediction results. In the seismic profile (Figure 7), the geostatistical inversion results from the Vp/Vs and Poisson's ratios show high consistency, demonstrating their validity in lithology discrimination (Figure 7b,c). The lithology profile characterized based on the Vp/Vs and Poisson's ratios shows good agreement with the log-predicted lithofacies at the well points and good continuity between the wells (Figure 7d).

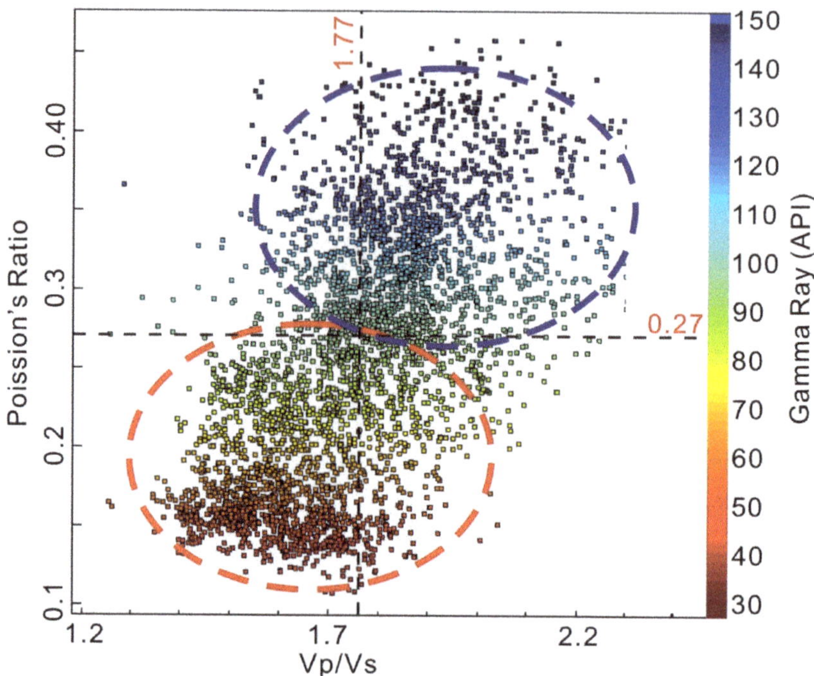


**Figure 6.** Cross plot of Poisson's ratio, the Vp/Vs ratio and gamma-ray values from W-1, W-2 W-3, W-4, and W-5 well.

The same geostatistical inversion process was applied to the Vp/Vs ratio and Poisson's ratio of a horizontal 75 ms time slice, with the high (red) and low (blue) values indicating sandstone and mudstone, respectively (Figure 8a,b). These two figures exhibit a high degree of consistency between the Vp/Vs and Poisson's ratio, illustrating the lithofacies distribution within the study area. The lateral spatial distribution pattern of the different lithologies within a specific interval was then calculated based on the geostatistical inversion of these two elastic parameters (Figure 8c). The lateral distribution of the sand body can be characterized by the accumulation of the sandstone thickness within a specific stratum (Figure 8d). The sand body distributions of LSTs from SQ1 to SQ4 are obtained to acquire the spatiotemporal sand body distribution pattern within the Pinghu Formation (Figure 9).

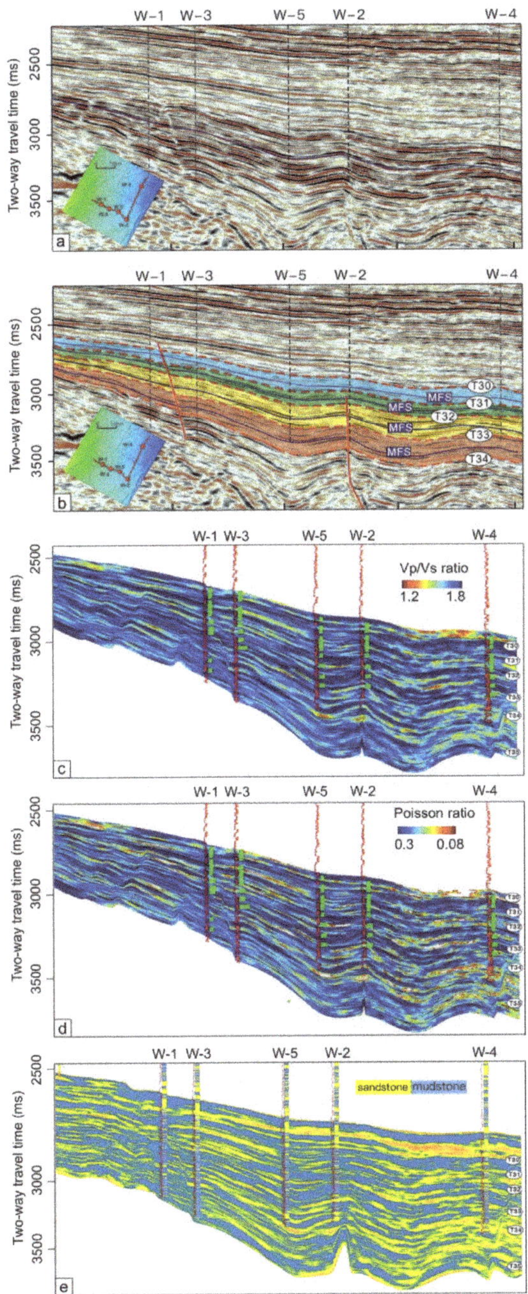

**Figure 7.** (**a**) The uninterpreted profile; (**b**) interpreted profile; the correlation profile inverted and (**c**) Vp/Vs; (**d**) Poisson's ratio; and (**e**) sand and mudstone profile of well W-4, W-2, W-5, W-3, an W-1.

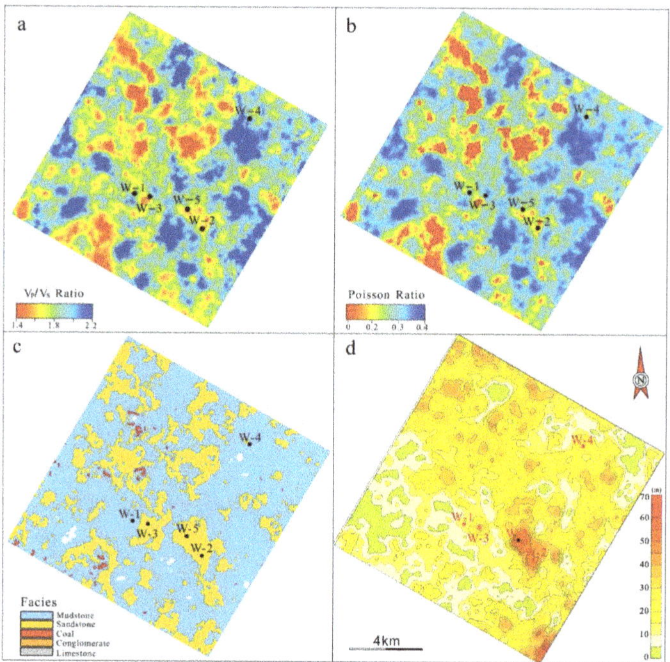

**Figure 8.** The inverted (**a**) Vp/Vs, (**b**) Poisson's ratio, (**c**) sand and shale section, and (**d**) the lateral distribution of sandstone in the LST stage of SQ4.

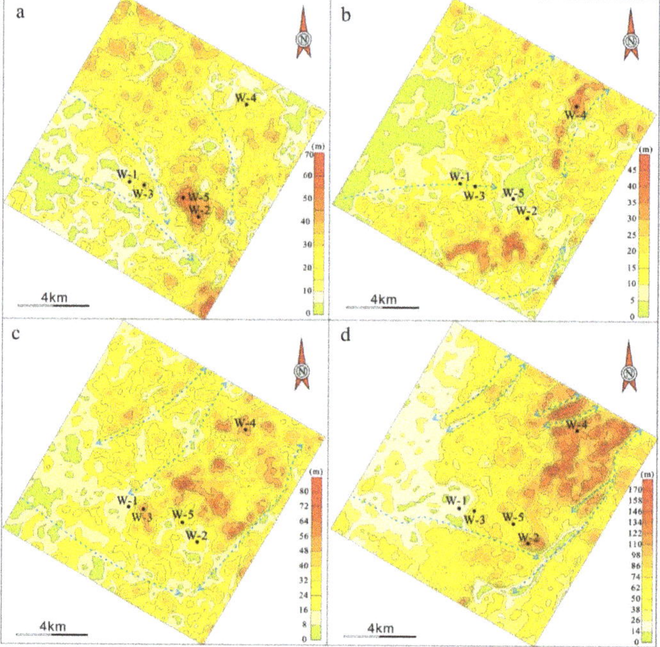

**Figure 9.** Lateral distribution of sand bodies within the LST stage of (**a**) SQ4, (**b**) SQ3, (**c**) SQ2, and (**d**) SQ1 based on the prestack seismic inversion technique. The blue dashed lines on the map indicate the possible dire.

### 4.3. Drillcore Analysis

The sedimentological analysis in this study is based on eight high-quality drill cores, which provide information on the depositional facies through the Pinghu Formation within the Xihu Sag. The core descriptions include the color of the fresh rock, grain size, sedimentary structures, the presence and type of clasts and bioclasts, deformation features, fracture characteristics (including width and orientation), and the presence of hydrocarbon and coal seams. Detailed drill core analyses from W-3, W-4, and W-5 from lower the Pinghu Formation are presented to demonstrate the procedure of the sedimentary facies analyses (Figures 10–12).

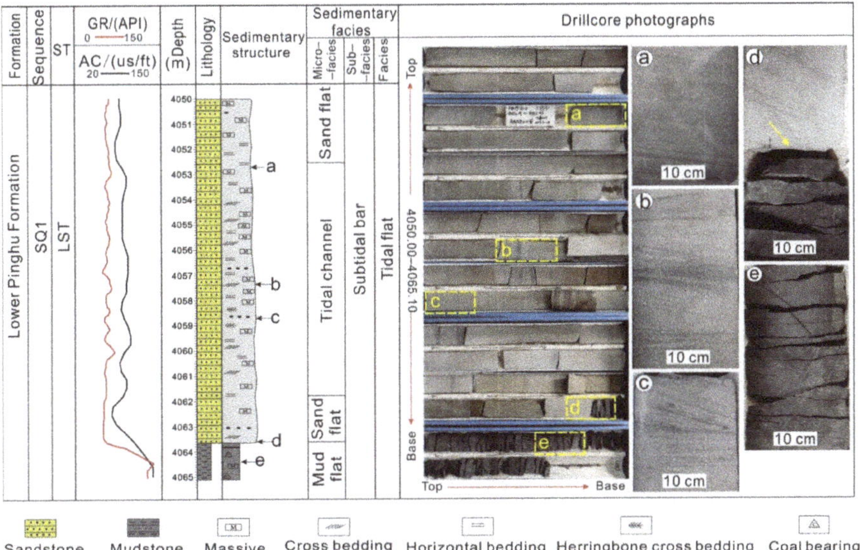


**Figure 10.** Well log, drill core, and facies interpretations for borehole W-3 from 4065–4050 m in the Xihu Depression. (**a**) 4052.68 m, grey-white fine sandstone with developed mainly parallel bedding with organic matter enriched within laminae; (**b**) 4057.26 m, grey fine sandstone with oblique bedding; (**c**) 4058.72 m, developed herringbone crossbedding; (**d**) 4063.48 m, the scouring surface and contact surface of grey sandstone and black mudstone; (**e**) 4064.36 m, black laminated mudstone with organic granules.

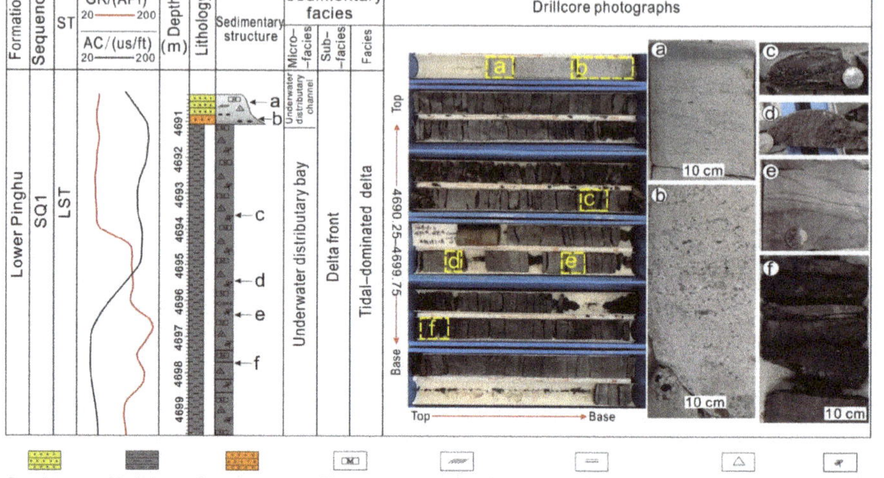

Sandstone  Mudstone  Conglomerate  Massive  Cross bedding  Horizontal bedding  Coal bearing  Plant fossil

**Figure 11.** Well log, drill core, and facies interpretations for borehole W-4 in the Xihu Depression. Core photographs at (**a**) 4690.50, grey massive mudstone with developed oblique bedding. Oil stains can be observed; (**b**) 4690.97 m, conglomerate with the size of the gravel decreasing upwardly; (**c**) 4693.59 m coal-bearing mudstone; (**d**) 4695.46 m, abundant plant fossils in mudstone; (**e**) 4696.38 siltstone interbedded black mudstone with developed ripple cross-lamination; (**f**) 4696.38 coal seam (1 cm thick) bearing mudstone.

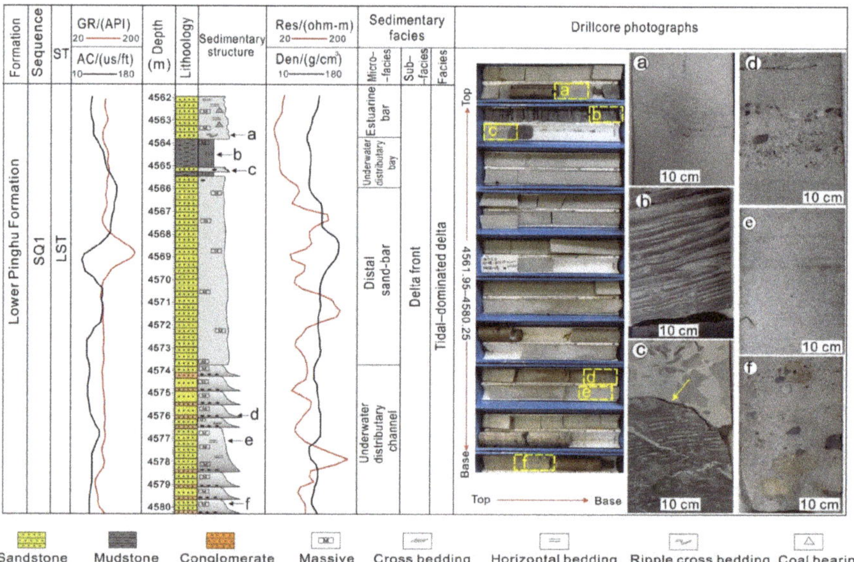

Sandstone  Mudstone  Conglomerate  Massive  Cross bedding  Horizontal bedding  Ripple cross bedding  Coal bearing

**Figure 12.** Well log, drill core, and facies interpretations for borehole W-5 in the Xihu Depression. Core photos (**a**) 4563.64 m, massive sandstone with deformed structure; (**b**) 4564.5 m, interbedded siltstone and mudstone with ripple cross lamination; (**c**) 4565.17 m, contacting surface of mudstone and sandstone. Mudstone breccia is observed in sandstone with a fracture developed in the mudstone; (**d**) 4576.00 m, scouring surface, the contacting surface of conglomerate and sandstone; (**e**) 4577.10 m, grey massive sandstone with faint parallel bedding; (**f**) 4579.86, conglomerate-bearing sandstone. The rounding and sorting of the gravel are good. Size ranges from 0.5 cm to 3 cm.

## 5. Results

### 5.1. Lithologies

The three main lithologies in the Pinghu Formation are sandstone (60%), siltstone (30%), and conglomerate (10%). Obvious lithological transitions can be observed in the drill cores; sedimentary structures were identified from hand specimens.

The 4050.0–4065.1 m interval in the W-3 drill core consists primarily of siltstone (45%), sandstone (40%), and mudstone (15%). The mudstone and sandstone interval correlates well with the high and low GR interval, respectively (Figure 10a). The scour structures are common in these lithologies (Figure 10d; yellow arrow), with an obvious lithological transition from mudstone to sandstone. The siltstone is gray-brown and characterized by parallel bedding that is usually intercalated with gray sandstone (Figure 10a). The observed changes in the bedding angle indicate frequent changes in the paleocurrent direction (Figure 10b) [71]. The sandstone commonly exhibits wavy and herringbone crossbedding structures, which are indicative of a high-energy environment and bidirectional water current forces (Figure 10c). The massive mudstone that was deposited in this interval is black, which is indicative of a reducing environment. Coal seams and oil stains are also commonly observed in this interval (Figure 10d,e).

The 4690–4670 m interval in the W-4 drill core consists primarily of mudstone (90%, ~9 m), sandstone, and conglomerate-bearing sandstone (~10%, ~1 m). The decrease in GR in the well log corresponds well with the transition of the drill core from mudstone to sandstone (Figure 11). The massive mudstone in this interval is dark gray to black, with locally wavy and lenticular bedding structures (Figure 11e). Scour surfaces are also observed in this interval (Figure 11e). Three light-to-dark cycles are identified from the base to the top of this interval. Oil stains (Figure 11f) and coal seams are commonly observed throughout this interval (Figure 11c), and large amounts of plant fossils are also observed (Figure 11d). The sandstone is a massive structure that possesses oblique crossbedding (Figure 11a). The interval above the mudstone contains abundant mud breccia and conglomerate, with the conglomerate particles distributed along the bedding orientation (Figure 11b).

The 4561.9–4580.3 m interval in the W-5 drill core consists primarily of conglomerate-bearing sandstone (25%), sandstone (60%), and mudstone (15%). The conglomerate-bearing sandstone is characterized by massive bedding (Figure 12a), ripple (Figure 12b), and trough crossbedding structures. Scour surfaces are a commonly observed feature along the lithological transition surfaces (Figure 12c). Mud breccia with fragments in the 0.5–10 cm range, is commonly observed above the scour surface, indicating a strong hydrodynamic environment. The conglomerate in the sandstone consists of 0.5–3.0 cm diameter clasts (Figure 12a,c,f), with the clast orientations largely consistent with the orientation of the rippled bedding (Figure 12d). The sandstone is gray and dominated by massive structures (Figure 12e). The mudstone possesses a laminated structure, consisting of gray siltstone that is intercalated with black mudstone (Figure 12b,c). The oblique and discontinuous laminae are indicative of frequent changes occurring within a high-energy hydrodynamic environment (Figure 12c).

### 5.2. Sand Body Prediction

A quantitative description of the sand bodies identified via prestack seismic inversion allows us to gain further insight into the hydrocarbon potential of the study area. The distribution pattern of the sand body that was deposited during the LST of SQ1 possesses obvious directional properties (Figure 8d). The thickness of this sand body decreases from ~150 to ~50 m along the NE–SW orientation. The general distribution pattern of the sand body that was deposited during the LST of SQ2 is similar to that of SQ1, with the thickness of the sand body possessing a decreasing pattern in the NE–SW orientation and a maximum thickness of ~90 m in the middle of the study area (Figure 8c). The sand body deposited during the LST of SQ3 is generally thin, with a maximum thickness of ~50 m. The sand body in SQ3 is oriented in a SE–NW direction and possesses a completely different distribution pattern compared with the sand bodies within SQ1 and SQ2, which exhibit a weak SE–NW orientation. However, a portion of the sand body in the eastern sector of the block possesses a NE–SW distribution pattern, similar to the sand body in SQ1 (Figure 8b). The sand body deposited during the LST of SQ4 distributed a NW–SE orientation, with the thickness of the sand body increasing from ~30 m to 70 m from west to east.

### 5.3. Sedimentary Facies Association

Two sedimentary facies associations, fluvial–delta and tidal flat, are identified on the analysis of the drill cores, well logs and seismic data.

*River delta facies association:* Well-log analysis of this sedimentary facies' association highlights predominantly sandstone (70%), siltstone (20%), and mudstone (10%) lithologies. The sandstones and siltstones are typically gray with massive bedding and locally developed oblique (Figure 11a), wavy (Figure 12a), and parallel (Figure 12e) bedding. Conglomerate-bearing sandstone is commonly observed, with conglomerate layers that are usually distributed orientally along with the bedding (Figures 10b and 11d,f).

*Interpretation:* The oblique and wavy bedding structures indicate a high-energy fluvial environment. The observed changes in the crossbedding orientation indicate frequent changes in the paleocurrent direction. The developed scouring surface along the lithological transition reflects a high-energy water column in a tidal channel and incised valley. The distribution of the conglomerate is linked to the fluvial influx of continental material. The relatively high sorting and roundness of this conglomerate-bearing sandstone indicate a high-energy, erosive fluvial current with a high sedimentation rate.

*Tidal flat facies association:* Well-log analysis of this sedimentary facies' association indicates that it comprises mudstone (60%), siltstone (30%), and sandstone (10%). The mudstone is dominated by massive bedding, oblique bedding, and herringbone crossbedding structures (Figure 10a–c) and is black in this interval. Large amounts of plant fossils (Figure 11d), oil stains, and coaly shale (Figure 11f) are commonly observed within the thick (~10 m) mudstone intervals (Figure 11). Ripple cross-lamination structures are developed along the lithological transition surface, and the gray sandstone and siltstone layers usually contain clay-rich laminae (Figure 11e). The sandstone that was deposited above the mudstone commonly contains abundant mudstone breccia and conglomerate, with clast orientations distributed irregularly (Figure 12c). A scouring surface is commonly observed along the contact between the sandstone and mudstone (Figure 10d). Interbedded sandstone, mudstone, and crossbedded sandstone are commonly observed in the transitional interval (Figure 10d). Siltstone is commonly developed adjacent to the black mudstone (Figure 10e) and possesses a dark gray color due to its organic carbon content (Figure 10a).

*Interpretation:* The oblique and wavy bedding structures indicate a high-energy fluvial environment. The well-preserved plant fossils and organic matter nodules within the black mudstone indicate anoxic bottom-water conditions during the highstand stage and terrestrial material input. The laminated mudstone is gray and dark gray in color with limited burrowing; the laminae is ~1–2 mm in thickness. These features reflect muddy tidal flat facies. The wavy bedding, orientally distributed conglomerate, and large amounts of muddy breccia and bidirectional crossbedding structures indicate a high-energy environment in the tidal channel. More recent crossbedding structures, typically cut older structures, can be observed. Trough and herringbone crossbedding indicate bidirectional flow on the tidal channel planforms [72,73]. The mudstones, with gray siltstone layers, bidirectional sand laminae and bedding, occasional flaser and lenticular bedding, and local burrows, are indicative of tidal flat deposits in relatively low-energy environments.

### 5.4. Sand Body and Sedimentary Facies Distribution

The Pinghu Formation was deposited in a transitional setting, with tidal and fluvial activity dominating the depositional environment and tidal and deltaic riverine facies being the main sand body deposits [33,65,67]. Sand body development is associated closely with the sedimentary facies distribution in a given sedimentary environment [74]. The spatial distribution of sand bodies across the study area during the Pinghu stage evolved from a NE–SW to NW–SE orientation. The spatiotemporal distribution of the sedimentary facies demonstrates a depositional environment change from tidal to fluvial facies (Figures 13 and 14). During SQ1 and SQ2, NE–SW-oriented tidal channel and subtidal bar facies were deposited across the eastern part of the study area (Figure 14d), with these facies representing the predominant sand bodies during SQ1 and SQ2 (50–100 m in thickness). A small-scale proximal bar (~20 m in thickness) was deposited along the western side of the A block. The proximal bar was deposited along the western part during SQ2 and increased gradually in both width and thickness to the east, with an incised channel beginning to form immediately from the west (Figure 14c). The tidal channel and subtidal bar facies developed in the northeastern part of the study area and decreased significantly in both continuity and thickness relative to that during SQ1 (Figure 14c). Incised channels and proximal bars formed along the southern and northwestern parts of the study area, with the tidal channel and subtidal bar distribution decreasing in both size and thickness to the northeast (Figure 14b). Incised channel, fluvial–delta, and proximal bar facies were distributed throughout most of the study area, and small-scale residual tidal deposits were identified along the northern boundary of the A block (Figure 14a). The sedimentary evolution from the SQ1 to the SQ4 sequence in the Pinghu Formation demonstrates the change in the hydrodynamic conditions over the sedimentary facies' distribution [75–77].

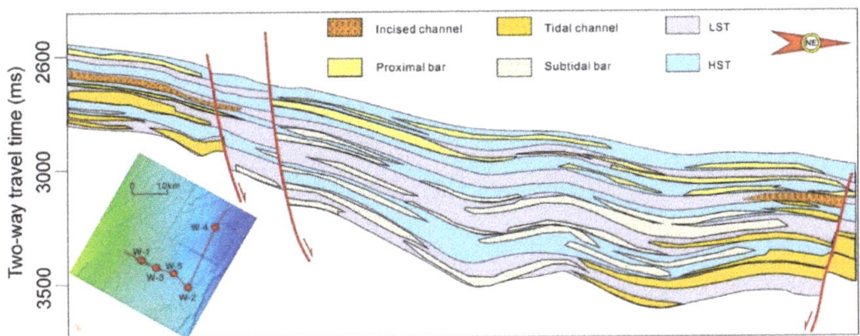


**Figure 13.** Horizontal distribution of sedimentary facies' profile of well W-4, W-2, W-5, W-3, and W-1 of the Pinghu Formation.

The tidal currents, flood tidal currents, and ebb tidal currents often flow along the shoreline, such as in the tidal currents of the Jiangsu Coastal zone [78]. The Xihu Sag is a NE–SW-oriented depression that lies along the Haijiao, Hupijiao, and Diaoyudao uplifts. In the case for the Xihu Sag, the direction of the tidal currents maybe parallels with the fault belts, which are in NE–SW direction. In a tectonic subsidence belt such as the Xihu Sag, the thickness of the sand body can reach several hundred meters [78–80]. Comparing with the tidal sand bodies developed at the Jiangsu coastal zone, east China, the tidal facies' sand body in the study area has a similar sand body thicknesses (~200 m in LST; Figure 9d). The obvious NE–SW distribution of the sand body probably indicates the direction of the tidal current (Figure 9d). The obvious weakening of the NE–SW orientation of the sand body is consistent with the gradually dropping sea level [59] and the associated decrease in tidal currents from SQ1 to SQ4 (Figure 9).

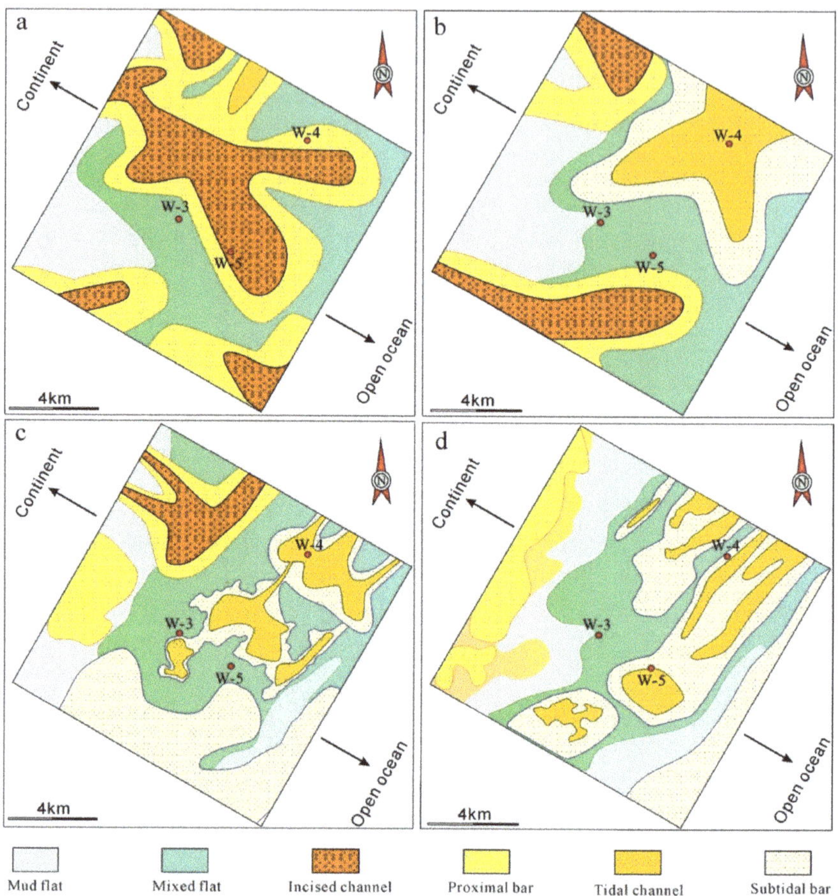



**Figure 14.** Horizontal distribution of sedimentary facies of the Pinghu Formation during (**a**) SQ4, (**b**) SQ3, (**c**) SQ2, and (**d**) SQ1 stage. The arrows indicate the relative location to the study area.

## 6. Discussion

### 6.1. Factors Controlling the Sand Body Distribution

On a basinal scale, the slope-break zone formed by syndepositional tectonic activities in rift-subsidence basins could exert significant control over the sequence stratigraphic framework and the spatial distribution of the sand body [57,81–83]. Within a third-order sequence, the sand body distribution pattern related intimately to the local tectonic activity as well [84]. The Pinghu Formation was deposited during the middle-to-late Eocene, in between the Yuquan and Pinghu movements, when the tectonic activity was relatively weak [49–51]. Still, the existing fault activities that occurred during the sedimentary period of the Pinghu Formation not only created the space for the accumulation of the sediments, but also controlled the development of the sand body significantly [33,85]. In A block, the faults belts are all in the NE–NNE direction; the enrichment sites of the sand body often occurred at the slope-break zone, which were usually distributed in the same direction as the fault belts [85], i.e., NE–NNE-oriented. However, the sand body distribution pattern from SQ1 to SQ4 changed from NE–SW-oriented (SQ1–SQ2) to NW–SE-oriented (SQ3–SQ4). The significant orientation change in the sand body distribution pattern indicates that the relatively stable landscape created by tectonic activities is not the dominant factor that controls the spatial distribution of the sand body.

The Pinghu Formation was deposited in a marginal-marine setting according to evidence from sedimentological, paleontological, and geochemical research [41,86,87]. In addition to the tectonic activities, the hydrodynamic processes could exert significant influence over the spatial distribution of the sedimentary facies [88] and the sand body developed within [89–91]. The Pinghu Formation was deposited in a freshwater–brackish water environment owing to its connectivity to the open ocean and the constant freshwater input from the continent. The hydrodynamic conditions rely on the interplay of riverine forces and tidal forces from the continent and open ocean, respectively [82]. The significant sea level drop could have led to the seaward shift of the coastline [92], which is further supported by the decreasing trend in the paleosalinity proxies (B/Ga and Sr/Ba) (Figure 15) [93–95], resulting in the weakening of the tidal activity and the simultaneous strengthening of riverine forces from SQ1 to SQ4. The gentle shelf topography of the Xihu Sag could have strengthened the effect of the tidal currents simultaneously [96]. The sea-level drop in the Xihu Sag [59,66,94] and the accompanied change in hydrodynamic driving forces exerted considerable influence on the development of the sedimentary facies and sand body.

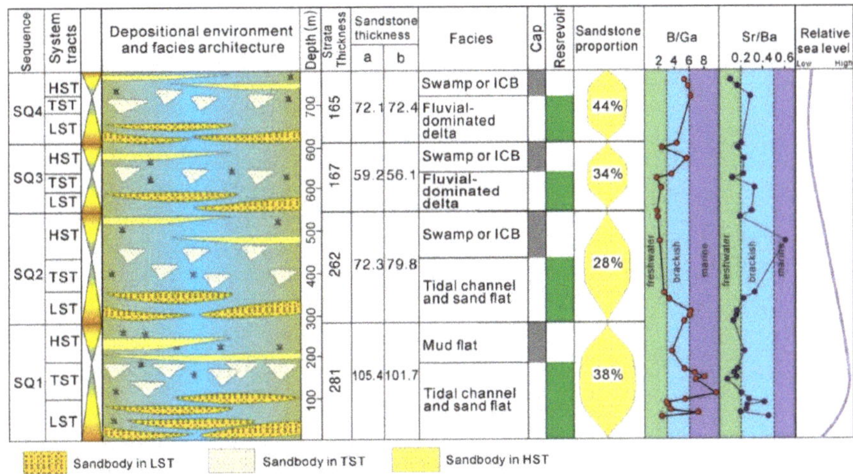

**Figure 15.** Synthetic chart of the Pinghu Formation in the Xihu Sag, showing the sedimentary environment and facies architecture, potential reservoir, sea interval, and vertical distribution of the sand body. ICB = inter-channel bay. The sandstone thickness, a, represents the calculated sandstone based on drill core data; b represents the sandstone thickness calculated based on the prestack inversion process. The threshold of B/Ga and Sr/Ba was modified from [95].

The gradual drop in sea level could affect the distribution of the sand body in three aspects: (1) the distribution of the different sand bodies of different sedimentary facies; (2) the thickness of the sand bodies [97–99]; and (3) the orientation of the sand body distribution pattern. During the SQ1 to SQ2 stage, the strong tidal activity was the dominant hydrodynamic force across most of the study area (~60%). This strong tidal hydrodynamic force transported large amounts of sediment to the eastern area, resulting in very thick sandstone deposits (~200 m). The orientation of the sand body distribution reflects a strong NE–SW-oriented tidal hydrodynamic force. These tidal activities weakened as the sea level fell from SQ1 to SQ4, with decreases in both the thickness and proportion of sandstone within SQ2 and SQ3, as well as the continuity and coverage area of the tidal facies' sandstone. Fluvial facies sand bodies expanded from the southern and western sides of the A block during SQ2 and SQ3 and covered almost the entire study area during SQ4, with fluvial activity dominating the sand body distribution within the study area during the low-sea level stage. Both the absolute thickness and the proportion of sandstone increased from

SQ3 to SQ4, which indicates that the fluvial system became the dominant hydrodynamic force owing to the significant drop in sea level, with more sediment transported into the study area during these latter sequences (Figure 13).

According to the relative orientation to the shoreline, the NE–SW paleocurrent direction was due to tidal activity, whereas the NW–SE paleocurrent direction was indicative of continental riverine inputs (Figure 14) [100,101]. The gradual change in paleocurrent direction from SQ1 to SQ4 indicates the evolution of hydrodynamics from a tidal- to a fluvial-dominated system across the study area. The orientation of the tidal current was usually parallel to the shoreline [100,102], leading to the formation of the lobe-shaped tidal channel sand body with a long axis parallel with the shoreline in the Xihu Sag [101]. The fluvial activities came from the continent to the open ocean in a direction perpendicular to the shoreline and/or to that of tidal current, thus forming the sand body with a long axis perpendicular to that formed from tidal currents. This distinction provides us with an insight into identifying the sand bodies derived from tidal and fluvial activities according to the geometric character of the sand bodies predicted with the prestack inversion technique (Section 5.2). The change in orientation of the sand body distribution from the NE–SW to NW–SE direction further supports the interpretation of significant control from the hydrodynamic conditions over the sand body distribution.

*6.2. Reservoir Characteristics*

The sedimentary evolution and sandstone distribution are illustrated in Figure 15 to highlight the favorable reservoir intervals in the Pinghu Formation across the study area. SQ1 contains the largest sandstone deposits (up to 200 m thick; 38% of the thickness of the whole sequence), which are found mainly in the northeastern portion of the study area. The average porosity and permeability of the sandstone are 11.1% and 3.3 darcies, respectively, in W-3, and 8.2% and 0.12 darcies, respectively, in W-2, which are favorable reservoir properties. However, the low permeabilities may impede the connectivity of the reservoirs in this interval. The sand body distribution pattern in SQ2 is similar to that in SQ1, although both the sand body scale and thickness are significantly smaller. The sandstone in SQ2 also possesses favorable reservoir properties, with porosities and permeabilities of 11.6% and 12.7 darcies, 18.8% and 803 darcies, and 11.87% and 2.9 darcies in W-2, W-3, and W-4, respectively. Fluvial facies sand bodies dominate SQ3 and SQ4, although both the sandstone scales and thicknesses are significantly smaller than in the SQ1 sandstone. The porosity and permeability of the SQ3 sandstone are 10.3% and 3.5 darcies, respectively, in W-4, and 16.2% and 34.5 darcies, respectively, in W-3. The fluvial sandstone in SQ4 possesses a porosity and permeability of 15.6% and 168 darcies in W-3, respectively, with this sandstone exhibiting high-quality reservoir properties owing to its high porosity and permeability. The SQ1 to SQ4 sandstone units in the Pinghu Formation all exhibit high-quality reservoir properties. However, the high permeability of the SQ4 sandstone may allow diffusion of hydrocarbons within these reservoirs; the low permeability of the SQ1 sandstone may indicate poor connectivity of the reservoirs in this sequence.

*6.3. Source Rock Characteristics*

Significant sea level changes can affect both the sand body distribution (i.e., reservoir) and source rock quality [94,103]. Previous organic geochemical studies have demonstrated that the mudstone within the Pinghu Formation has good to excellent source rock properties [44,51,104].

The total organic carbon (TOC) content in the Pinghu Formation generally exhibits a decreasing trend from SQ1 to SQ4. The TOC values in SQ1, SQ2, and SQ3 are 3.48%, 2.16%, and 1.81%, respectively, in W-3, and 3%, 1.22%, and 2.97%, respectively, in W-4, whereas the average TOC value in SQ4 is <1% [94]. The hydrogen index (HI) for the SQ1, SQ2, and SQ3 samples are 155, 145, and 110 mg HC/g, and 154, 130, and 195 mg HC/g in W-3 and W-4, respectively, with these high HI values indicating a moderately high hydrocarbon-generating potential within the Pinghu Formation [105]. The average vitrinite reflectance values for SQ1, SQ2, and SQ3 are 0.69, 0.58, and 0.54, and 0.77, 0.78, and 0.73 in W-3 and W-4, respectively, with maximum burial temperatures of 440 °C, 436 °C, and 432 °C, and 449 °C, 440 °C, and 429 °C in W-3 and W-4, respectively. These features indicate that the source rock is relatively immature and mostly within the oil-generating window [106,107]. The organic-rich mudstones comprise ~70% of the Pinghu Formation by volume, with the SQ1 mudstone being a better source rock than those in the other sequences due to its generally higher TOC and HI and moderate maturity.

The gradual drop in sea level during the deposition of the Pinghu Formation may have affected the quality of the source rock based on two key factors: (1) the water column salinity, which influences both the biomass and type of organic matter [93,94] and (2) the redox condition of the benthic environment, which influences the preservation of organic materials [68,108,109]. The B/Ga and Sr/Ba ratios exhibit a generally decreasing bottom-up trend through the Pinghu Formation, with average (range) values of 0.28 (0.05–1.98) and 3.37 (2.11–9.02), respectively (Figure 15). The threshold B/Ga and Sr/Ba values indicate that the water column salinity decreased from high-brackish (~20–30 psu) to fresh/low-brackish (~5–15 psu) conditions [68,93,95]. A generally brackish water column will create an environment that favors enhanced primary production, which includes various phytoplankton, such as blue algae, green algae, and diatoms [94]. This is further supported by the microfossil evidence such as dinoflagellates that were deposited mainly in the brackish-marine environment discovered within the study interval. The relatively high base sea level during SQ1 may create a more reducing environment that favors the preservation of organic matter in mudstone [109]. Furthermore, seawater intrusions occurred during the early stage of the Pinghu Formation, leading to the development of coal seams in SQ1 and SQ2, which have been proven to be an excellent hydrocarbon source rock for the Xihu Sag [44,51,110,111].

## 7. Conclusions

For a better understanding of the complex sand body distribution within the Pinghu Formation, the stochastic inversion method was used to prestack seismic data from A block in the Baochu slope zone in the Xihu Sag. The sand body prediction results based on the prestack seismic data inversion process show high consistency with the well log-derived sandstone thicknesses, indicating that prestack seismic inversion is a powerful approach for sand body prediction in petroliferous provinces, even with complex sand body distribution patterns. The application of this technique to other petroliferous provinces could be of great significance for future hydrocarbon exploration industries.

The temporal-spatial distribution pattern of the sedimentary facies established based on the sand body prediction results and petrological analysis exhibits a strong dependence on the weakening of the tidal activities and the strengthening of fluvial activities, resulting from the sea level dropping during the Pinghu stage. The evolution of the hydrological forces resulting from the sea level drop led to the change in the sedimentary facies of the sand body with different physical properties. The sand body of transitional facies developed during SQ2 and SQ3, which acquired better physical properties and are the favorable reservoirs of the Pinghu Formation.

The gradually dropping sea level also caused the simultaneous change in paleosalinity and paleoredox conditions and the type of organic matter transported into the basin, which in combination, affected the development of high-quality source rocks within the Pinghu Formation significantly. Therefore, the sea level change played a significant role in the development of both favorable reservoirs and high-quality source rocks in the Xihu Sag and probably in other petroliferous provinces developed in marginal-marine environments distributed all over the world.

**Author Contributions:** Conceptualization, W.W. and Z.C.; methodology, Z.C.; software, S.Z.; investigation, Z.C.; resources, Y.L.; writing—original draft preparation, Z.C.; writing—review and editing, W.W., J.Z. and S.C.; funding acquisition, Y.L. All authors have read and agreed to the published version of the manuscript.

**Funding:** This research was funded by the National Science Foundation of China (42072137) and the National Science and Technology Major Project of China (34000000-19-ZC0613-0010 and 34000000-21-ZC0613-0050).

**Acknowledgments:** We thank Hui Jian for assistance with drafting figures. We thank SINOPEC Shanghai Offshore Oil and Gas Company, Shanghai for providing a valuable opportunity for core description, sampling, and geological data.

**Conflicts of Interest:** The authors declare no conflict of interest.

# References


1. Taner, M.T.; Sheriff, R.E. Application of amplitude, frequency, and other attributes to stratigraphic and hydrocarbon determination. In *Applications to Hydrocarbon Exploration*; Payton, C.E., Ed.; American Association of Petroleum Geologists: Tulsa, OK, USA, 1977; pp. 301–327.
2. Liu, L.W.; Hao, X.G.; Zhao, W.; Guo, S.Y.; Guo, S.P. Application of multi-attribute analysis to the identification of river palaeochannel in the thin interbeded reservoir. *J. Daqing Pet. Inst.* **2006**, *30*, 4. (In Chinese)
3. Brown, A.R. *Interpretation of Three-Dimensional Seismic Data*; Society of Exploration Geophysicists and American Association of Petroleum Geologists: Tulsa, OK, USA, 2011; p. 247.
4. Pigott, J.D.; Kang, M.H.; Han, H.C. First Order Seismic Attributes for Clastic Seismic Facies Interpretation: Examples from the East China Sea. *J. Asian Earth Sci.* **2013**, *10*, 34–54. [CrossRef]
5. Wang, K.Y.; Zhou, Y.; Chen, Y.Q.; Liu, J.R.; Zhang, W.; Bao, B.Q. Prediction of reservoir thickness based on spectral decomposition and seismic multi-attribute. *Prog. Geophys.* **2014**, *29*, 1271–1276.
6. Huang, W.; Wang, J.; Chen, H.; Yang, J.; Wu, J.; Ma, N.; Meng, Z.; Jing, Y.; Xu, F.; Ning, B.; et al. Application of multi-wave and multi-component seismic data in the description on shallow-buried unconsolidated sand bodies: Example of block J of the Orinoco heavy oil belt in Venezuela. *J. Pet. Sci. Eng.* **2021**, *205*, 108786. [CrossRef]
7. Lindseth, R.O. Synthetic sonic logs—A process for stratigraphic interpretation. *Geophysics* **1979**, *44*, 3–26. [CrossRef]
8. Morozov, I.B.; Ma, J. Accurate poststack acoustic-impedance inversion by well-log calibration. *Geophysics* **2009**, *74*, R59–R67. [CrossRef]
9. Li, X.T. Application of Post-Stack Seismic Inversion in Discription of the Reservoir Distribution. In *Applied Mechanics and Materials*; Trans Tech Publications, Ltd.: Bäch, Switzerland, 2014; Volume 477, pp. 1088–1091.
10. Karim, S.U.; Islam, M.S.; Hossain, M.M.; Islam, M.A. Seismic reservoir characterization using model based post-stack seismic inversion: In case of fenchuganj gas field, Bangladesh. *J. Jpn. Pet. Inst.* **2016**, *59*, 283–292. [CrossRef]
11. Das, B.; Chatterjee, R.; Singha, D.K.; Kumar, R. Post-stack seismic inversion and attribute analysis in shallow offshore of Krishna-Godavari basin, India. *J. Geol. Soc. India* **2017**, *90*, 32–40. [CrossRef]
12. Hampson, D.P.; Brian, H.R.; Brad, B. Simultaneous inversion of pre-stack seismic data. In *SEG Technical Program Expanded Abstracts 2005*; Society of Exploration Geophysicists: Houston, TX, USA, 2005; pp. 1633–1637.
13. Li, L.G.; Gan, L.D.; Du, W.H.; Dai, X.F. Pre-stack seismic inversion applied to reservoir prediction and natural gas detection in SLG gas field. *Nat. Gas GeoSci.* **2008**, *19*, 261–265.
14. Wang, W.; Zhang, X.; Cui, X.D.; Zhang, X.Z. Application of pre-stack inversion technology to tracing sand body: A case study from Huagang Formation in Pinghu Oil and Gas Field. *Lithol. Reserv.* **2010**, *22* (Suppl. S1), 69–73. (In Chinese)
15. Li, H.C.; Li, Z.D.; Lu, G.D.; Yu, Y. Application of reconstruction inversion in prediction of tight reservoir in faulted basin: Take Beizhong oilfield in Hailar basin as an example. *Prog. Geophys.* **2020**, *35*, 1391–1399.
16. Shi, S.; Feng, J.; Liu, Z.; Qi, Y.; Duan, P.; Han, Q.; Zuo, J. Seismic inversion-based prediction of the elastic parameters of rocks surrounding roadways: A case study from the Shijiazhuang mining area of North China. *Acta Geophys.* **2021**, *69*, 2219–2230. [CrossRef]
17. Vail, P.R.; Mitchum, R.M., Jr.; Thompson, S., III. Seismic stratigraphy and global changes of sea level: Part 4. Global cycles of relative changes of sea level: Section 2. Application of seismic reflection configuration to stratigraphic interpretation. In *Seismic Stratigraphy—Applications to Hydrocarbon Exploration*; American Association of Petroleum Geologists: Tulsa, OK, USA, 1977; pp. 63–81.
18. Verma, S.; Matt, S. The early Paleozoic structures and its influence on the Permian strata, Midland Basin: Insights from multi-attribute seismic analysis. *J. Nat. Gas Sci. Eng.* **2020**, *82*, 103521. [CrossRef]
19. Buist, C.; Bedle, H.; Rine, M.; Pigott, J. Enhancing Paleoreef reservoir characterization through machine learning and multi-attribute seismic analysis: Silurian reef examples from the Michigan Basin. *Geosciences* **2021**, *11*, 142. [CrossRef]
20. Bouchaala, F.; Guennou, C. Estimation of viscoelastic attenuation of real seismic data by use of ray tracing software: Application to the detection of gas hydrates and free gas. *C. R. Géosci.* **2012**, *344*, 57–66. [CrossRef]
21. Matsushima, J.; Ali, M.Y.; Bouchaala, F. A novel method for separating intrinsic and scattering attenuation for zero-offset vertical seismic profiling data. *Geophys. J. Int.* **2017**, *211*, 1655–1668. [CrossRef]
22. Wu, J.; Li, F. Prediction of oil-bearing single sandbody by 3D geological modeling combined with seismic inversion. *Pet. Explor. Dev.* **2009**, *36*, 623–627.

23. Bosch, M.; Mukerji, T.; Gonzalez, E.F. Seismic inversion for reservoir properties combining statistical rock physics and geostatistics: A review. *Geophysics* **2010**, *75*, 75A165–75A176. [CrossRef]

24. Alvarez, P.; Bolívar, F.; Di Luca, M.; Salinas, T. Multiattribute rotation scheme: A tool for reservoir property prediction from seismic inversion attributes. *Interpretation* **2015**, *3*, SAE9–SAE18. [CrossRef]

25. Sokolov, A.; Schulte, B.; Shalaby, H.; van der Molen, M. Seismic inversion for reservoir characterization. In *Applied Techniques to Integrated Oil and Gas Reservoir Characterization*; Elsevier: Amsterdam, The Netherlands, 2021; pp. 329–351.

26. Yuan, S.J. Progress of pre-stack inversion and application in exploration of the lithological reservoirs. *Prog. Geophys.* **2007**, *22*, 879–886. (In Chinese)

27. Yu, P. Prestack seismic inversion and its application. *Shandong Ind. Technol.* **2014**, *10*, 72. (In Chinese)

28. Zhang, R.; Sen, M.K.; Srinivasan, S. A prestack basis pursuit seismic inversion. *Geophysics* **2013**, *78*, R1–R11. [CrossRef]

29. Zhou, L.; Zhong, F.; Yan, J.; Zhong, K.; Wu, Y.; Wu, X.; Lu, P.; Zhang, W.; Liu, Y. Prestack inversion identification of organic reef gas reservoirs of Permian Changxing Formation in Damaoping area, Sichuan Basin, SW China. *Pet. Explor. Dev.* **2020**, *47*, 89–100. [CrossRef]

30. Jiang, Y.; Diao, H.; Zeng, W. Coal source rock conditions and hydrocarbon generation model of Pinghu Formation in Xihu Depresion, East China Sea Basin. *Bull. Geol. Sci. Technol.* **2020**, *39*, 30–39. (In Chinese)

31. Kang, S.; Shao, L.; Qin, L.; Li, S.; Liu, J.; Shen, W.; Chen, X.; Eriksson, K.A.; Zhou, Q. Hydrocarbon Generation Potential and Depositional Setting of Eocene Oil-Prone Coaly Source Rocks in the Xihu Sag, East China Sea Shelf Basin. *ACS Omega* **2020**, *5*, 32267–32285. [CrossRef]

32. Yang, F.L.; Xu, X.; Zhao, W.F.; Sun, Z. Petroleum accumulations and inversion structures in the Xihu Depression, East China Sea Basin. *J. Pet. Geol.* **2011**, *34*, 429–440. [CrossRef]

33. Abbas, A.; Zhu, H.; Zeng, Z.; Zhou, X. Sedimentary facies analysis using sequence stratigraphy and seismic sedimentology in the Paleogene Pinghu Formation, Xihu Depression, East China Sea Shelf Basin. *Mar. Pet. Geol.* **2018**, *93*, 287–297. [CrossRef]

34. Yuan, J.; Lu, Y.; Li, Z. Physical Characteristics and Influencing Factors of Deep Reservoirs in Xihu Sag. *Offshore Oil* **2019**, *39*, 16–21. (In Chinese)

35. Nio, S.D.; Yang, C.S. Sea-level fluctuations and the geometric variability of tide-dominated sandbodies. *Sediment. Geol.* **1991**, *70*, 161–193. [CrossRef]

36. Flood, Y.S.; Hampson, G.J. Quantitative analysis of the dimensions and distribution of channelized fluvial sandbodies within a large outcrop dataset: Upper Cretaceous Blackhawk Formation, Wasatch Plateau, Central Utah, USA. *J. Sediment. Res.* **2015**, *85*, 315–336. [CrossRef]

37. Ding, F.; Xie, C.; Zhou, X.; Jiang, C.; Li, K.; Wan, L.; Zhang, P.; Niu, H. Defining stratigraphic oil and gas plays by modifying structural plays: A case study from the Xihu Sag, east China Sea Shelf Basin. *Energy Geosci.* **2021**, *2*, 41–51. [CrossRef]

38. Ye, J.; Gu, H. A Study on Hydrocarbon Generation History and Expulsion History of Source Rock in the Zhedong Central Anticline Belt, Xihu Depression. *Mar. Geol. Quat. Geol.* **2001**, *21*, 75–80. (In Chinese)

39. Dai, L.; Li, S.; Lou, D.; Liu, X.; Suo, Y.; Yu, S. Numerical modeling of Late Miocene tectonic inversion in the Xihu Sag, East China Sea Shelf Basin, China. *J. Asian Earth Sci.* **2014**, *86*, 25–37. [CrossRef]

40. Chen, Y.Z.; Xu, Z.X.; Xu, G.S.; Xu, F.H.; Liu, J.S. Coupling relationship between abnormal overpressure and hydrocarbon accumulation in a central overturned structural belt, Xihu Sag, East China Sea Basin. *Oil Gas Geol.* **2017**, *38*, 570–581.

41. Ye, J.; Qing, H.; Bend, S.L.; Gu, H. Petroleum systems in the offshore Xihu Basin on the continental shelf of the East China Sea. *AAPG Bull.* **2007**, *91*, 1167–1188. [CrossRef]

42. Wang, Q.; Li, S.; Guo, L.; Suo, Y.; Dai, L. Analogue modelling and mechanism of tectonic inversion of the Xihu Sag, East China Sea Shelf Basin. *J. Asian Earth Sci.* **2017**, *139*, 129–141. [CrossRef]

43. Zhou, X.H.; Jiang, Y.M.; Tang, X.J. Tectonic setting, prototype basin evolution and exploration enlightenment of Xihu Sag in East China Sea Basin. *China Offshore Oil Gas* **2019**, *31*, 1–10. (In Chinese)

44. Zhu, X.; Chen, J.; Li, W.; Pei, L.; Liu, K.; Chen, X.; Zhang, T. Hydrocarbon generation potential of Paleogene coals and organic rich mudstones in Xihu sag, East China Sea Shelf basin, offshore eastern China. *J. Pet. Sci. Eng.* **2020**, *184*, 106450. [CrossRef]

45. Gao, W.; Sun, P.; Zhao, H.; Yang, Y. Study of dep reservoirs characters and main control factors of Huagang Formation in Xihu sag, East China Sea. *J. Chengdu Univ. Technol.* **2016**, *43*, 396–404. (In Chinese)

46. Diao, H.; Liu, J.; Hou, D.; Jiang, Y.; Zhang, T.; Zeng, W. Coal-bearing source rocks formed in the transitional stage from faulting to depression nearshore China—A case from the Pinghu Formation in the Xihu Sag. *Mar. Geol. Quat. Geol.* **2019**, *39*, 102–114.

47. Wang, L.; Yang, R.; Sun, Z.; Wang, L.; Guo, J.; Chen, M. Overpressure: Origin, Prediction, and Its Impact in the Xihu Sag, Eastern China Sea. *Energies* **2022**, *15*, 2519. [CrossRef]

48. Zhu, Y.; Li, Y.; Zhou, J.; Gu, S. Geochemical characteristics of Tertiary coal-bearing source rocks in Xihu depression, East China Sea basin. *Mar. Pet. Geol.* **2012**, *35*, 154–165. [CrossRef]

49. Yuan, W.; Xu, X.; Zhou, X. The new geochronology results of the Pinghu Formation in Xihu Depression: Evidence from the SHRIMP Zircon U-Pb ages of the volcanic rocks. *Geol. J. China Univ.* **2014**, *20*, 407–414. (In Chinese)

50. Su, A.; Chen, H.; Chen, X.; He, C.; Liu, H.; Li, Q.; Wang, C. The characteristics of low permeability reservoirs, gas origin, generation and charge in the central and western Xihu depression, East China Sea Basin. *J. Nat. Gas Sci. Eng.* **2018**, *53*, 94–109. [CrossRef]

51. Wang, Y.; Qin, Y.; Yang, L.; Liu, S.; Elsworth, D.; Zhang, R. Organic geochemical and petrographic characteristics of the coal measure source rocks of Pinghu Formation in the Xihu Sag of the East China Sea Shelf Basin: Implications for coal measure gas potential. *Acta Geol. Sin.* **2020**, *94*, 364–375. [CrossRef]
52. Quan, Y.; Chen, Z.; Jiang, Y.; Diao, H.; Xie, X.; Lu, Y.; Du, X.; Liu, X. Hydrocarbon generation potential, geochemical characteristics, and accumulation contribution of coal-bearing source rocks in the Xihu Sag, East China Sea Shelf Basin. *Mar. Pet. Geol.* **2022**, *136*, 105465. [CrossRef]
53. Van Wagoner, J.C.; Posamentier, H.W.; Mitchum, R.M.J.; Vail, P.R.; Sarg, J.F.; Loutit, T.S.; Hardenbol, J. *An Overview of the Fundamentals of Sequence Stratigraphy and Key Definitions*; Society of Economic Paleontologists and Mineralogists: Tulsa, OK, USA, 1988; Volume 42, pp. 39–45.
54. Catuneanu, O.; Abreu, V.; Bhattacharya, J.P.; Blum, M.D.; Dalrymple, R.W.; Eriksson, P.G.; Fielding, C.R.; Fisher, W.L.; Galloway, W.E.; Gibling, M.R.; et al. Towards the standardization of sequence stratigraphy. *Earth-Sci. Rev.* **2009**, *92*, 1–33. [CrossRef]
55. Brown, L.F., Jr.; Fisher, W.L. Seismic-stratigraphic interpretation of depositional systems: Examples from brazilian rift and pull-apart basins: Section 2. Application of seismic reflection configuration to stratigraphic interpretation. In *Seismic Stratigraphy—Applications to Hydrocarbon Exploration*; American Association of Petroleum Geologists: Tulsa, OK, USA, 1977; pp. 213–248.
56. Emery, D.; Myers, K. (Eds.) *Sequence Stratigraphy*; John Wiley & Sons: New York, NY, USA, 2009; p. 297.
57. Wei, W.; Lu, Y.; Xing, F.; Liu, Z.; Pan, L.; Algeo, T.J. Sedimentary facies associations and sequence stratigraphy of source and reservoir rocks of the lacustrine Eocene Niubao Formation (Lunpola Basin, central Tibet). *Mar. Pet. Geol.* **2017**, *86*, 1273–1290. [CrossRef]
58. Kadkhodaie, A.; Rezaee, R. Intelligent sequence stratigraphy through a wavelet-based decomposition of well log data. *J. Nat. Gas Sci. Eng.* **2017**, *40*, 38–50. [CrossRef]
59. Zhang, J.; Pas, D.; Krijgsman, W.; Wei, W.; Du, X.; Zhang, C.; Liu, J.; Lu, Y. Astronomical forcing of the Paleogene coal-bearing hydrocarbon source rocks of the East China Sea Shelf Basin. *Sediment. Geol.* **2020**, *406*, 105715. [CrossRef]
60. Catuneanu, O. *Principles of Sequence Stratigraphy*; Elsevier: Amsterdam, The Netherlands, 2006; p. 375.
61. Carroll, A.R.; Bohacs, K.M. Stratigraphic classification of ancient lakes: Balancing tectonic and climatic controls. *Geology* **1999**, *27*, 99–102. [CrossRef]
62. Bohacs, K.M.; Carroll, A.R.; Neal, J.E.; Mankiewicz, P.J. Lake-basin type, source potential, and hydrocarbon character: An integrated sequence-stratigraphic-geochemical framework. In *Lake Basins through Space and Time*; Gierlowski-Kordesch, E., Kelts, K.R., Eds.; AAPG Studies in Geology; AAPG: Tulsa, OK, USA, 2000; Volume 46, pp. 3–34.
63. Zecchin, M.; Mellere, D.; Roda, C. Sequence stratigraphy and architectural variability in growth fault-bounded basin fills: A review of Plio-Pleistocene stratal units of the Crotone Basin, southern Italy. *J. Geol. Soc.* **2006**, *163*, 471–486. [CrossRef]
64. Yu, C.; Chen, J.; Du, Y.; Zhang, Y. Sequence stratigraphic division of Pinghu Formation in Xihu Sag, East China Sea. *Mar. Geol. Quat. Geol.* **2007**, *27*, 85–90. (In Chinese)
65. Zhang, J.; Xu, F.; Zhong, T.; Zhang, T.; Yu, Y. Sequence stratigraphic model and sedimentary evolution of Pinghu Formation Huagang Formation in Xihu Sag, East China Sea shelf basin. *Mar. Geol. Quat. Geol.* **2012**, *1*, 35–41. (In Chinese) [CrossRef]
66. Chen, L. Evolution of sedimentary environment of Pinghu Formation in Xihu Sag, East China Sea. *Mar. Geol. Quat. Geol.* **1998**, *4*, 69–78, (In Chinese with English Abstract).
67. Zhou, R. Study on Sequence Stratigraphy and Sedimentary Facies of Low Porosity and Permeability Reservoirs of Pinghu Formation Huagang Formation in Xihu Sag. Ph.D. Thesis, Chengdu University of Technology, Chengdu, China, 2015. (In Chinese)
68. Xie, G.; Shen, Y.; Liu, S.; Hao, W. Trace and rare earth element (REE) characteristics of mudstones from Eocene Pinghu Formation and Oligocene Huagang Formation in Xihu Sag, East China Sea Basin: Implications for provenance, depositional conditions and paleoclimate. *Mar. Pet. Geol.* **2018**, *92*, 20–36. [CrossRef]
69. Russell, B.; Hampson, D. The old and the new in seismic inversion. *CSEG Rec.* **2006**, *31*, 5–11.
70. Schuster, G.T. *Seismic Inversion*; Society of Exploration Geophysicists: Houston, TX, USA, 2017; p. 344.
71. Collinson, J. *Sedimentary Structures*; Dunedin Academic Press Ltd.: Edinburgh, UK, 2019.
72. Chakrabarti, A. Sedimentary structures of tidal flats: A journey from coast to inner estuarine region of eastern India. *J. Earth Syst. Sci.* **2015**, *114*, 353–368. [CrossRef]
73. Ainsworth, R.B.; Hasiotis, S.T.; Amos, K.J.; Krapf, C.B.; Payenberg, T.H.; Sandstrom, M.L.; Vakarelov, B.K.; Lang, S.C. Tidal signatures in an intracratonic playa lake. *Geology* **2012**, *40*, 607–610. [CrossRef]
74. Miall, A.D. *The Geology of Fluvial Deposits: Sedimentary Facies, Basin Analysis, and Petroleum Geology*; Springer: Berlin/Heidelberg, Germany, 2013; p. 575.
75. Jiang, Z.; Liang, S.; Zhang, Y.; Zhang, S.; Qin, L.; Wei, X. Sedimentary hydrodynamic study of sand bodies in the upper subsection of the 4th member of the Paleogene Shahejie Formation the eastern Dongying Depression, China. *Pet. Sci.* **2014**, *11*, 189–199. [CrossRef]
76. Chen, J.Q.; Pang, X.Q.; Chen, D.X. Sedimentary facies and lithologic characters as main factors controlling hydrocarbon accumulations and their critical conditions. *J. Palaeogeogr.* **2015**, *4*, 413–429. [CrossRef]
77. Xia, S.; Liu, Z.; Liu, J.; Chang, Y.; Li, P.; Gao, N.; Ye, D.; Wu, G.; Yu, L.; Qu, L.; et al. The controlling factors of modern facies distributions in a half-graben lacustrine rift basin: A case study from Hulun Lake, Northeastern China. *Geol. J.* **2018**, *53*, 977–991. [CrossRef]

78. Li, C.X.; Zhang, J.Q.; Deng, B. Holocene regression and the tidal radial sand ridge system formation in the Jiangsu coastal zone, east China. *Mar. Geol.* **2001**, *173*, 97–120. [CrossRef]
79. Yang, C.S. Active, moribund and buried tidal sand ridges in the East China Sea and the southern Yellow Sea. *Mar. Geol.* **1989**, *88*, 97–116. [CrossRef]
80. Yang, Z.G.; Lin, H.M. *Chinese Quaternary Stratigraphy and Its Correlation with Stratigraphy of Asian-Pacific Region*; Geological Publishing House: Beijing, China, 1996; p. 207.
81. Feng, Y. Controlling effect of faulted lake basins, valleys and structural slope breaks on sand bodies. *Chin. J. Pet.* **2006**, *1*, 13–16. (In Chinese)
82. Deng, H.; Wang, R.; Xiao, Y.; Guo, J.; Xie, X. Tectono-sequence stratigraphic analysis in continental faulted basins. *Earth Sci. Front.* **2008**, *15*, 1–7. [CrossRef]
83. Feng, Y.; Jiang, S.; Hu, S.; Li, S.; Lin, C.; Xie, X. Sequence stratigraphy and importance of syndepositional structural slope-break for architecture of Paleogene syn-rift lacustrine strata, Bohai Bay Basin, E. China. *Mar. Pet. Geol.* **2016**, *69*, 183–204. [CrossRef]
84. Jiang, S.; Henriksen, S.; Wang, H.; Lu, Y.; Ren, J.; Cai, D.; Feng, Y.; Weimer, P. Sequence-stratigraphic architectures and sand-body distribution in Cenozoic rifted lacustrine basins, east China. *AAPG Bull.* **2013**, *97*, 1447–1475. [CrossRef]
85. Li, J.; Hou, G.; Qin, L.; Xie, J.; Jiang, X. Effect of Sand Body Enrichment Under the Restriction of a Tectonic Transfer Zone: A Case Study on the Pinghu Formation in the Kongqueting Region on the Pinghu Slope. *J. Ocean Univ. China* **2021**, *20*, 765–776. [CrossRef]
86. Zhang, J.; Lu, Y.; Krijgsman, W.; Liu, J.; Li, X.; Du, X.; Wang, C.; Liu, X.; Feng, L.; Wei, W.; et al. Source to sink transport in the Oligocene Huagang Formation of the Xihu Depression, East China Sea Shelf Basin. *Mar. Pet. Geol.* **2018**, *98*, 733–745. [CrossRef]
87. Zhu, X.; Chen, J.; Zhang, C.; Wang, Y.; Liu, K.; Zhang, T. Effects of evaporative fractionation on diamondoid hydrocarbons in condensates from the Xihu Sag, East China Sea Shelf Basin. *Mar. Pet. Geol.* **2021**, *126*, 104929. [CrossRef]
88. Maniscalco, R.; Casciano, C.I.; Distefano, S.; Grossi, F.; Di Stefano, A. Facies analysis in the Second Cycle Messinian evaporites predating the early Pliocene reflooding: The Balza Soletta section (Corvillo Basin, central Sicily). *Ital. J. Geosci.* **2019**, *138*, 301–316. [CrossRef]
89. Hu, M.Y.; Ke, L.; Liang, J.S. The characteristics and pattern of sedimentary facies of Huagang Formation in Xihu depression. *J. Oil Gas Technol.* **2010**, *32*, 1–5.
90. Liu, C. Study on sedimentary facies for Pinghu Formation in Pinghu oil and gas field in East China Sea Basin. *Offshore Oil* **2010**, *30*, 9–13.
91. Chang, J.H.; Hsu, H.H.; Su, C.C.; Liu, C.S.; Hung, H.T.; Chiu, S.D. Tectono-sedimentary control on modern sand deposition on the forebulge of the Western Taiwan Foreland Basin. *Mar. Pet. Geol.* **2015**, *66*, 970–977. [CrossRef]
92. Distefano, S.; Gamberi, F.; Baldassini, N.; Di Stefano, A. Quaternary evolution of coastal plain in response to sea-level changes: Example from south-east Sicily (Southern Italy). *Water* **2021**, *13*, 1524. [CrossRef]
93. Wei, W.; Algeo, T.J.; Lu, Y.; Lu, Y.; Liu, H.; Zhang, S.; Peng, L.; Zhang, J.; Chen, L. Identifying marine incursions into the Paleogene Bohai Bay Basin lake system in northeastern China. *Int. J. Coal Geol.* **2018**, *200*, 1–17. [CrossRef]
94. Chen, Q.; Pu, R.; Xue, X.; Han, M.; Wang, Y.; Cheng, X.; Wu, H. Controlling Effect of Wave-Dominated Delta Sedimentary Facies on Unconventional Reservoirs: A Case Study of Pinghu Tectonic Belt in Xihu Sag, East China Sea Basin. *Geofluids* **2022**, *2022*, 8163011. [CrossRef]
95. Wei, W.; Algeo, T.J. Elemental proxies for paleosalinity analysis of ancient shales and mudrocks. *Geochim. Cosmochim. Acta* **2020**, *287*, 341–366. [CrossRef]
96. Li, S.; Yu, X.; Steel, R.; Zhu, X.; Li, S.; Cao, B.; Hou, G. Change from tide-influenced deltas in a regression-dominated set of sequences to tide-dominated estuaries in a transgression-dominated sequence set, East China Sea Shelf Basin. *Sedimentology* **2018**, *65*, 2312–2338. [CrossRef]
97. Blum, M.D.; Törnqvist, T.E. Fluvial responses to climate and sea-level change: A review and look forward. *Sedimentology* **2000**, *47*, 2–48. [CrossRef]
98. Satur, N.; Hurst, A.; Cronin, B.T.; Kelling, G.; Gürbüz, K. Sand body geometry in a sand-rich, deep-water clastic system, Miocene Cingöz Formation of southern Turkey. *Mar. Pet. Geol.* **2000**, *17*, 239–252. [CrossRef]
99. Seidler, L.; Steel, R. Pinch-out style and position of tidally influenced strata in a regressive–transgressive wave-dominated deltaic sandbody, Twentymile Sandstone, Mesaverde Group, NW Colorado. *Sedimentology* **2001**, *48*, 399–414. [CrossRef]
100. Stevenson, J.C.; Ward, L.G.; Kearney, M.S. Sediment transport and trapping in marsh systems: Implications of tidal flux studies. *Mar. Geol.* **1988**, *80*, 37–59. [CrossRef]
101. Chen, L.L. Genesis analysis of transgressive tidal channel sand bodies of Pinghu Formation in Xihu Sag, East China Sea. *Offshore Oil* **2000**, *2*, 15–21. (In Chinese)
102. Distefano, S.; Gamberi, F.; Borzì, L.; Di Stefano, A. Quaternary Coastal Landscape Evolution and Sea-Level Rise: An Example from South-East Sicily. *Geosciences* **2021**, *11*, 506. [CrossRef]
103. Arthur, M.A.; Sageman, B.B. Sea-level control on source-rock development: Perspectives from the Holocene Black Sea, the mid-Cretaceous Western Interior Basin of North America, and the Late Devonian Appalachian Basin. In *The Deposition of Organic-Carbon-Rich Sediments: Models, Mechanisms, and Consequences*; Society for Sedimentary Geology: Broken Arrow, OK, USA, 2005; pp. 35–59.

104. Cheng, X.; Hou, D.; Zhou, X.; Liu, J.; Diao, H.; Jiang, Y.; Yu, Z. Organic geochemistry and kinetics for natural gas generation from mudstone and coal in the Xihu Sag, East China Sea Shelf Basin, China. *Mar. Pet. Geol.* **2020**, *118*, 104405. [CrossRef]
105. Tissot, B.P.; Welte, D.H. *Petroleum Formation and Occurrence*; Springer: New York, NY, USA, 2013; p. 540.
106. Anders, D. Geochemical exploration methods. In *Source and Migration Processes and Evaluation Techniques. AAPG Treatise of Petroleum Geology, Handbook of Petroleum Geology*; Merrill, R.K., Ed.; AAPG: Tulsa, OK, USA, 1991; pp. 89–95.
107. Peters, K.E.; Moldowan, J.M. *The Biomarker Guide, Interpreting Molecular Fossils in Petroleum and Ancient Sediments*; Prentice Hall: Englewood, NJ, USA, 1993; p. 363.
108. Cheng, X.; Hou, D.; Zhou, X.; Liu, J.; Diao, H.; Wei, L. Geochemical Characterization of the Eocene Coal-Bearing Source Rocks, Xihu Sag, East China Sea Shelf Basin, China: Implications for Origin and Depositional Environment of Organic Matter and Hydrocarbon Potential. *Minerals* **2021**, *11*, 909. [CrossRef]
109. Shen, Y.; Qin, Y.; Cui, M.; Xie, G.; Guo, Y.; Qu, Z.; Yang, T.; Yang, L. Geochemical Characteristics and Sedimentary Control of Pinghu Formation (Eocene) Coal-bearing Source Rocks in Xihu Depression, East China Sea Basin. *Acta Geol. Sin.* **2021**, *95*, 91–104. [CrossRef]
110. Xiong, B.; Wang, C.; Zhang, J.; Zhang, X. The distribution and exploration implications of coal beds of Pinghu Formation, Paleologene in Xihu Sag. *Offshore Oil* **2007**, *27*, 27–33.
111. Su, A.; Chen, H.; Zhao, J.X.; Zhang, T.W.; Feng, Y.X.; Wang, C. Natural gas washing induces condensate formation from coal measures in the Pinghu Slope Belt of the Xihu Depression, East China Sea Basin: Insights from fluid inclusion, geochemistry, and rock gold-tube pyrolysis. *Mar. Pet. Geol.* **2020**, *118*, 104450. [CrossRef]

*Article*

# Depositional and Diagenetic Controls on Reservoir Quality of Callovian-Oxfordian Stage on the Right Bank of Amu Darya

Yuzhong Xing [1], Hongjun Wang [1], Liangjie Zhang [1], Muwei Cheng [2], Haidong Shi [1], Chunqiu Guo [1], Pengyu Chen [1] and Wei Yu [3,*]

[1]   Research Institute of Petroleum Exploration & Development, PetroChina, Beijing 100083, China
[2]   International Corporation (Turkmenistan), CNPC, Beijing 100080, China
[3]   SimTech LLC, Katy, TX 77494, USA
*   Correspondence: yuwei127@gmail.com

**Abstract:** Based on the detailed analysis of sedimentology, diagenesis, and petrophysics, this study characterized the Middle-Lower Jurassic Callovian-Oxfordian carbonate reservoirs of 68 key wells in the Amu Darya Basin and assessed the controlling factors on the quality of the target intervals. We identified 15 types of sedimentary facies developed in seven sedimentary environments using sedimentary facies analysis, such as evaporative platform, restricted platform, open platform, platform margin, platform fore-edge upslope, platform fore-edge downslope, and basin facies. The target intervals went through multiple diagenetic stages, including the syndiagenetic stage, early diagenetic stage, and middle diagenetic stage, all of which had a significant impact on the reservoir quality. Main diagenetic processes include dissolution and fracturing which improve the reservoir quality as well as cementation, compaction, and pressure solution that reduce the reservoir quality. By analyzing the reservoir quality, we identified nine fluid flow units and five types of reservoir facies. Among them, the dissolved grain-dominated reservoir facies is of the highest quality and is the best storage and flow body, while the microporous mud-dominated reservoir facies of platform fore-edge downslope and open marine facies is of the lowest quality and could not become the flow unit unless it was developed by fracturing.

**Keywords:** carbonate reservoir; reservoir quality; depositional model; diagenetic process; flow unit; Callovian-Oxfordian Stage

**Citation:** Xing, Y.; Wang, H.; Zhang, L.; Cheng, M.; Shi, H.; Guo, C.; Chen, P.; Yu, W. Depositional and Diagenetic Controls on Reservoir Quality of Callovian-Oxfordian Stage on the Right Bank of Amu Darya. *Energies* **2022**, *15*, 6923. https://doi.org/10.3390/en15196923

Academic Editors: Xingguang Xu, Kun Xie and Yang Yang

Received: 30 August 2022
Accepted: 19 September 2022
Published: 21 September 2022

**Publisher's Note:** MDPI stays neutral with regard to jurisdictional claims in published maps and institutional affiliations.

## 1. Introduction

As the largest sedimentary basin in Central Asia [1], the Amu Darya Basin is rich in oil and natural gas resources [2], thanks to its superior source-reservoir-cap assemblages and reservoir forming conditions. Recently, many petroleum geologists have conducted detailed research on the sedimentary facies and reservoir characters [3–7] of Callovian-Oxfordian carbonate sedimentation, the identification of reservoirs of reefs and shoal facies, and the evolution and distribution patterns [8–12] of pores and fractures using coring, well logging, seismic data, and geological modeling.

Most of these studies focus on describing and identifying a certain aspect of sedimentary facies or reservoir characteristics; the main factor affecting reservoir quality have not yet been comprehensively investigated. As the reservoir quality in carbonate formation is mainly controlled by comprehensive factors such as sedimentary facies characteristics (e.g., rock structure, textures, and mineral composition), and various post-depositional diagenesis and fracturing [13,14], the inherent geological properties of the rocks result in the static and dynamic heterogeneity of the reservoir [15,16]. Therefore, studying the controlling factors of sedimentation and diagenesis on reservoir quality is a substantial step for hydrocarbon exploration and production in carbonate reservoirs.

This study is the first to analyze the factors affecting the reservoir quality and establish the quantitative standards of different reservoir facies based on the study of the sedimentology and diagenesis of the Callovian-Oxfordian carbonate formation in the Amu Darya Basin. The main purpose was to determine the various characteristics and controlling factors of the Callovian-Oxfordian reservoir quality by establishing a sedimentary model and a diagenesis model for the target area. The research results can be used to guide the fine geological modeling, the preparation of oil and gas reservoir development plans, and the well placement.

## 2. Geological Setting

### 2.1. Geographical Information and History of Petroleum Exploration and Production

The Amu Darya basin is located in southern Turan plate, occupying the western part of Uzbekistan, eastern and central parts of Turkmenistan, north Afghanistan, and north-eastern part of Iran; the total area of the basin is 437,319 km$^2$.

The basin's main tectonic elements include: Central Karakum arch, North Amu Darya Sub-basin, Bahardok monocline, Kopet Dag Foredeep, and Murgab Sub-basin (Figure 1A). The basin's structural style was finally shaped during the Miocene-Quaternary period. The Alpine compression led to a reactivation of existing and formation of new regional faults. At the post-rift stage, the basin was steadily subsiding during a very long period of time and a thick sedimentary sequence was deposited. The most important hydrocarbon traps formed during the period were the Callovian-Oxfordian reefs [17].

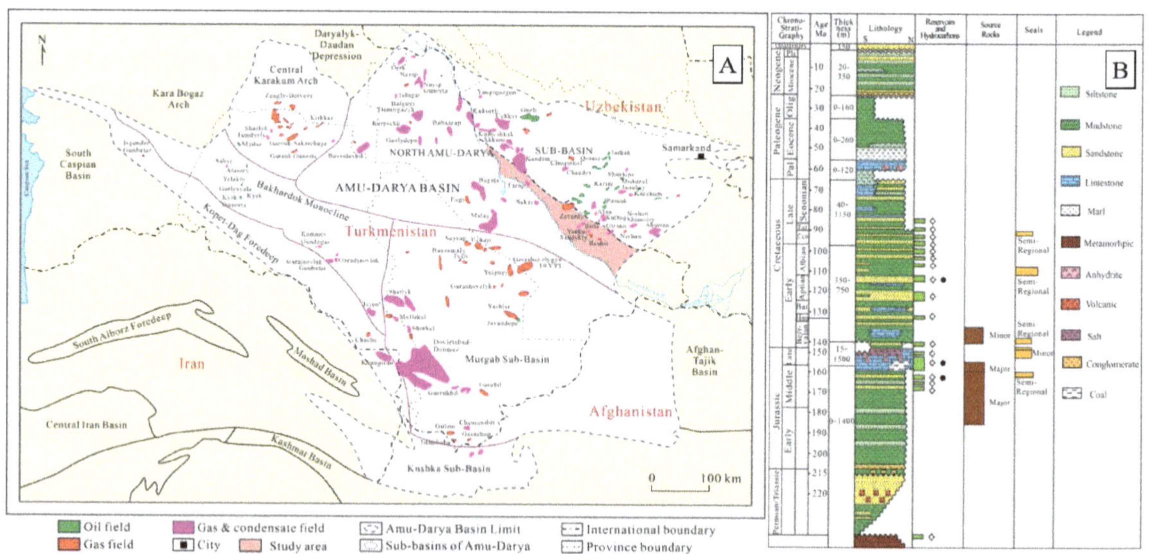


**Figure 1.** (**A**) Map of the Amu Darya Basin and adjacent areas, the study area is indicated by red shade; (**B**) Stratigraphic column and major petroleum system of the basin.

The basement of the Amu Darya Basin is composed of Paleozoic igneous rocks and metamorphic rocks with widely varying burial depths. The Permian, Triassic continental facies, and Lower Jurassic marine-continental alternating facies coal-bearing clastic rock formation is well developed above the basement. The middle-upper Jurassic and Cretaceous formations are marine carbonate rock and evaporite formations, with highly favorable source-reservoir-cap assemblages and hydrocarbon migration or accumulation conditions (Figure 1B).

Sedimentary cover overlying the "Hercynian" basement of the Turan plate comprises three major mega sequences: the Permian-Lower Triassic "Transitional Complex" probably deposited in a back arc environment during the closure of the Paleotethys ocean; the

Upper Triassic-Oligocene "Platform Complex" formed on a Tethyan passive margin; the uppermost Oligocene-Quaternary molasse deposited during the Alpine compressional phase caused by the collision of the Arabian and Indian plates with the Eurasian continent ("the Himalayan Orogeny"). The Himalayan Orogeny is responsible for the formation of the majority of structures in which the present-day hydrocarbon accumulations are trapped.

The exploration and development of the Amu Darya Basin has gone through three stages: ① Regional reconnaissance and pre-exploration stage (1929–1964): two sets of main gas-bearing strata, namely, the post-salt Cretaceous sandstone and the sub-salt lower Jurassic carbonate rock, were determined; ② Shallow exploration and development stage (1965–1996): the main discovered gas reservoirs were in the post-salt Cretaceous and the shallow sub-salt Jurassic structural traps; ③ Deep exploration and development stage (1997–present): the main discovered gas reservoirs were in the deep sub-salt carbonate rocks.

### 2.2. Major Geological Characteristics of Petroleum System

The basin's two main source rocks are Lower-Middle Jurassic shales and coals with humic organic matter, and Callovian-Oxfordian deep-water marine black shales with sapropelic kerogen. Both source rocks are in the gas-generation window over much of the Amu Darya Basin [1].

Proven hydrocarbon reservoirs occur in the Lower-Middle Jurassic (clastics), Upper Jurassic (mainly carbonates), Cretaceous (mainly clastics), and Paleocene (carbonates). Callovian-Oxfordian carbonates including reefal ones, and the Hauterivian Shatlyk sandstones are the most important reservoirs in the basin.

The Gaurdak salt (Kimmeridgian-Tithonian) is the most important regional seal, providing a caprock for Callovian-Oxfordian carbonates.

### 2.3. Location of Study Area and Main Target Strata

The study area (the right bank of Amu Darya) is located near the border of Turkmenistan and Uzbekistan in the northeast of the Amu Darya Basin (Figure 1A). The total area of the study area is 14,314 km$^2$. The main target strata for exploration and development are the Middle and Upper Jurassic carbonate rocks. The Amu Darya Right Bank shows a structural pattern of being higher in the east, west, north, and lower in the middle and south; the faults in the west of the study area are mainly NW-striking slip faults, while those in the east are mainly reverse faults (Figure 2A). Being the main oil and gas production strata, the Callovian–Oxfordian is dominated by thick marine reef limestone, grainstone, and tight limestone interbedded with thin mudstone, with a thickness is 300–450 m. The sub-salt carbonate rocks are classified as Callovian XVI, XVa2, XVz, and Xva1 intervals, and Oxfordian XVhp, XVm, XVp, and Xvac (Figure 2B). Hereinto, the Xvac is interbedded with light gray limestone and gray white anhydrite; the XVp is massive limestone; the XVm is mainly reef limestone and bioclastic limestone mixed with micrite; the upper part of the XVhp is dark gray argillaceous limestone, and the lower part Is brownish-gray tight limestone; the xVa1 is gray limestone with relatively developed organisms and a slightly coarser texture; the XVz is mainly dark gray tight limestone; the xVa2 is gray limestone with crystal powder; the XVI layer is tight bedded limestone. The main production intervals of natural gas are XVp, XVm, XVhp, xVa1, and so on. The reservoir rocks are mostly reef and shoal facies limestone in the sedimentary facies belts such as platform fore-edge upslope, platform margins, and open platforms. Furthermore, the tectonic fractures of different scales in the study area greatly enhanced the accumulation, migration, and flow capacity of oil and gas.

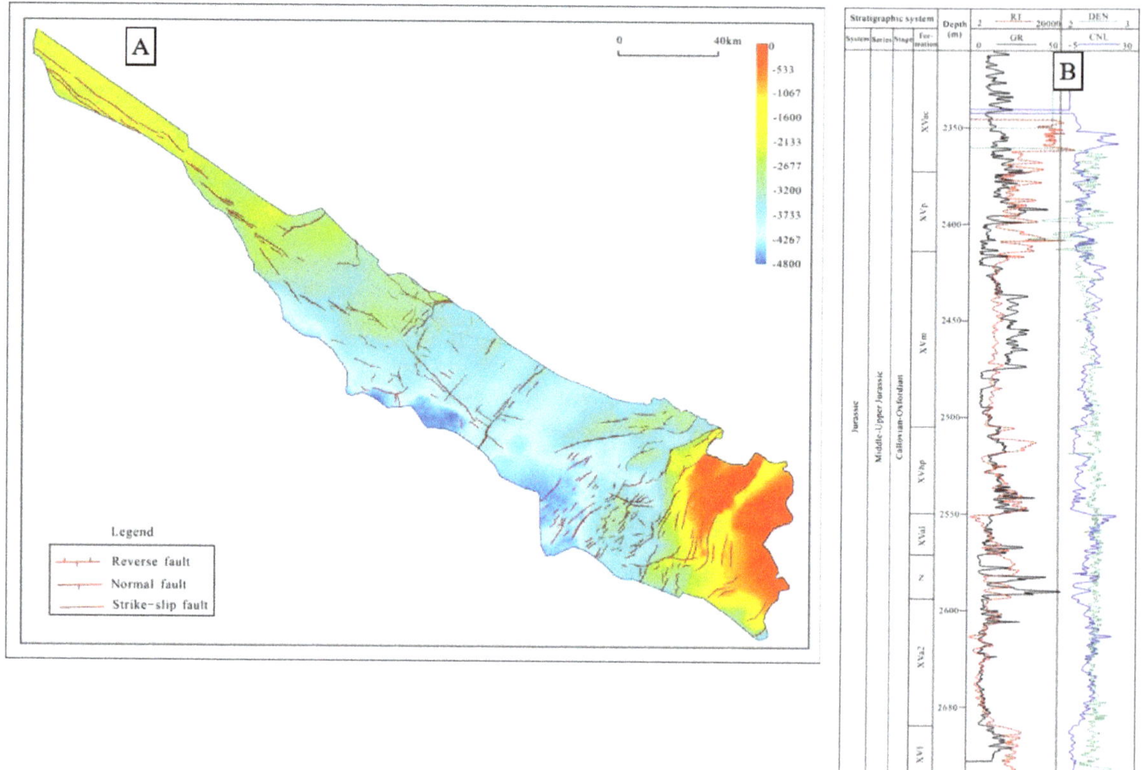

**Figure 2.** (**A**) Structure map (carbonate top) of the study area. (**B**) Target intervals and the log characteristics of Callovian–Oxfordian.

## 3. Data and Methods

In this study, the 600 m coring data and thin sections analysis data of 68 wells in the Callovian-Oxfordian formation were used, mainly including the porosity and permeability analysis data of 65 wells (13,450 samples), core photos of 62 wells, thin sections of 19 wells, and scanning electron microscope data of 13 wells. Thin sections of 15 wells (594 samples) were stained to accurately identify pore types and diagenetic process.

### 3.1. Petrography

The Dunham classification system [18] as modified by Embry and Klovan [19] was mainly used for sedimentary facies analysis, with reference made to the carbonate sedimentary facies description method proposed by Flügel in 2010 [20]. The vertical division of sedimentary facies and stratigraphic correlation were mainly based on core data and petrophysical logging (mainly using natural gamma-ray GR, sonic differential time DT and density logging RHOB).

### 3.2. Reservoir Quality

The flow unit method was used to quantitatively evaluate the reservoir quality using the flow unit index. The flow unit method was proposed by Amaefule et al. [21] by correcting the equation proposed by Kozeny [22] and Carman [23], and flow zone indicator (FZI) values were used to define different flow units [15,24–26]. The input data include porosity ($\phi_e$) and permeability (K) measured on core plugs [27].

In this study, the porosity and permeability obtained by testing 13,450 core plug samples in the target intervals are used to evaluate reservoir quality, whereas density

(RHOB), sonic (DT), and resistivity (LLD and LLS) logs are mainly used to evaluate reservoir quality in uncored intervals.

$$0.0314\sqrt{\frac{k}{\Phi_e}} = (\frac{\phi_e}{1-\phi_e})(\frac{1}{\tau \times S_g \times \sqrt{F_s}}) \tag{1}$$

The left-hand side of the equation is the reservoir quality index *RQI* (μm), defined as follows:

$$RQI = 0.0314\sqrt{k/\phi_e} \tag{2}$$

where $\Phi_e$ is the effective porosity and $k$ is the permeability.

The normalized porosity or pore-matrix ratio (*PMR*) is defined as follows:

$$PMR = \frac{\phi_e}{1-\phi_e} \tag{3}$$

*FZI* (μm) is the flow zone indicator, defined as follows:

$$FZI = (\frac{1}{\tau \times S_g \times \sqrt{F_s}}) \tag{4}$$

By combining Equations (2)–(4), Equation (1) can be transformed into the following:

$$RQI = PMR \times FZI \tag{5}$$

Taking the logarithm on both sides of Equation (5) gives:

$$logRQI = logPMR + logFZI \tag{6}$$

## 4. Results

### 4.1. Facies Analysis

Since diagenetic events and reservoir quality are mainly controlled by depositional attributes, the sedimentary facies must first be investigated in details. Based on the core and thin sections, 15 sedimentary facies (F1–F15) were deposited in seven facies belts (FB1–FB1), such as evaporative platform, restricted platform, open platform, platform margin, platform fore-edge upslope, platform fore-edge downslope, and basin facies in the Oxford period are identified based on rock texture, grain types, sorting and comparative analysis with the standard facies models. See Table 1 and Figure 3 for sedimentary facies characteristics and representative photomicrographs of depositional facies. To provide a basis for our ongoing discussion on reservoir quality, only the facies belts are briefly described in this paper.

**Table 1.** Facies characteristics of the Callovian-Oxfordian stage.

| Facies Belts Name | Code | Facies Name | Code | Main Lithology | Texture | Sorting | Energy Level |
|---|---|---|---|---|---|---|---|
| Evaporative platform | FB1 | Sabkha | F1 | Gypsum | Mudstone | – | Low |
| | | Evaporation lagoon | F2 | Gypsum-bearing limestone | Mudstone | – | Low |
| Restricted platform | FB2 | Tidal flat | F3 | Algal limestone | Mud/packstone | Poor | Low |
| | | Restricted platform shoal | F4 | Micrite oolitic limestone | Pack/Wackstone | Well | High |
| | | Lagoon | F5 | Bioclastic micrite | Wack/packstone | Poor | Low |

**Table 1.** *Cont.*

| Facies Belts Name | Code | Facies Name | Code | Main Lithology | Texture | Sorting | Energy Level |
|---|---|---|---|---|---|---|---|
| Open platform | FB3 | Open platform shoal | F6 | Bioclastic limestone | Grain/packstone | Well | High |
| | | Inter-shoal | F7 | Bioclastic micrite | Wack/packstone | Poor | Low |
| platform Margin | FB4 | Platform margin shoal | F8 | Biosparitic calcirudite | Grainstone | Well | High |
| | | Organic reef | F9 | Reef (Bioclastic) limestone | Grainstone | Poor | High |
| Platform fore-edge upslope | FB5 | Shoal (upslope) | F10 | Bioclastic limestone | Grainstone | Medium | High |
| | | Bioherm | F11 | Micritic grainstone | Grain/packstone | Poor | High |
| | | Mud (upslope) | F12 | Micrite | Wack/mudstone | Poor | Low |
| Platform fore-edge downslope | FB6 | Lime-mud mound | F13 | Micrite boundstone | Wack/packstone | Poor | Low |
| | | Mud (downslope) | F14 | Micrite | Wack/mudstone | Poor | Low |
| Basin | FB7 | Basin mud | F15 | Argillaceous limestone mudstone | Mudstone | Poor | Very low |

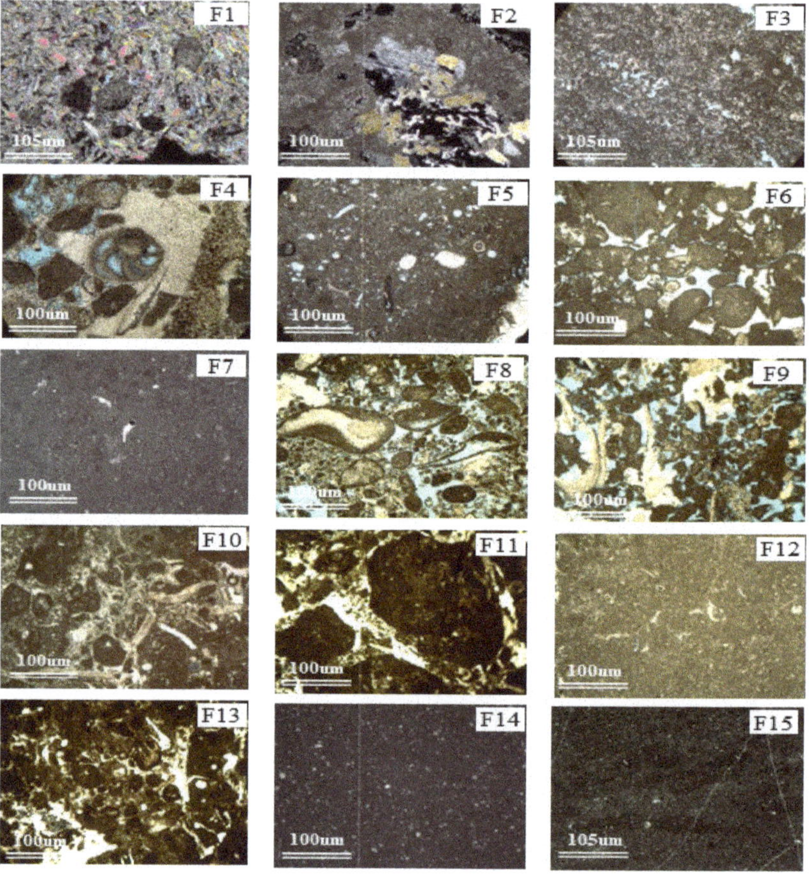

**Figure 3.** Photomicrographs of 15 facies identified in the Callovian-Oxfordian stage.

**(1) Evaporative platform:** The evaporative platform facies belt is formed under arid and hot climate conditions, including Sabkha facies (F1) dominated by gypsum and evaporative lagoon facies (F2) dominated by micrite and gypsum interlayer. The evaporation lagoon facies is mainly at the top of Callovian-Oxfordian strata (Figure 3).

**(2) Restricted platform:** The restricted platform facies belt has three sedimentary facies: tidal flat (F3), restricted platform shoal (F4), and lagoon (F5). The main lithology of tidal flat facies is algal limestone, with occasional dolomite; the main lithology of shoal facies is micrite oolitic limestone and calcarenite; the main lithology of lagoon facies is bioclastic micrite (Figure 3).

**(3) Open platform:** The open platform is composed of open platform shoal (F6) and inter-shoal (F7) facies. The main lithology of the open platform shoal facies is bioclastic limestone while the main lithology of the inter-shoal facies is bioclastic micrite (Figure 3).

**(4) Platform margin:** The platform margin facies is a transition zone between deep water deposits and shallow water deposits, which is the sedimentary environment with the strongest hydrodynamic force and can be classified into two sedimentary facies: platform margin shoal (F8) and organic reef (F9). The main lithology of the shoal facies at the platform margin consists of well-sorted bioclastic limestone, calcarenite, and oolitic limestone. The reef facies is distributed in groups and belts along the platform margin and intergrow vertically with bioclastic shoals. Its main lithology is rudist skeleton reef limestone and boundstone, with a small amount of coral reef limestone. The biological framework is mainly composed of rudist, coral, moss, and algae (Figure 3).

**(5) Platform fore-edge upslope:** The platform fore-edge upslope facies belt is divided into three sections: the shoal (F10), the bioherm (F11), and the mud facies of the platform fore-edge upslope (F12). The lithology of the shoal is dominated by calcarenite and bioclastic limestone. They consist of intraclasts or spherulites, and so on. The bioclasts are dominated by bivalves, followed by echinoderms, foraminifera, Crinoidea, and algae. The lithology of bioherm facies is mainly micritic grainstone composed of bioclasts, spherulites, clotted limestones, and algal nodules. The mud facies of the platform fore-edge upslope is mostly microcrystalline limestone, with a few foraminifera and pleopods (Figure 3).

**(6) Platform fore-edge downslope:** The platform fore-edge downslope facies belt is classified as lime-mud mound (F13) and mud (F14) facies. The lime-mud mound is only found in the middle and lower parts of the platform foreslope and is composed of wackestone, and a small amount of packstone. It is composed of very pure fine-grained lime-mud or lime-mud with spherulites and bioclasts, and the bioclasts are mainly rinoideaa, brachiopods, gastropods, etc. The mud facies of platform fore-edge downslope are dark thin-layer bioclastic microcrystalline limestone and spherulitic microcrystalline limestone deposited in deep water and a low-energy environment (Figure 3).

**(7) Basin:** Two types of basin facies are developed in the study area (F15): One is a deep Callovian sea basin beneath the platform foreslope with a deep and low-energy water body. Its main lithology consists of thin-layer dark microcrystalline limestone and marl rich in organic matter and argillaceous lamination, with a trace of bivalve and siliceous sponge spicules. The other type is the silled bay basin (deep lagoon), which was directly transformed from the early basin due to strong limitation of seawater circulation caused by a large-scale decline in sea level at the end of the Oxford period. The sediments contain few biological fossils, and the main lithology is mudstone with a high GR (Gamma-Ray) log response (Figure 3).

### 4.2. Diagenesis

Petrographic studies of core samples, core plugs and thin sections reveal that different sedimentary facies of the Callovian-Oxfordian strata have been subjected to complex and strong diagenetic processes, including cementation, mechanical compaction, chemical compaction (pressure solution), dissolution, fracturing, silicification, anhydrification, filling, dolomitization, and recrystallization. The first five were the most common (Figure 4). Considering the diagenetic process controls on reservoir quality (i.e., porosity and permeability), these five main diagenetic processes can be classified into two categories:

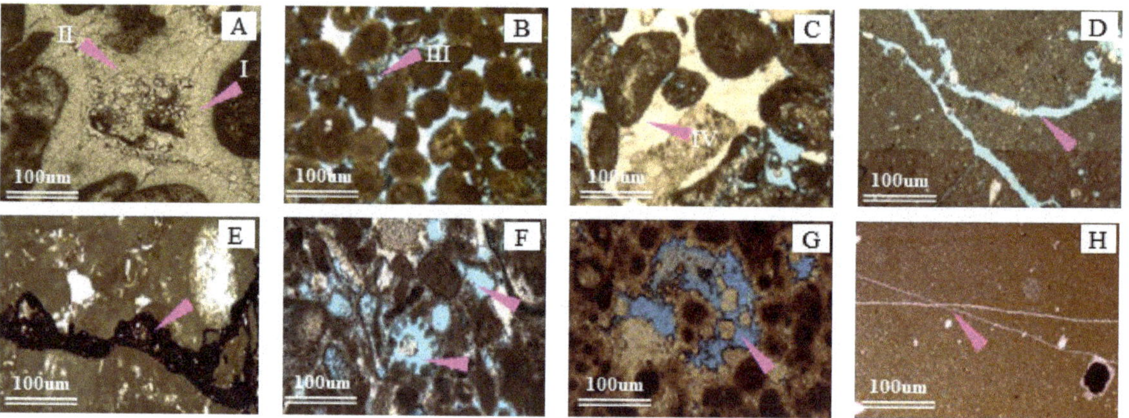

**Figure 4.** Photomicrographs of diagenetic features of the Callovian-Oxfordian stage. ((**A–C**) refer to cementation, where I is isopachous calcite cementation, II is equigranular calcite cementation, III is medium-coarse calcite filling, and IV is intergrowth calcite filling; (**D**) is the fracture formed by mechanical compaction; (**E**) is the stylolite formed by chemical compaction; (**F**) indicates mold pores, intragranular dissolution pores, and intergranular dissolution pores formed by dissolution; (**G**) refers to intergranular and intragranular dissolution pore; (**H**) refers to the shear fracture formed by fracturing).

### 4.2.1. Reservoir Quality Reducing Diagenetic Processes

(1)  Cementation

Cementation mainly occurs in a high-energy reef and shoal facies limestone where various primary pores form, and the cement is predominantly calcite. Based on the distribution patterns of cement, four-stage cementation can be identified in the Callovian-Oxfordian Stage [5,7]. Stage I: Quasi-syngenetic comb-shell-shaped calcite (0.05–0.15 mm) covered particles with a uniform thickness (of 0.05–0.15 mm) or growing along the pore wall of a biological reef skeleton pore (Figure 4A). Stage II: In the early stage of diagenesis, the equigranular (0.05–0.1 mm) sparry calcite filled in the pores between reef skeletons or bioclastic grains, growing along the pore edges, or forming a second-generation cement texture with the asaphopsoides isopachous calcite cement in Stage I (Figure 4A); Stage III: Medium-coarse grained (0.3–2.0 mm) equigranular sparry calcite filled in primary pores from late early diagenetic stage to early mid-diagenetic stage (Figure 4B). Stage IV: Intergrowth calcite filled in the remaining pores in the late mid-diagenetic stage (Figure 4C). Although the cementation process in the four stages significantly reduced the primary porosity and deteriorated the reservoir quality, the high-energy reef and shoal facies medium-thick calcite cementation process provided a rigid framework, thus reducing the impact of compaction on the quality of such reservoirs.

(2)  Compaction

Mechanical compaction and chemical compaction are common in the studied intervals. Mechanical compaction mainly occurs in reef limestone and shoal facies grainstone, which leads to deformation, directional alignment, and fracturing of fine bioclasts (Figure 4D), and has little impact on the primary pore and reservoir quality of loose, porous reef limestone, and shoal facies grainstone. Chemical compaction (pressure solution) mainly occurs in the sedimentary facies with high mud content, forming stylolites (Figure 4E). Since the soluble components are dissolved and migrated during chemical compaction (pressure solution) and most of them enter the pores in the form of cement, the stylolites are filled with residual clay or organic matters from pressure solution and closed. As a result, chemical compaction (pressure solution) is highly destructive to the reservoir quality.

4.2.2. Reservoir Quality Enhancing Diagenetic Processes

(1)  Dissolution

Dissolution includes freshwater leaching and dissolution during the syngenetic sedimentary period and burial dissolution during the middle and late diagenetic periods [4]. The platform reef shoal of the highstand system tract was intermittently exposed to the atmospheric environment and was selectively dissolved (the bioclastic skeleton and bioclasts, such as rudist, coral, foraminifera, bryozoa, and rhodophyta were preferentially dissolved) under the influence of freshwater leaching resulting in abundant intergranular dissolution pores, intragranular dissolution pores, and mold pores were developed in the grain-dominated shoal and reef facies (Figure 4F,G). The burial dissolution was mainly caused by the organic acid generated in the hydrocarbon generation process that dissolved the carbonate rock, thus further enlarging the pores [28]. Although the dissolved mold or dissolved pores were partially filled with later calcite cement, these two types of dissolution significantly improved the reservoir quality.

(2)  Fracturing

Fracturing is very common in the Callovian−Oxfordian formation, and the tectonic fractures formed are often partially filled with calcite and carbonized asphalt (Figure 4H). Microfractures are important for linking up pores and cavities in the reservoir and improving its performance; additionally, the tectonic fractures are prone to dissolution, which is very conducive to the formation of secondary dissolution pores and dissolution fractures and greatly improve the reservoir quality.

*4.3. Reservoir Quality*

Generally, the quality and heterogeneity of carbonate reservoirs are affected by geological contexts such as sedimentation, diagenesis, and porosity characteristics, which affect the FZI value. Consequently, the reservoir quality can be quantitatively evaluated using the FZI method [29–32].

4.3.1. Flow Unit

Ideally, on a log-log crossplot of reservoir quality index (RQI) versus normalized porosity (PMR), sample points with similar FZI values should be plotted on the same straight line with a slope of 1, while sample points with different FZI values should be plotted on another parallel straight line. Samples on the same straight line constitute an independent flow unit due to similar pore throat properties. When PMR = 1, the intercept of each straight line represents the average FZI value of a flow unit [15,24].

The flow units of the Callovian-Oxfordian formation were identified using the data from 68 key wells. Figure 5 is the porosity−permeability crossplot of the identified nine flow units (FUs) and the RQI of different flow units as a function of PMR are shown in Figure 6. As shown in Figure 5 and the statistical parameters of different flow units (Table 2), the average porosity gradually decreases with the increase in the number of identified FUs; the average permeability gradually increases as the number of identified FUs increases, and the FZI gradually increases. Consequently, as the number of identified FUs increases, the storage capacity of the interval gradually decreases, whereas the flow capacity and reservoir quality improve. Furthermore, by classifying FUs in key wells, the longitudinal distribution of different reservoir qualities can be better evaluated.

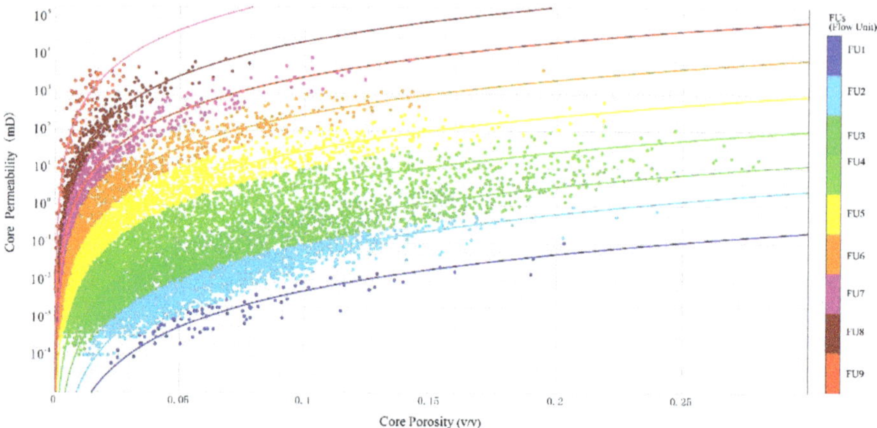

Figure 5. Porosity−permeability cross plots of different flow units.

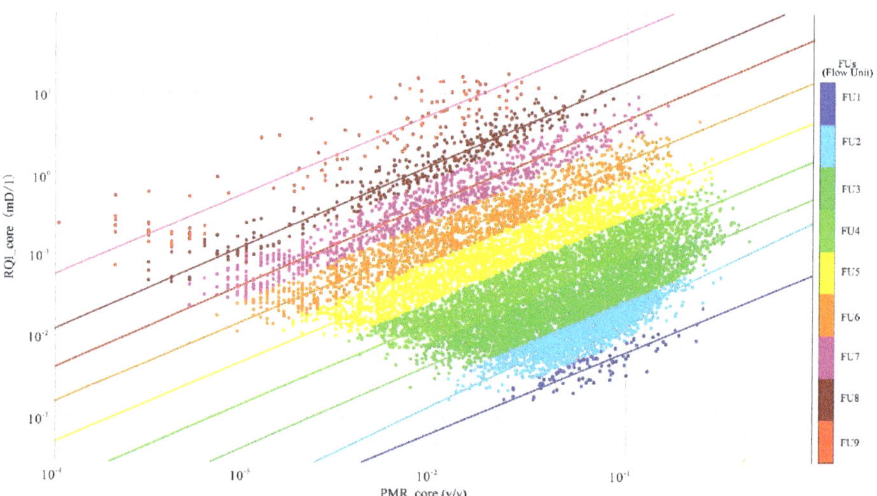

Figure 6. RQI−PMR cross plots of different flow units.

Table 2. Statistics of Parameters of Different Flow Units of the Callovian-Oxfordian stage.

|  |  | FU1 | FU2 | FU3 | FU4 | FU5 | FU6 | FU7 | FU8 | FU9 |
|---|---|---|---|---|---|---|---|---|---|---|
| Porosity (%) | Min | 2.3 | 1.2 | 0.7 | 0.4 | 0.2 | 0.1 | 0.1 | 0.00 | 0.00 |
|  | Max | 20.3 | 24.0 | 30.7 | 26.6 | 24.9 | 20.5 | 14.2 | 7.8 | 0.8 |
|  | Mean | 7.3 | 6.4 | 6.6 | 6.1 | 4.3 | 2.4 | 1.6 | 1.4 | 0.5 |
| Permeability (mD) | Min | 0.000 | 0.000 | 0.000 | 0.000 | 0.000 | 0.000 | 0.000 | 0.001 | 0.001 |
|  | Max | 0.103 | 0.925 | 18.469 | 162.00 | 769.80 | 4189.5 | 8801.3 | 7874.3 | 7405.8 |
|  | Mean | 0.004 | 0.026 | 0.406 | 4.330 | 18.528 | 44.772 | 118.20 | 296.04 | 662.93 |
| Reservior Quality Index (RQI) | Min | 0.002 | 0.002 | 0.003 | 0.005 | 0.01 | 0.011 | 0.023 | 0.050 | 0.073 |
|  | Max | 0.022 | 0.066 | 0.260 | 0.819 | 1.970 | 4.610 | 9.170 | 10.750 | 17.720 |
|  | Mean | 0.005 | 0.014 | 0.039 | 0.113 | 0.222 | 0.392 | 0.800 | 2.020 | 4.680 |
| Flow Zone Indicator (FZI) | Min | 0.029 | 0.090 | 0.282 | 0.892 | 2.819 | 8.927 | 28.210 | 89.407 | 283.47 |
|  | Max | 0.089 | 0.282 | 0.891 | 2.818 | 8.912 | 28.128 | 88.967 | 278.354 | 2381.5 |
|  | Mean | 0.066 | 0.193 | 0.526 | 1.644 | 5.073 | 16.168 | 48.338 | 143.885 | 685.95 |

### 4.3.2. Pore Types

The results of petrographic studies of cores and thin sections, as well as scanning electron microscopy (SEM) analysis, show that different pore types exist in the Callovian-Oxfordian formation (Figure 7), which can be classified into three types based on genesis:

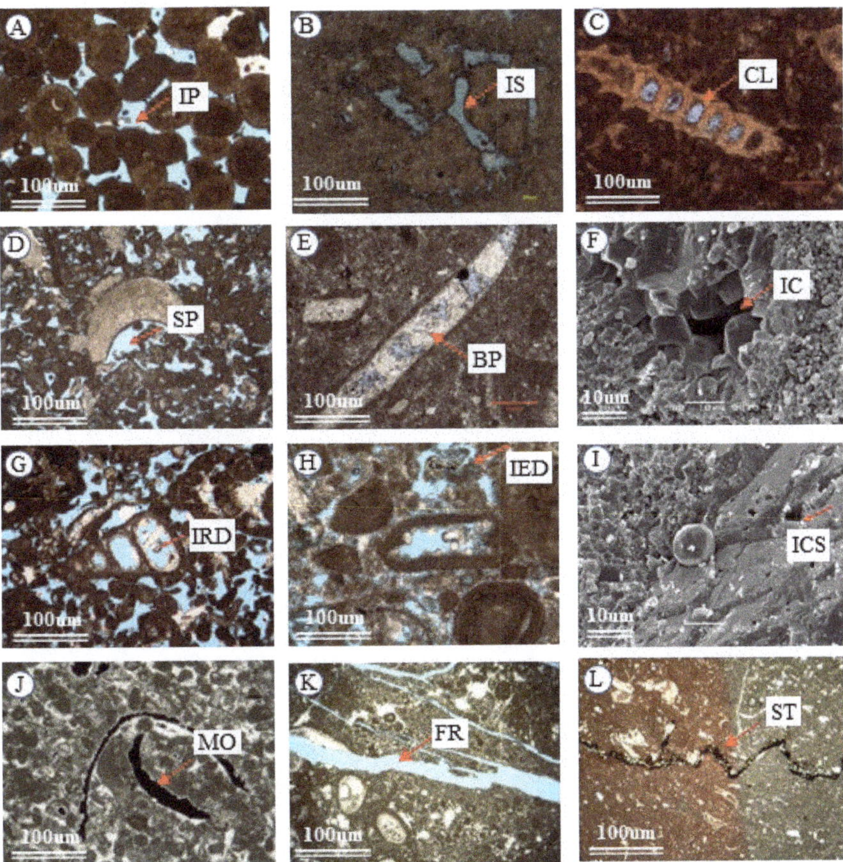


**Figure 7.** Photomicrographs and SEM images of various pore types of the Callovian-Oxfordian stage (**A–L**). They include interparticle (IP), intraskeletal (IS), coelomopore (CL), shadow pore (SP), bioboring pore (BP), intercrystalline (IC), intragranular dissolved pore (IRD), intergranular dissolved pore (IED), intracrystal solution pore (ICS), modic (MO), fracture (FR), and stylolite (ST) pore type.

(1) Primary pore types

Primary pores refer to the pores formed during sedimentation, the primary pores in the studied formaton include interparticle pores, intraskeletal pores, coelomopore, shadow pores, bioboring pores, and intercrystalline pores (Figure 7A–F). An interparticle pore is the dominant pore type of packstone-grainstone of shoal facies (Figure 7A), which are mainly the pore spaces between beach facies bioclasts, pelletoids, and oolities. In terms of strata, interparticle pores are the dominant pore type of XVm and XVa1 layers. The intraskeletal and coelomopore pores are mainly developed in the biohermal facies (XVm) limestone in the platform margin facies belt, with a few intercrystalline pores, which is a pore type in the dolomitized packstone-grainstone main facies belt (Figure 7F), and is mainly developed in the Xvac layer and exposed environments such as the top of Sabkha and shoal facies; bioboring pores and shelter pores are less visible in the target interval.

(2)    Secondary pore types

Secondary pores refer to the pores formed after sediment deposition. The target stratum mainly includes intragranular dissolved pores, intergranular dissolved pores, intracrystal dissolution pores, mold pores, fractures, stylolites, and vugs (Figure 7G–L). Due to atmospheric freshwater dissolution, various dissolution pores were formed and mainly distributed in reef limestone, calcarenite, calcirudite, bioclastic limestone, and other grainstones in the reef (shoal) facies; burial dissolution shaped the vugs which are distributed along the fractures, mainly near the fracture zone; bioboring pores were usually filled and then dissolved (Figure 7E). Additionally, tectonic fractures of different scales (Figure 7K) are common in the outcrops, cores, and thin sections. Tectonic fractures greatly improved the reservoir quality (especially, the permeability of the mud dominated facies belt).

(3)    Micropores

Petrographic and SEM analysis indicate that there are micropores (including round, subrounded, oblique to rhombic micrite interparticle pores in limestones) and micro molds in micrites are only visible under the microscope in low-energy, mud-dominated facies belts (e.g., mud facies on the platform fore-edge downslope) of the Callovian-Oxfordian stage (Figure 7I). Such micropores can only be identified by SEM images, not in the routine petrological studies.

## 5. Discussion

### 5.1. Sedimentary Model

The recognition of sedimentary facies helps to construct the sedimentary environment of the target intervals and to exhibit a sedimentary model. The sedimentary model helps guide the deployment of well locations during the early stages of oil and gas exploration and development, and to establish a more accurate fine geological model (especially the sedimentary facies model) during the middle and later stages to guide the efficient development of oil and gas reservoirs.

Through the analysis of sedimentary facies of Callovian-Oxfordian formation, 7 sedimentary facies belts and 15 sedimentary facies are identified; the conceptual sedimentary model is depicted in Figure 8. A relatively similar model was proposed in the Zagros area and other surrounding areas in the Middle East [33–35], which indicate a slightly sloping depositional setting and proposed a carbonate ramp conceptual model.

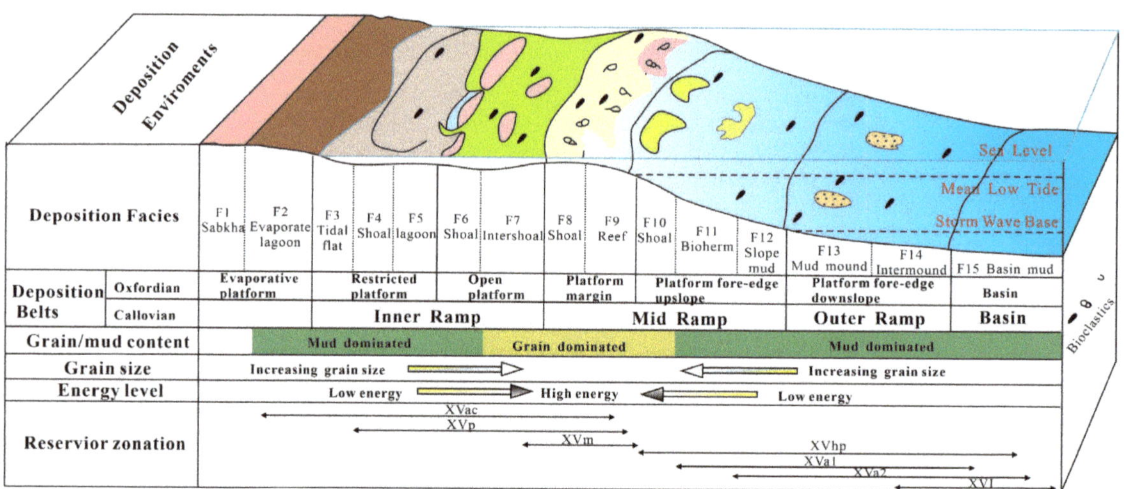


**Figure 8.** Conceptual depositional model presented for the Callovian-Oxfordian stage in the Amu Darya Basin.

The area under study in the Callovian period is a carbonate gentle slope sedimentary system composed of an inner ramp, mid ramp, outer ramp, and basin facies belts. The inner ramp is composed of an internal closed inner ramp (Sabkha, tidal flat, and lagoon) and an external restricted inner ramp (shoal and restricted subtidal zone). The shoal at the high point of the ancient landform serves as the barrier between the inner and the outer zones. The water body of the inner ramp exposed above the sea level is blocked and highly evaporative with gypsum-bearing micrite deposited. The restricted inner ramp has a shallow water body, with poor circulation, and is dominated by micrite deposits. The mid ramp is divided into two zones: an inner zone (open subtidal zone) and an outer zone (shoal and reef flat). The water body in the inner zone has low-energy but good circulation, with prosperous organisms and low-energy bioclastic shoals deposited. The water body in the outer zone is highly energetic, with high-energy shoals or reef–shoal complexes developed. The outer ramp is divided into two zones: an inner zone and an outer zone. The inner zone is located above the lowest storm surface and is dominated by storm action induced packstone deposition. The outer zone is in a low-energy environment, with wackstone and lime-mud mounds developed. The basin facies is located below the lowest storm surface, with marlstone, calcareous mudstone, and mudstone deposited.

In the early Oxfordian period, under regional transgression, the outer zone of the mid ramp and outer ramp in the Callovian period were gradually submerged, and the inner ramp—mid ramp gradually developed into an edged shelf-type carbonate platform (this model is similar to Wilson's platform model). The evaporative platform facies belt is located in the supratidal low-energy zone, which is often exposed above the sea surface and has poor water circulation. It can be classified into two sedimentary facies: Sabkha and evaporative lagoon. The restricted platform facies zone is located in the intertidal-subtidal low-energy zone, with tidal flat facies, low-energy shoal facies, and lagoon facies developed. The water body in the open platform facies belt has good circulation, with the intra-platform bioclastic shoals and inter-shoal facies developed. The platform margin facies zone is located in the subtidal high-energy zone, with reef facies and high-energy bioclastic shoal facies developed. The upslope of platform front is located below the wave base, with point reefs developed locally. The reef shoals and the inter-reef shoals sedimentary landform are quite different. The downslope of platform front and the basin facies have quiet water bodies and are characterized by low-energy micrite, wackstone, and mudstone deposition.

*5.2. Diagenesis History*

By establishing and interpreting the diagenetic sequence based on the petrographic characteristics of diagenesis and its relationship, the main diagenesis process of the Callovian-Oxfordian stage can be summarized as follows (Figure 9):

(1)  Syngenetic diagenensis: In the sedimentary environment of biological reef and shoal facies with high-energy, the porosity was reduced and the reservoir performance deteriorated due to edged asaphopsoides isopachous calcite cementation and equigranular calcite cementation. The main pore types at this stage are the remaining primary pores like interparticle pores, intraskeletal and coelomopore pores.

(2)  Early diagenesis: This stage is characterized by diagenetic processes such as mechanical compaction and chemical compaction (pressure solution), equigranular calcite cementation, and intergrowth calcite cementation. The reservoir is dominated by the remaining primary pores, as the number of primary pores was severely reduced.

(3)  Early middle diagenesis: At this stage, intergrowth calcite cementation occurred, and chemical compaction was further strengthened, but local strong diagenetic fracturing and dissolution improved the reservoir quality. The reservoir space is still dominated by remaining primary pores, with a few secondary pores.

(4)  Late middle diagenesis: The main diageneses are dissolution and fracturing, as well as secondary pores such as intergranular dissolution pores and intragranular

dissolution pores, mold pores, intercrystalline dissolution pores, vugs, and fractures were widely developed.

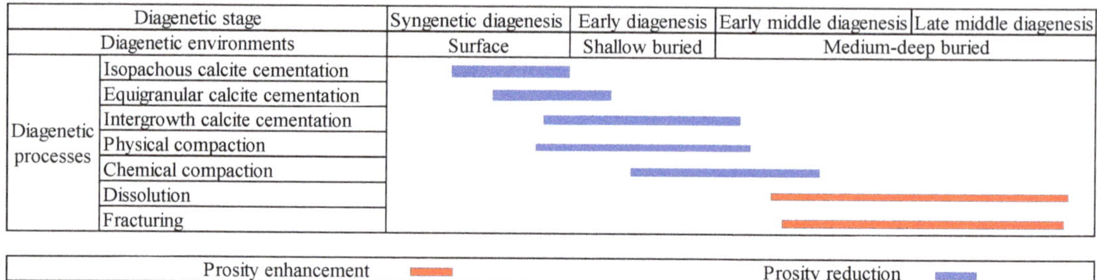

| Diagenetic stage | | Syngenetic diagenesis | Early diagenesis | Early middle diagenesis | Late middle diagenesis |
|---|---|---|---|---|---|
| Diagenetic environments | | Surface | Shallow buried | Medium-deep buried | |
| Diagenetic processes | Isopachous calcite cementation | | | | |
| | Equigranular calcite cementation | | | | |
| | Intergrowth calcite cementation | | | | |
| | Physical compaction | | | | |
| | Chemical compaction | | | | |
| | Dissolution | | | | |
| | Fracturing | | | | |

| Prosity enhancement | Prosity reduction |
|---|---|

**Figure 9.** Paragenetic sequence of diagenetic processes in the Callovian-Oxfordian stage in the studied area. Effects of diagenetic processes on porosity are shown as blue (for porosity reducing), red (porosity enhancing reducing).

These results have been similarly reported in previous studies. As Lee et al. [36] noted, the mechanical compaction can reduce porosities to 30% in carbonates, which occurred the depth shallower than 750 m and before major chemical compaction. Tavakoli et al. [37,38] pointed out the diagenetic heterogeneity and explained the similar influence on reservoir performance in different sedimentary environments.

*5.3. Controlling Factors of the Reservoir Quality*

Using a comprehensive analysis of petrographic and petrophysical data, we evaluated the quality of the Callovian-Oxfordian reservoir on the Right Bank of Amu Darya. To establish a reliable relationship between physical property analysis data (porosity and permeability) and lithofacies data, five reservoir types (RF1–RF5) were identified based on facies characteristics (such as rock texture), dominant diagenesis, and pore types; the distribution and average values for porosity, permeability, RQI, porosity to matrix ratio (PMR), and FZI of different reservoir facies (RFs) were recorded (Figure 10, Table 3).

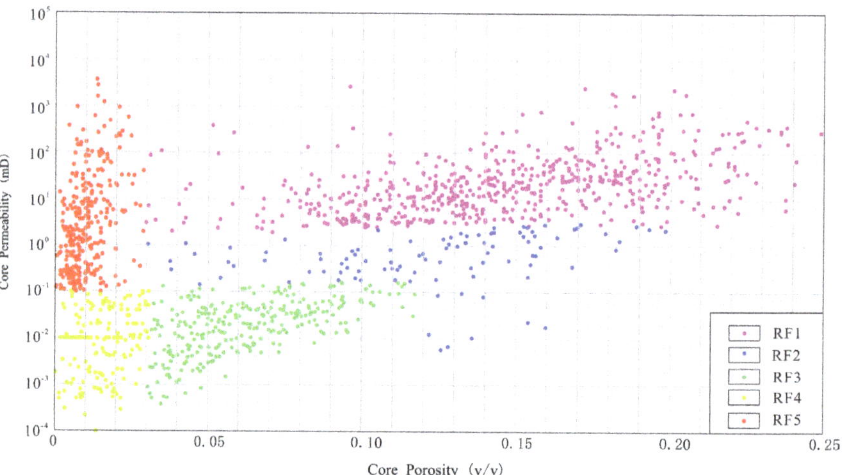


**Figure 10.** Porosity−Permeability cross plots of reservoir facies of the Callovian−Oxfordian stage.

As depicted in Figure 10 and Table 3, From RF1 to RF4, the average porosity, permeability, and RQI gradually decrease, indicating that the reservoir storage capacity and flow capacity gradually decrease, and the reservoir quality gradually deteriorates. An exception occurs in the statistical values of RF5, wherein its porosity and PMR are lower than those of RF1, RF2, RF3, and RF4, indicating that RF5 has a low storage capacity (even no storage capacity at all), whereas its permeability and RQI are higher than those of RF2, RF3, and RF4. This indicates that its flow capacity is higher than those of the three reservoir facies.

Based on the statistics presented above, we analyzed the sedimentological characteristics of different reservoir facies (RF1–RF5) of the Callovian-Oxfordian formation, their corresponding relation with FUs, and their distribution characteristics (Figures 11 and 12).

**Table 3.** Statistical parameters calculated for reservoir facies (RFS) of the Callovian−Oxfordian stage.

|  |  | RF1 | RF2 | RF3 | RF4 | RF5 |
|---|---|---|---|---|---|---|
| Porosity (%) | Min | 2.9 | 2.9 | 3.0 | 0.0 | 0.0 |
|  | Max | 25.0 | 19.7 | 11.8 | 3.2 | 2.8 |
|  | Mean | 14.3 | 11.2 | 5.9 | 1.3 | 1.1 |
| Permeability (mD) | Min | 1.900 | 0.004 | 0.004 | 0.0001 | 0.009 |
|  | Max | 2701.400 | 3.200 | 0.170 | 0.121 | 3985.800 |
|  | Mean | 22.500 | 0.500 | 0.016 | 0.006 | 1.820 |
| Reservoir quality Index(RQI) | Min | 0.007 | 0.007 | 0.002 | 0.002 | 0.064 |
|  | Max | 0.235 | 0.235 | 0.070 | 0.535 | 17.719 |
|  | Mean | 0.483 | 0.078 | 0.017 | 0.024 | 0.363 |
| Ratio of Porosity to Matrix (PMR) | Min | 0.030 | 0.030 | 0.031 | 0.000 | 0.000 |
|  | Max | 0.389 | 0.261 | 0.137 | 0.033 | 0.032 |
|  | Mean | 0.127 | 0.103 | 0.064 | 0.014 | 0.016 |
| Flow zone indicator (FZI) | Min | 0.164 | 0.038 | 0.028 | 0.070 | 2.200 |
|  | Max | 343.653 | 7.542 | 2.010 | 2303.910 | 2381.515 |
|  | Mean | 4.535 | 0.864 | 0.282 | 2.556 | 26.837 |

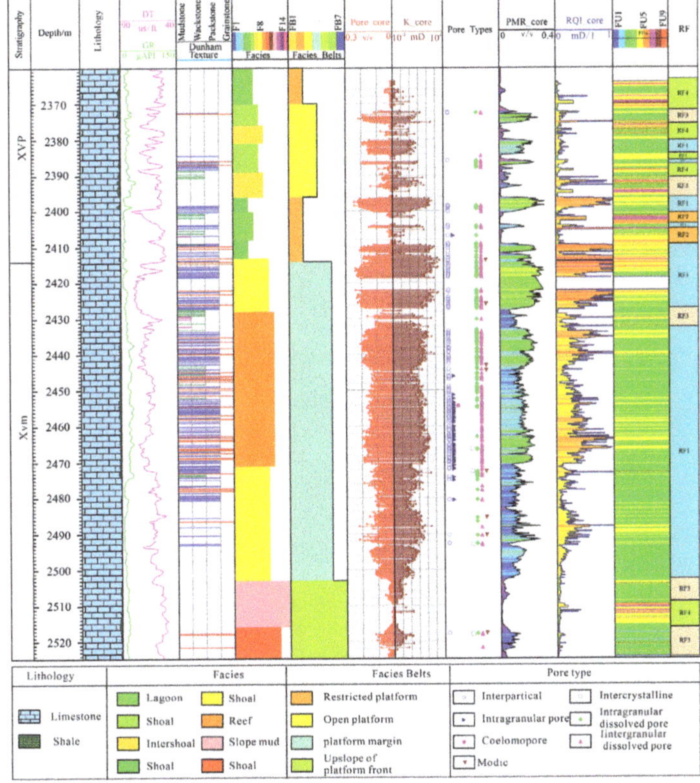

**Figure 11.** Correlation between different parameters including depositional, diagenetic, porosity, permeability, and flow units in the Callovian-Oxfordian stage of well S1.

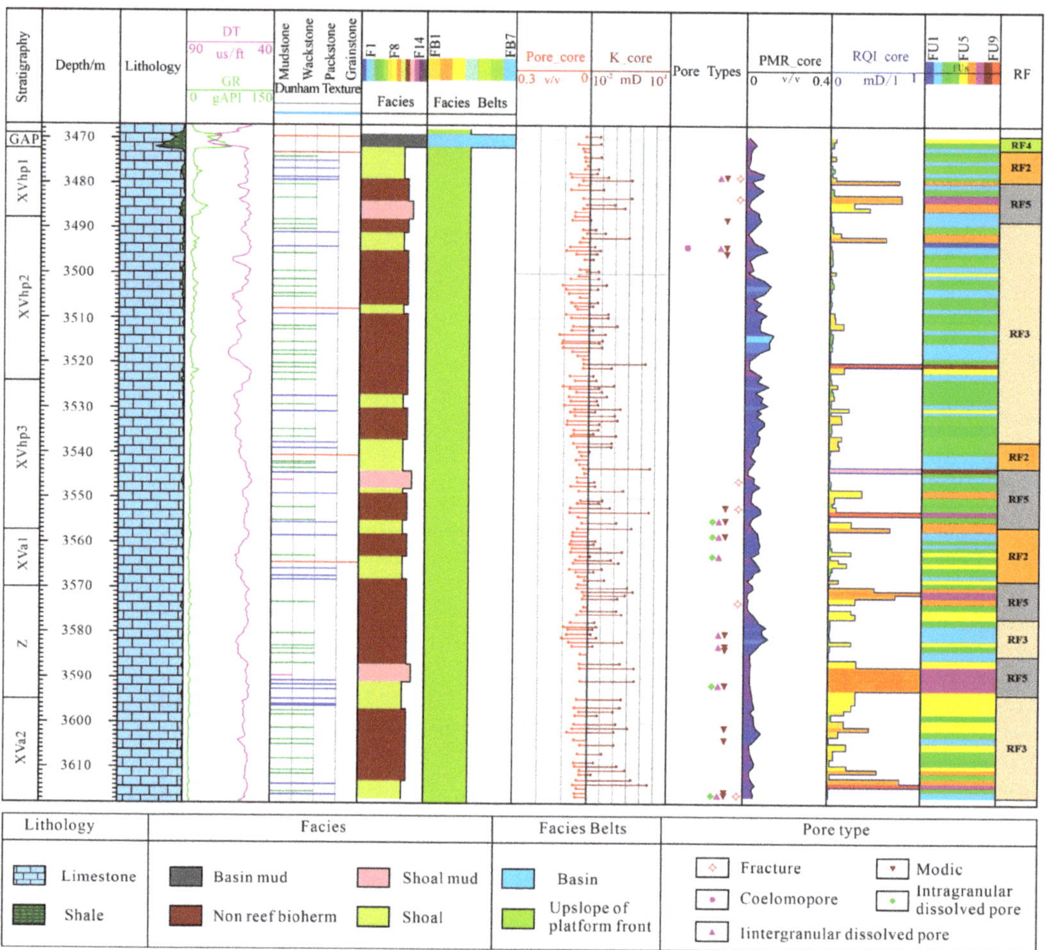





**Figure 12.** Correlation between different parameters including depositional, diagenetic, porosity, permeability, and flow units in the Callovian-Oxfordian stage of well C1.

### 5.3.1. Dissolved Grain-Dominated Reservoir Facies—RF1

This reservoir facies (RF1) is mainly distributed in the high-energy shoal and biohermal facies in the inner ramp and mid ramp facies belts, and the main lithology is grainstone or packstone with bioclasts, oolites, and so on. Since this reservoir facies is usually developed near the sequence boundary and has undergone strong dissolution, it has pore types such as intragranular dissolution pores, intergranular dissolution pores, and mold pores formed by atmospheric dissolution (Figure 11). According to the statistical results of FZI, porosity, and permeability (Table 3, Figures 10 and 11), RF1 mainly corresponds to three flow units, namely, FU4, FU5, and FU6, with a high storage capacity and flow capacity, an average porosity of 14.3% and an average permeability of 22.5 mD. This kind of reservoir facies is mainly distributed in the XVm layer of the Callovian-Oxfordian formation, which has a thickness of about 30 m in general. When it is at the platform margin, the thickness can reach 70–80 m (Figure 11), followed by XVp and XVa1 layers, with a thickness of 5–8 m. RF1 mainly distributed in the western part of the study area and has good lateral continuity. The reservoir facies is characterized by good connectivity, abundant reserves, high single-well and stable production.

### 5.3.2. Compacted/Cemented Grain-Dominated Reservoir Facies—RF2

The reservoir facies (RF2) is mainly composed of packstone and grainstone deposited on a restricted platform, open platform, and platform fore-edge upslope. Generally, it is far away from the sequence boundary and has been subjected to strong cementation and mechanical compaction, with weak dissolution. The remaining interparticle pores are the most common pore type. Although the reservoir facies is dominated by grainstone with high primary porosity, the pore throat radius is small and the permeability is low due to strong mechanical compaction. In addition, as the calcite cement plays an important blocking role in the pore space and pore throat size, the quality of the RF2 reservoir is lower than that of RF1 (Table 3), which is characterized by a medium FZI, porosity, and permeability (Table 3 and Figure 10). The average porosity is 11.2%, the average permeability is 0.5 mD and is mainly composed of flow units FU3 and FU4. Generally, RF2 has a higher storage capacity, but its flow capacity is lower than that of RF1. It is mainly distributed in XVp and XVhp layers with a small thickness and sporadic distribution. For example, the RF2 thickness of Well S1 is 5.4 m (Figure 11) and that of Well C1 is 12.4 m (Figure 12). RF2 mainly distributed in the western part of the study area, it has poor lateral continuity and planar distribution.

### 5.3.3. Packstone/Wackstone Dominated Reservoir Facies—RF3

The reservoir facies (RF3) mainly consists of bioherm (F11), mud mound (F13), and inter-shoal (F7) facies of open platform deposited in the upslope and downslope of the platform front-edge. Since the reservoir facies is less affected by the dissolution of atmospheric fresh water, only a few dissolution pores are developed. Although the occasionally developed mold pores increased porosity, they had little impact on permeability since most of the mold pores are isolated pores, leaving the porosity and permeability of packstone and wackstone relatively low (the average porosity is 5.9%, and the average permeability is 0.016 mD). The RF3 mostly associated with FU2 and FU3 (Tables 2 and 3). In the open platform or platform margin facies belt, the thickness of a single layer of the RF3 is small, usually 3–5 m (Figure 11); in the slope facies at platform front-edge, the thickness of a single layer of the RF3 is relatively large, usually 20–50 m (Figure 12). RF3 mainly distributed in the middle part of the study area and it has poor lateral continuity.

### 5.3.4. Microporous Mudstone-Dominated Reservoir Facies—RF4

The reservoir facies (RF4) is widely distributed, especially in the central and eastern part of the basin, including mudstone-dominated sedimentary facies in the inner ramp, mid ramp, and outer ramp, and are mainly distributed in the F12–F15 facies, and the XVhp, XVz, and XVI layers. It is difficult to observe the macropore type of these facies on cores and only a few micropores are visible by microscopic or SEM image. Additionally, due to diagenetic processes such as fine-grained dolomitization and anhydrite cementation, the porosity and permeability of these mud-dominated facies are usually further reduced, making it difficult to significantly improve the reservoir quality even if there is dissolution. Consequently, the RF4 is characterized by low porosity (average porosity of 1.3%) and low permeability (average permeability of 0.006 mD) and has the lowest reservoir quality and poor lateral continuity. The thickness of a single layer is generally 2–3 m and is mainly composed of flow units FU1 and FU2 (Figures 10–12). Due to the poor storage and flow capacity, the RF4 is mainly an interlayer that blocks the fluid flow in the Callovian-Oxfordian stage.

### 5.3.5. Fracturing Mudstone-Dominated Reservoir Facies—RF5

RF5 is mainly developed in sedimentary facies belts with high mud content and low-energy environments, such as slope mud (F12) on platform fore-edge upslope and slope mud (F14) on platform fore-edge upslope. The main lithology is mudstone. The most prominent characteristic of RF5 is that the permeability of rocks is significantly increased due to the fracturing in the late middle diagenetic stages (Figure 9), thus greatly enhancing the flow capacity of the reservoir. The average porosity is 1.1% and the average permeability is 1.82 mD. It is mainly distributed in the XVhp, Z, and XVI layers of the Callovian-Oxford formation. When compaction and cementation dominate and there is no fracturing, the RF5 corresponds to flow unit FU1 with a low reservoir quality, and is mainly an interlayer; when both fracturing and dissolution occur, RF5 corresponds to FU2; when many open microfractures are developed, the permeability of this RF is greatly improved, and the corresponding FUs are FU8 and FU9, which mainly form flow channels for fluid (Figures 11 and 12). RF5 mainly distributed in the fracture development area, especially in the eastern part of the study area which is strongly affected by tectonic movement. The lateral continuity mainly depends on the development of natural fractures.

## 6. Conclusions

Through a comprehensive evaluation of sedimentary facies, diagenesis, and reservoir quality of the Callovian-Oxfordian formation of the Middle and Lower Jurassic series on the right bank of the Amu Darya River, the following main conclusions are drawn:

(1) Through facies analysis, we believe that the Callovian period is a gentle slope carbonate platform sedimentary model, which mainly includes four sedimentary facies belts, namely, the inner ramp, mid ramp, outer ramp, and basin. The Oxfordian period is an edged shelf-type carbonate platform sedimentary model, which mainly includes 15 sedimentary facies deposited in seven sedimentary facies belts, namely, the evaporative platform, restricted platform, open platform, platform margin, platform fore-edge upslope, platform fore-edge downslope, and basin facies.

(2) The diagenesis fields of carbonate rocks in the Callovian-Oxfordian stage include atmospheric freshwater, shallow burial, and medium-deep burial environments. The main diagenetic processes include four-stage cementation, mechanical compaction, chemical compaction (pressure solution), dissolution, and fracturing, among which the dissolution and fracturing are the main processes that improve reservoir quality.

(3) The main primary and secondary pore types in the target formation were identified; nine FUs were determined using the flow unit index method; five reservoir facies, as well as their reservoir quality and variation characteristics, were identified according to rock texture and flow characteristics, diagenesis, and pore types.

(4) The reservoir quality evaluation revealed that the reservoir facies with the highest reservoir quality is RF1, which is the highest storage capacity and has strong flow capacity. Its corresponding flow units are FU4, FU5, and FU6. The reservoir facies with the second highest reservoir quality is RF2, and its corresponding flow units are FU3 and FU4, which are less distributed in the target intervals. RF4 has the lowest reservoir quality, and its corresponding flow units are FU1 and FU2, which mainly form an interlayer that blocks the fluid flow. Despite its very low storage capacity, RF5 serves mainly as a flow channel for the fluid in the Callovian-Oxfordian Stage due to the development of micro-fractures.

**Author Contributions:** Data curation, P.C.; Investigation, M.C.; Methodology, L.Z.; Software, H.S.; Validation, C.G.; Writing—original draft, Y.X.; Writing—review & editing, H.W. and W.Y. All authors have read and agreed to the published version of the manuscript.

**Funding:** This work was supported by PetroChina Science and Technology Major Project (2021DJ3301). The authors would like to thank the management of RIPED for authorizing the presentation of this work.



## References

1. Ulmishek, G.F. *Petroleum Geology and Resources of the Amu Darya Basin, Turkmenistan, Uzbekistan, Afghanistan and Iran*; USGS: Reston, VA, USA, 2004; pp. 1–38.
2. Xu, S.; Wang, S.; Sun, X. Petroleum geology and resource potential. *Oil Forum* **2007**, *6*, 31–38.
3. Zhang, B.; Zheng, R.; Liu, H. Characteristics of Carbonate Reservoir in Callovian-Oxfordian of Samandepe Gas field Turkmenistan. *Acta Geol. Sin.* **2010**, *84*, 117–125.

4.  Zheng, R.; Liu, H.; Wu, L. Geochemical characteristics and diagenetic fluid of the Callovian-Oxfordian carbonate reservoirs in Amu Darya basin. *Acta Petrol. Sin.* **2012**, *28*, 961–970.

5.  Wen, H.; Gong, B.; Zheng, R. Deposition and Diagenetic System of Carbonate in Callovian-Oxfordian of Samandepe Gas field, Turkmenistan. *J. Jilin Univ. Earth Sci. Ed.* **2012**, *42*, 991–1002.

6.  Liu, S.; Zheng, R.; Yan, W. Characteristics of Oxfordian carbonate reservoir in Agayry area, Amu Darya Basin. *Lithol. Reserv.* **2012**, *24*, 57–63.

7.  Dong, X.; Zheng, R.; Wu, L. Diagenesis and porosity evolution of carbonate reservoirs in Samandepe Gas Field, Turkmenistan. *Lithol. Reserv.* **2010**, *22*, 54–61.

8.  Xu, W.; Zheng, R.; Fei, H. The sedimentary facies of Callovian-Oxfordian Stage in Amu Darya basin, Turkmenistan. *Geol. China* **2012**, *39*, 954–964.

9.  Xu, W.; Zheng, R.; Fei, H. Characteristics and timing of fractures in the Callovian-Oxfordian boundary of the right bank of the Amu Darya River Turkmenistan. *Nat. Gas Ind.* **2012**, *32*, 33–38.

10.  Wang, L.; Zhang, Y.; Wu, L. Characteristics and identification of bioherms in the Amu Darya Right Bank Block, Turkmenistan. *Nat. Gas Ind.* **2010**, *30*, 30–33.

11.  Zeng, Z.; Chen, F.; Chen, Z. Forward Modeling of Callovian-Oxfordian Reefs of Upper Jurassic in AS Block of Amu Darya Right Bank in Turkmenistan. *Xinjiang Pet. Geol.* **2011**, *32*, 201–203.

12.  Zhang, W.; Liu, X.; Zhang, M. Forward Seismic Modeling and its Response of Point reef Reservoir in Amu Darya Area of Turkmenistan. *J. Oil Gas Technol.* **2011**, *33*, 72–75.

13.  Lucia, F.J. *Carbonate Reservoir Characterization: An Integrated Approach*; Springer Science & Business Media: Berlin/Heidelberg, Germany, 2007.

14.  Ahr, W.M. *Geology of Carbonate Reservoirs: The Identification, Description and Characterization of Hydrocarbon Reservoirs in Carbonate Rocks*; John Wiley & Sons: Hoboken, NJ, USA, 2011.

15.  Mehrabi, H.; Ranjbar-Karami, R.; Roshani-Nejad, M. Reservoir rock typing and zonation in sequence stratigraphic framework of the Cretaceous Dariyan Formation, Persian Gulf. *Carbonates Evaporites* **2019**, *34*, 1833–1853. [CrossRef]

16.  Mehrabi, H.; Bahrehvar, M.; Rahimpour-Bonab, H. Porosity evolution in sequence stratigraphic framework: A case from Cretaceous carbonate reservoir in the Persian Gulf, southern Iran. *J. Petrol. Sci. Eng.* **2021**, *196*, 107699. [CrossRef]

17.  Lv, G.; Liu, H.; Zhang, B.; Deng, M.; Zhang, X. *Exploration and Development of Large Sub-Salt Carbonate Gas Fields on Right Bank of Amu Darya*; Science Press: Beijing, China, 2013.

18.  Dunham, R.J. Classification of carbonate rocks according to their depositional texture. In *Classification of Carbonate Rocks*; Ham, W.E., Ed.; American Association of Petroleum Geologists Memoir: Tulsa, OK, USA, 1962; Volume 1, pp. 108–121.

19.  Embry, A.F.; Klovan, J.E. A late devonian reef tract on northeastern banks Island, NWT. *Can. Pet. Geol. Bull.* **1971**, *19*, 730–781.

20.  Flügel, E. *Microfacies of Carbonate Rocks: Analysis, Interpretation and Application*; Springer Science & Business Media: Berlin/Heidelberg, Germany, 2010.

21.  Amaefule, J.O.; Altnubay, M.; Tiab, D.; Kersey, D.G.; Keeland, D.K. Enhanced reservoir description: Using core and log data to identify hydraulic (flow) units and predict permeability in un-cored intervals/wells. *Soc. Pet. Eng.* **1993**, *26436*, 1–16.

22.  Kozeny, J. *Ber Kapillare Leitung des Wassers im Boden, Sitzungsberichte*; Royal Academy of Science Vienna: Vienna, Austria, 1927; Volume 136, pp. 271–306.

23.  Carman, P.C. Fluid flow through granular beds. *Trans. Inst. Chem. Eng.* **1937**, *15*, 150–166. [CrossRef]

24.  Gomes, J.S.; Riberio, M.T.; Strohmenger, C.J.; Negahban, S.; Kalam, M.Z. *Carbonate Reservoir Rock Typing the Link between Geology and SCAL*; SPE: Abu Dhabi, United Arab Emirates, 2008; p. 118284.

25.  Rahimpour-Bonab, H.; Mehrabi, H.; Navidtalab, A.; Izadi-Mazidi, E. Flow unit distribution and reservoir modeling in Cretaceous carbonates of the Sarvak Formation, Abteymour Oilfield, Dezful embayment, SW IRAN. *J. Pet. Geol.* **2012**, *35*, 1–24. [CrossRef]

26.  Mehrabi, H.; Rahimpour-Bonab, H.; Enayati-Bidgoli, A.H.; Esrafili-Dizaji, B. Impact of contrasting paleoclimate on carbonate reservoir architecture: Cases from arid Permo-Triassic and humid Cretaceous platforms in the south and southwestern Iran. *J. Pet. Sci. Eng.* **2015**, *126*, 262–283. [CrossRef]

27.  Tiab, D.; Donaldson, E.C. *Petrophysics: Theory and Practice of Measuring Reservoir Rock and Fluid Transport Properties*, 2nd ed.; Gulf Professional Publishing, Elsevier: Amsterdam, The Netherlands, 2004.

28.  Deng, Y.; Wang, Q.; Cheng, X.; Wu, L.; Zhao, C.; Chen, R. Origin and distribution laws of $H_2S$ in carbonate gas reservoirs in the Right Bank Block of the Amu Darya River. *Nat. Gas Ind.* **2011**, *31*, 21–23.

29.  Burrowes, A.; Moss, A.; Sirju, C.; Pritchard, T. *Improved Permeability Prediction in Heterogeneous Carbonate Formations*; Society of Petroleum Engineers: Barcelona, Spain, 2010; p. 131606.

30.  Enayati-Bidgoli, A.H.; Rahimpour-Bonab, H.; Mehrabi, H. Flow unit characterization in the Permian-Triassic carbonate reservoir succession at South Pars Gas field, of shore Iran. *J. Pet. Geol.* **2014**, *37*, 205–230. [CrossRef]

31.  Sfdari, E.; Kadkhodaie-Ilkhchi, A.; Rahimpour-Bonab, H.; Soltani, B. A hybrid approach for lithofacies characterization in the framework of sequence stratigraphy: A case study from the South Pars gas field, the Persian Gulf basin. *J. Pet. Sci. Eng.* **2014**, *121*, 87–102. [CrossRef]

32.  Nabawy, B.S. Impacts of the pore- and petro-fabrics on porosity exponent and lithology factor of Archie's equation for carbonate rocks. *J. Afr. Earth Sci.* **2015**, *108*, 101–114. [CrossRef]

33. Jamalian, M.; Adabi, M.H.; Moussavi, M.R.; Sadeghi, A.; Baghbani, D.; Ariyafar, B. Facies characteristic and paleoenvironmental reconstruction of the lower cretaceous fahliyan formation in the Kuh-e-siah area, Zagros basin, southern Iran. *Facies* **2011**, *57*, 101–122. [CrossRef]
34. Wolpert, P.; Bartenbach, M.; Suess, P.; Rausch, R.; Aigner, T.; Le Nindre, Y.M. Facies analysis and sequence stratigraphy of the uppermost Jurassic–Lower Cretaceous Sulaiy Formation in outcrops of central Saudi Arabia. *GeoArabia* **2015**, *20*, 67–122. [CrossRef]
35. Noori, H.; Mehrabi, H.; Rahimpour-Bonab, H.; Faghih, A. Tectono-sedimentary controls on Lower Cretaceous carbonate platforms of the central Zagros, Iran: An example of rift-basin carbonate systems. *Mar. Pet. Geol.* **2019**, *110*, 91–111. [CrossRef]
36. Lee, E.Y.; Kominz, M.; Reuning, L.; Gallagher, S.J.; Takayanagi, H.; Ishiwa, T.; Knierzinger, W.; Wagreich, M. Quantitative compaction trends of Miocene to Holocene carbonates off the west coast of Australia. *Aust. J. Earth Sci.* **2021**, *68*, 1149–1161. [CrossRef]
37. Tavakoli, V.; Rahimpour-Bonab, H.; Esrafili-Dizaji, B. Diagenetic controlled reservoir quality of South Pars gas field, an integrated approach. *Comptes. Rendus Geosci.* **2011**, *343*, 55–71. [CrossRef]
38. Enayati-Bidgoli, A.; Navidtalab, A. Effects of progressive dolomitization on reservoir evolution: A case from the Permian–Triassic gas reservoirs of the Persian Gulf, offshore Iran. *Mar. Pet. Geol.* **2020**, *119*, 104480. [CrossRef]

*Article*

# Laboratory Evaluation and Field Application of a Gas-Soluble Plugging Agent: Development of Bottom Water Plugging Fracturing Technology

**Aiguo Hu** [1,2,3], **Kezhi Li** [3], **Yunhui Feng** [1,2], **Hucheng Fu** [3] **and Ying Zhong** [1,2,*]

[1]   State Key Laboratory of Oil and Gas Reservoir Geology and Exploitation, Chengdu University of Technology, Chengdu 610059, China
[2]   College of Energy, Chengdu University of Technology, Chengdu 610059, China
[3]   Research Institute of Engineering and Technique, Sinopec Huabei Sub-Company, Zhengzhou 450006, China
*   Correspondence: zhongying8911@163.com

**Abstract:** The currently reported bottom water sealing materials and fracturing technologies can hardly simultaneously achieve the high production and low water cut of gas reservoirs due to the complexity of various formation conditions. Therefore, without controlling the fracturing scale and injection volume, a kind of polylactide polymer water plugging material with a density of 1.15–2.0 g/cm$^3$ is developed, which can be used to seal the bottom water of a gas–water differential layer by contact solidification with water and automatic degradation with natural gas. This technology can not only fully release the production capacity of the gas reservoir but also effectively control water production and realize the efficient fracturing development of the target gas reservoir. Laboratory test results show that the smart plugging agent has a bottom water plugging rate of 100%, and the low-density plugging agent has a dissolution rate of 96.7% in methane gas at 90 °C for 4 h and a dissolution rate of 97.6% in methane gas at 60 °C for 6 h, showing remarkable gas degradation performance. In addition, settlement experiments show that the presence of a proppant can increase the settlement rate of a plugging agent up to many times (up to 21 times) in both water and guanidine gum solution. According to the actual conditions of well J66-8-3, a single-well water plugging fracturing scheme was prepared by optimizing the length of fracture, plugging agent dosage, and plugging agent sinking time, and a post-evaluation method was proposed. It has guiding significance to the development of similar gas reservoirs.

**Keywords:** gas-soluble plugging agent; bottom water; water plugging fracturing technology; unconventional oil and gas resources

**Citation:** Hu, A.; Li, K.; Feng, Y.; Fu, H.; Zhong, Y. Laboratory Evaluation and Field Application of a Gas-Soluble Plugging Agent: Development of Bottom Water Plugging Fracturing Technology. *Energies* **2022**, *15*, 6761. https://doi.org/10.3390/en15186761

Academic Editor: Paul W. J. Glover

Received: 25 July 2022
Accepted: 6 September 2022
Published: 15 September 2022



## 1. Introduction

The discovery of unconventional oil and gas resources not only increased the potential of global oil and gas resources but also changed the world energy structure and helped to solve short-term energy crises [1–3]. As an important reservoir stimulation technology, hydraulic fracturing ensures the stable production of oil and gas fields [4–6].

At present, most unconventional oil and gas reservoirs in China are characterized by high water content and low permeability [7,8]. The separation between water and oil and gas layers is thin, and the pressure difference between reservoirs is small [9,10]. Thus, the high water production caused by hydraulic fracturing becomes another significant problem [11]. Therefore, conventional fracturing methods were mainly used to control the height of the fracture [12,13], avoid connecting the water layer, and carry out the small-scale and low-displacement fracturing technology [14]. These methods can control the height of fractures to a certain extent and avoid direct contact between water and oil layers. However, after fracturing, the fractures are shorter and the reservoir permeability is lower. In addition, the barrier between the gas and water layers is thin, and the stress difference between the reservoir and the barrier layer is small. Even if the scale and displacement are reduced in the fracturing process, controlling the propagation of fractures is extremely difficult, and pressing through the water layer is still easy.

In recent years, fracturing technology has made breakthrough progress with the advances in intelligent materials and computer technology (including intelligent optimization of fracturing parameters, intelligent monitoring of fracturing fractures, and so on) [15–17]. Many bottom water sealing materials have been applied in the field, and the water production rate of fractured oil and gas wells has been effectively controlled. Xu prepared a new expansion particle profile control agent, and laboratory test results attest to the strong plugging ability of the plugging agent [18]. Moreover, physical simulation tests and field applications were used to verify the plugging ability and plugging strength of the expandant bottom water fracturing technology on pores to ensure the performance of the expandant bottom water fracturing fluid. Through the dynamic simulation and data analysis of a field oil well, it has been previously found that there is a point-like or segmented water producing section in the horizontal section of the well, which produces water from the edge and bottom. On this basis, a highly selective gel was developed, which entered deep into the reservoir to plug the fracture, causing the pressure in the tube to slowly rise. The well maintained fluid production at 27 t/d after fracturing, increased oil production from 4 to 20 t/d, and reduced the water cut from 75% to 25% [19]. Zhang et al. studied segmented water search and water plugging technology [20]. First, a horizontal well was segmented reasonably, and then the water layer was determined using the designed intelligent equipment (due to the different properties of oil and water and the different temperature of different positions, water can be found intelligently by capturing the change in water pressure and temperature). Then, the water was cut off from the determined layers. This technology can effectively and intelligently selectively shut off water, which provides a more effective method for delaying the influx of bottom water and further improving oil recovery. Zhao et al. reported a new precipitator for fracturing and measured the settling velocity, plugging capacity, and conductivity of the new precipitator barrier and compared these with those of six other commonly used settling agents [21]. Filtration tests showed that the new settling agent could form a barrier and has a strong plugging effect on the fracturing fluid. Qi et al. developed a foam gel system that uses polyacrylamide as the main agent and contains the in situ heating system of nitrite and ammonium [22]. The foam gel has good water plugging ability due to volume expansion, and its performance can be controlled by adjusting the pH value to meet the different configuration requirements of different reservoirs.

However, the currently reported bottom water sealing materials so far have different advantages and disadvantages due to the complexity of various formation conditions, summarized in Table 1.

**Table 1.** Summary of partial plugging materials.

| References | Plugging Materials | Advantages | Points That Can Be Improved |
| --- | --- | --- | --- |
| Xu, 2016 [18] | Swellable particle plugging agent | Good expansion performance, strong water blocking ability | blocks water but also gas; cannot achieve 100% plugging rate |
| Liu et al., 2020 [23] | Highly selective gel | Shell microspheres, coalescence, good hydration expansion effect | No selective |
| Li et al., 2019 [24] | Gel | Selective expansion | Hard to combine the water control and fracturing stimulations and bottom water |
| Li et al., 2019 [19] | Hydrogels | Highly selective water shutoff | Hard to combine the water control and fracturing stimulations and bottom water |

**Table 1.** *Cont.*

| References | Plugging Materials | Advantages | Points That Can Be Improved |
|---|---|---|---|
| Guo et al., 2017 [25] | Foamed-gel system | Good expansion effect and good water blocking effect | Hard to combine the water control and fracturing stimulations and bottom water |
| Zhang et al., 2018 [26] | Acrylamide monomer polymer gel | The plugging rate is over 96.5% in high permeability sandpack | Hard to combine the water control and fracturing stimulations and bottom water |
| Qi et al., 2018 [22] | Double crosslinked gel | High efficiency plugging and selective water shutoff; the volumetric expansion ratio exceeds 130% | Hard to combine the water control and fracturing stimulations and bottom water |
| Liu, 2014 [27] | Gel | High strength and adjustable forming gel time | Hard to combine the water control and fracturing stimulations and bottom water |
| Xie et al., 2022 [28] | Composite gel | Plugging ratio of up to 99.5% | Hard to combine the water control and fracturing stimulations and bottom water |
| Zhao et al., 2019 [21] | Precipitation | Unifying the water control between fracturing stimulations and bottom water | No selective |
| Zhang et al., 2019 [29] | Ion precipitation | In situ reaction, remarkable effect of water shutoff | No selective |

To sum up, the main problems are as follows:

(1) Unifying the water control between fracturing stimulations and bottom water is difficult [30,31]. Without fracturing, there is low or no production, and fractures communicate with water layers after fracturing.

(2) Simply controlling the fracture height cannot solve the problems of water control and bottom water simulation [32]. For instance, the small-scale fracture control mode still has high water production after fracturing.

(3) The existing traditional fracturing plugging agents may block water but also gas and hinder gas production during water control [33].

Therefore, we put forward in this work a new bottom water plugging fracturing material to try to solve the above problems. Without controlling the fracturing scale and injection volume, a water-blocking material that solidifies in contact with water and automatically degrades in contact with natural gas was developed to block the bottom water of gas–water differential layers. While fully releasing the production capacity of gas reservoirs, the technology can effectively control water and increase gas production and achieve the efficient fracturing development of gas reservoirs in a target area. Experimental test results show that the bottom water sealing rate of this intelligent plugging agent reaches 100% and that it has great degradation performance in the presence of natural gas. The construction was improved according to the water plugging fracturing field test, and a post-fracturing evaluation method was put forward.

## 2. Experimental Materials and Methodology

### 2.1. Experimental Materials

Pyrophosphoryl chloride ($Cl_4O_3P_2$, Mn = 251.76) was obtained from Chongqing Chuandong Chemical Co., Chongqing, China, and p-hydroxybenzaldehyde ($C_7H_6O_2$, Mn = 122.12), toluene ($C_7H_8$, Mn = 92.14), and sodium borohydride ($NaBH_4$, Mn = 37.83) were purchased from Qianyan Science and Technology Co., Hong Kong. Lactide ($C_6H_8O_4$, Mn = 144) was obtained from Suzhou Qihang Biotechnology Co., Suzhou, China, stannous isooctylate ($C_{16}H_{30}O_4Sn$, Mn = 405.10) was obtained from Hubei Chunshuo Chemical Co., Zhijiang, China, and proppant (20/40 mesh ceramsite) was purchased from Guanghan Huitong Plastic Co., Deyang, China.

### 2.2. Flowchart of Laboratory Experiment

The synthesis and performance testing of the gas-soluble plugging agent is the key to verify the performance of the plugging agent. The first is the synthesis of the plugging agent, and then the synthetic plugging agent needs curing performance test, solubility test, and settlement experiment, respectively. An initial diverting capacity test of proppant is required prior to the curing performance test. Various laboratory experiments of plugging agent are shown in Scheme 1.

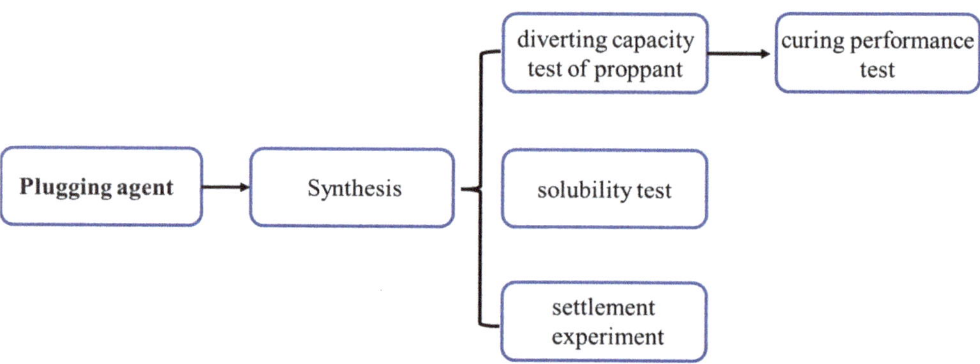


**Scheme 1.** Flowchart of laboratory experiment.

### 2.3. Synthesis of the Plugging Agent

The gas-soluble plugging agent is a type of plugging agent that is soluble in natural gas and is injected into the formation with the proppant. It is solidified by water under the formation temperature condition. This can reduce the water-phase permeability of artificial fractures and increase the shielding ability and pressure resistance strength between gas and water layers to control the bottom water from flowing up along artificial fractures and leading to a large amount of water flowing out of gas wells. At the same time, the plugging agent dissolves when it comes into contact with natural gas, which restores the conductivity of proppant fractures and plays a dual role in controlling the bottom water of gas reservoirs and ensuring the normal production of natural gas. For this purpose, the laboratory synthesized a plugging agent material that solidifies in water and dissolves in gas. The 1 mol pyrophosphoryl chloride and 4 mol p-hydroxybenzaldehyde react in toluene to produce intermediate with aldehyde group. The aldehyde group is directly reduced to hydroxyl group by adding sodium borohydride. Finally, a highly branched poly (lactide) polymer was synthesized with lactide catalyzed by stannous isooctylate. The main synthesis route is shown in Scheme 2.

The plugging agent was synthesized using solution polymerization, and the reactor is shown in Figure 1.

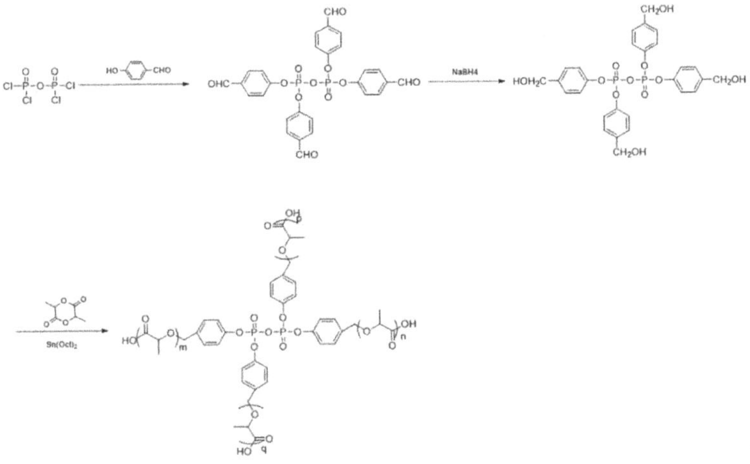

**Scheme 2.** Synthesis of plugging agent.

**Figure 1.** Plugging agent synthesis reactor.

### 2.4. Diverting Capacity Test of Proppant

The main purpose of the experiment is to study the curing performance of the plugging agent system. Thus, the initial permeability of the fracture should be measured first. The experiment of diverting capacity test of proppant was based on methods previously reported by researchers, and the specific experimental steps are as follows:

(1) Prepare the diversion chamber (Figure 2) for the experimental test without filling the proppant, and move it to the experimental test platform to ensure that the diversion chamber is placed horizontally to ensure uniform stress loading. Install the displacement sensor on the test platform, open the hydraulic press to load a closing stress of 1 MPa to ensure that the parts fit tightly, and set the data on the displacement meter test panel to zero.

(2) Load the closing stress of the hydraulic press under a predetermined pressure, and record the displacement gauge data to obtain the diversion chamber deformation under the predetermined pressure. Then, record the average value as h1.

(3) After the pressure relief, add a mixture of the plugging agent and a ceramic particle to the diversion chamber (Figure 1) under a predetermined sand paving concentration, and record the original sand thickness h2 using vernier calipers. Move the diversion chamber to the experimental test platform and pressurize to 1 MPa. Load the displacement sensor and set it to zero to record the height of the four corners.

(4) Connect the differential pressure sensor and inlet and outlet pipeline of the diversion chamber to ensure that the valves are closed. Open the displacement pump to suppress the pressure of the diversion chamber. Control the displacement of the displacement pump at 30 mL/min and suppress the pressure of the diversion chamber to 2 MPa. During this period, check whether there is leakage in each pipeline.

(5) After ensuring that each line is well sealed, open the vacuum pump to evacuate the line for more than 30 min. Then, open the displacement pump to saturate the fracture and line with the test fluid (simulating formation water).

(6) Test the proppant conductivity step by step under pressure. When the proppant is pressured to the predetermined closing stress, use the vernier caliper to record the height of the four corners and the data of the displacement gauge to obtain the total deformation h3 under the predetermined closing stress, thus obtaining the proppant thickness = h2 − h3 + h1 in the diversion chamber under the predetermined closing pressure.

(7) Open the conductivity test program, input the parameters of the experiment, and then switch to the experimental test window to start the diversion capacity test.

(8) Adjust the advection pump to the set test flow rate, test the permeability within 30 min, and record the fluid quality, diversion chamber pressure, and flow pressure difference along the fracture.

(9) After the test, close the advection pump, open the drain valve of the hydraulic press, unload the oil pressure, remove the pipeline of the diversion chamber, and take out the proppant in the diversion chamber and the diversion chamber.

(10) According to the permeability calculation formula in the "Recommended Method for Short-term Conductivity Evaluation," solve for the original permeability $K$ under predetermined pressure and sand paving concentration (Equation (1)):

$$K = \frac{99.998\mu QL}{A\Delta P} \tag{1}$$

where $K$ is the proppant pack permeability ($\mu m^2$), $\mu$ is the viscosity of the liquid at the experimental temperature (mPa·s), $Q$ represents the flow rate ($cm^3/s$), $L$ represents the length between the pressure measuring holes (cm), $A$ is the cross-flow area ($cm^2$), and $\Delta P$ is the pressure difference (KPa).

**Figure 2.** Diversion chamber before and after the placement of the plugging agent and proppant.

*2.5. Curing Performance Test*

Curing with water is one of the properties of the plugging agent. The following steps were completed to further explore the curing ability of the plugging agent.

(1) Fill the diversion chamber with the plugging agent and proppant.

(2) Connect the appropriate pressure sensor and import and export pipeline of the bypass chamber to ensure that each valve is in a closed state. Open the displacement pump to suppress the pressure of the diversion chamber. Control the displacement of the displacement pump at 30 mL/min and suppress the pressure of the diversion chamber to 2 MPa. During this period, check whether there is leakage in each pipeline.

(3)  After ensuring that the pipelines are well sealed, open the vacuum pump to evacuate the pipelines for more than 30 min. Then, open the displacement pump to saturate the cracks and pipelines with the test fluid (distilled water).

(4)  Heat the diversion chamber and set the formation temperature (60 °C or 90 °C).

(5)  Open the hydraulic pressurizer to load the closure stress of the diversion chamber, and stop pressurizing when the pressure rises to the initial pressure of the experimental design, ensuring that the pressure stability time is more than 30 min. Continue heating for n hours (heating time determined by melting and consolidation tests).

(6)  Check the plugging agent breakthrough pressure step by step. When the outlet end of the continuous liquid outflow shows that the plugging agent has been broken, the highest pressure is the highest breakthrough pressure (maximum plugging strength) of the plugging agent.

After the plugging agent breakthrough, continue forward flooding to simulate formation water. Then, measure the water-phase permeability $K$ within 30 min after the breakthrough to obtain the temporary plugging rate, as shown in Equation (2):

$$Z = \frac{K - K_1}{K} \times 100\% \tag{2}$$

where $Z$ represents the temporary plugging rate (%), $K$ represents the original water-phase permeability ($\mu m^2$), and $K_1$ represents the post-plugging permeability ($\mu m^2$).

(7)  If the displacement pressure reaches 20 MPa and still does not break, shut off the pump.

(8)  After the test is completed, close the advection pump, unload the oil pressure, remove the pipeline of the diversion chamber, and take out and observe the diversion chamber.

### 2.6. Solubility Test

The gas solubility experiment is a key experiment to evaluate whether a plugging agent can be automatically degraded by gas. The evaluation methods and steps adopted in this study are as follows:

(1)  Add the liquid (fracturing fluid) into the plugging agent, stir it, and then put it into an oven for curing. Take the solidified plugging agent as the basic sample.

(2)  Take a proper amount of the solidified sample and put it into a sealable glass bottle filled with distilled water.

(3)  Drive the distilled water in the glass bottle with methane gas so that the solidified sample is completely immersed in methane gas.

(4)  Place the glass bottle in a water bath and set it to the formation temperature to heat.

(5)  Take out the plugging agent every 1 h and weigh it to calculate the dissolution. Repeat steps 2–4 to obtain the dissolution of each plugging agent every hour until the plugging agent weight changes almost no more.

(6)  Calculate and analyze the dissolution of the different plugging agents at different formation temperatures and times.

Note: proppant was not included in the plugging agent mentioned above because it is not possible to calculate the plugging agent dissolution rate accurately after the addition of proppant. Therefore, pure plugging agent was used.

### 2.7. Settlement Experiment of the Plugging Agent

The proppant settlement experiment was used for reference to simulate the underground fracture situation, and the settlement law of the water plugging material alone in the fracturing fluid and mixed with proppant was measured.

(1)  Measure the plugging agent and proppant with different volume ratios.

(2)  Shake well the plugging agent and proppant to mix them evenly.

(3)  Add water (or a guanidine gum solution) into the mixture of the plugging agent and proppant and simulate the mixing of the two in the sand mixing truck.

(4)    Add the mixture of the plugging agent and proppant to a graduated cylinder filled with water (or a guanidine gum solution) and calculate the settling rate.

*2.8. Preparation of Fracturing Fluid*

The guanidine gum base fluid prepared for laboratory testing and the fracturing fluid formulation used in the well J66-8-3 are shown below:

Base liquid: 0.42% guanidine gum (HPG) +1.0% anti-expansion agent +0.1% fungicide +0.5% foaming agent +0.2% drainage aid +0.2% $Na_2CO_3$. Cross-linking agent: cross-linking agent A:B = 100:6, the cross-linking ratio is 100:0.3.

The preparation method is carried out according to Chinese oil and gas standard SY/T 5107-2016 (water-based fracturing fluid performance evaluation method).

## 3. Results and Discussion

### 3.1. Research on Plugging Agent

In order to prove the successful synthesis of the polymer plugging agent, the final product poly (lactide) was analyzed by mass spectrometry.

As can be seen from Figure 3, the mass number corresponding to the peak with the largest mass number of the mass spectrum (that is, the molecular ion peak) is also greater than the molecular mass of various monomers and intermediates, thus confirming the successful synthesis of the polymer.

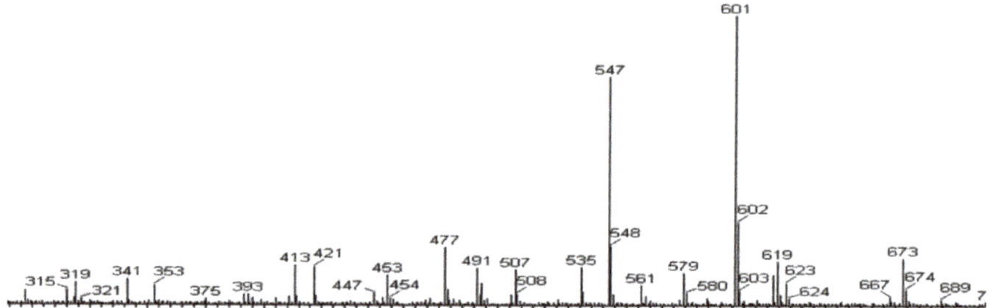

**Figure 3.** Mass spectrometry of polymers.

### 3.2. Performance of the Water Plugging Material

### 3.2.1. Consolidation Test

- Diversion capacity of the proppant.

Measuring the original permeability of fractures is necessary to study the curing performance of the plugging agent. A 20/40-mesh ceramsite with a bulk density of 1.65 $g/cm^3$ (apparent density of 3.0 $g/cm^3$) was used to test the permeability and diversion capacity of the proppant (pure proppant and its placement in the diversion chamber are shown in Figure 4). These provide the calculation basis for the plugging efficiency of the plugging agent. Because temperature has almost no effect on the diversion capacity of pure proppant, the test temperature was the normal temperature.

The permeability and diversion capacity of the proppant at different closing pressures are shown in Figure 5. It can be found that the formation permeability and proppant conductivity both decrease with the increase in closure pressure. This is because the fracture tends to heal with the increase in formation closure pressure, resulting in smaller fractures and lower permeability, which affects proppant conductivity. From the test results, we found that the permeability of the 20/40-mesh ceramsite is 68.9 D at a 45 MPa closure pressure.

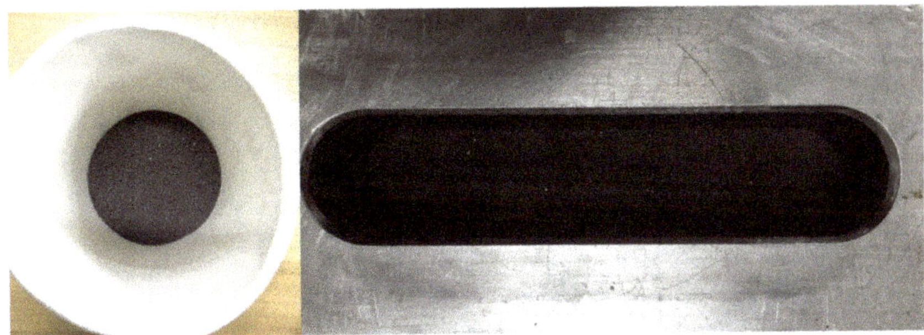

**Figure 4.** Pure proppant and its placement in the diversion chamber.

- Plugging performance evaluation.

Different groups were tested to determine the cure and plugging ability of the plugging agent and proppant mixture at different densities. The densities and ratios of the plugging agent to the proppant are shown in Table 2.

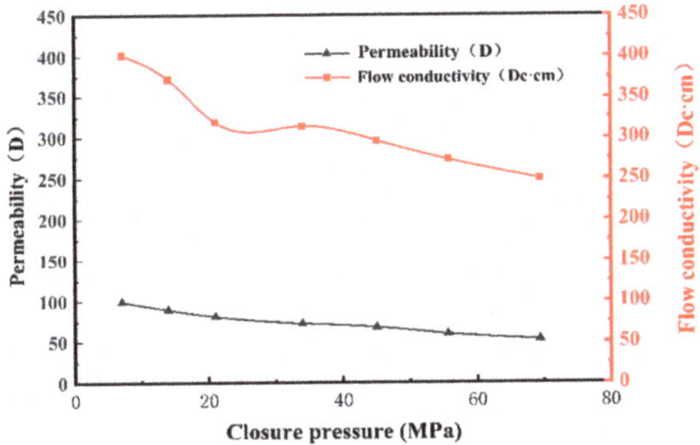


**Figure 5.** Permeability and diversion capacity of pure proppant at different closure pressures.

**Table 2.** Parameters of the different experimental groups.

| Group | Density | Volume Ratio of the Plugging Agent to the Proppant | Test Temperature (°C) |
|---|---|---|---|
| 1 | Low density, 1.15 g/cm$^3$ | 1:1 | 90 |
| 2 | Low density, 1.15 g/cm$^3$ | 1:2 | 90 |
| 3 | High density, 1.56 g/cm$^3$ | 1:0.5 | 60 |
| 4 | High density, 1.56 g/cm$^3$ | 1:1 | 60 |
| 5 | High density, 1.74 g/cm$^3$ | 1:0.5 | 60 |
| 6 | High density, 1.74 g/cm$^3$ | 1:1 | 60 |
| 7 | High density, 1.90 g/cm$^3$ | Only the plugging agent | 60 |
| 8 | High density, 1.90 g/cm$^3$ | 1:0.5 | 60 |
| 9 | High density, 1.90 g/cm$^3$ | 1:1 | 60 |
| 10 | High density, 2.00 g/cm$^3$ | Only the plugging agent | 60 |
| 11 | High density, 2.00 g/cm$^3$ | Only the plugging agent | 90 |
| 12 | High density, 2.00 g/cm$^3$ | 1:0.5 | 60 |

A total of 50 mL of the plugging agent and proppant were measured in various volume ratios. Then, reagents were poured into the beaker, stirred evenly, and weighed 32.25 g, with a sand paving concentration of 5 kg/m$^2$. A guanidine gum solution and glue breaker (guanidine gum concentration: 0.42%; glue breaker dosage: 0.04%) were added into the plugging agent and then put into the diversion chamber. Then, the curing ability of the plugging agent was tested. The bonding procedure followed after the solidification of the plugging agent. Under the pressure difference of 1 MPa, distilled water was used for displacement. The plugging curve and optical images of the plugging agent and proppant after the test are shown in Figures 6 and 7.

Observation and experimental data showed a liquid outflow at the outlet end of group 1 (the average flow rate was 0.2 mL/min), and no droplet appeared after 13 min. The displacement pressure difference was continuously increased to 5, 10, 15, and 20 MPa without breakthrough, and the plugging performance was good. A large amount of liquid was released from the outlets of groups 4, 6, and 9 (the average flow rates were 22, 15, and 25 mL/min, respectively). This indicates that high-density plugging agents (densities of 1.56, 1.74, and 1.90 g/cm$^3$) cannot achieve plugging under a 1:1 ratio. Obviously, the higher the density of the plugging agent, the more volume fraction of plugging agent is required to achieve effective plugging. However, no droplet appeared at the outlet ends of groups 8 and 13. The displacement pressure difference was continuously increased to 5, 10, and 13 MPa without breakthrough. When the displacement pressure difference of group 12 reached 15 MPa, the liquid was released from the outlet (the maximum flow rate was 15.62 mL/min). In addition, in group 8, when the displacement pressure difference reached 20 MPa, the liquid was released from the outlet (the highest flow rate was 4.19 mL/min), and the plugging performance was good. In other groups, the displacement pressure was continuously increased to 25 MPa, but no breakthrough was achieved, and the plugging efficiency reached 100%. The experimental results confirm the excellent plugging ability of the plugging agent. The optical photos of Figure 7 confirm that the plugging agent is a pale-yellow particle, distinct from the black proppant. The mixture becomes more yellow with the increase in the volume fraction of plugging agent.

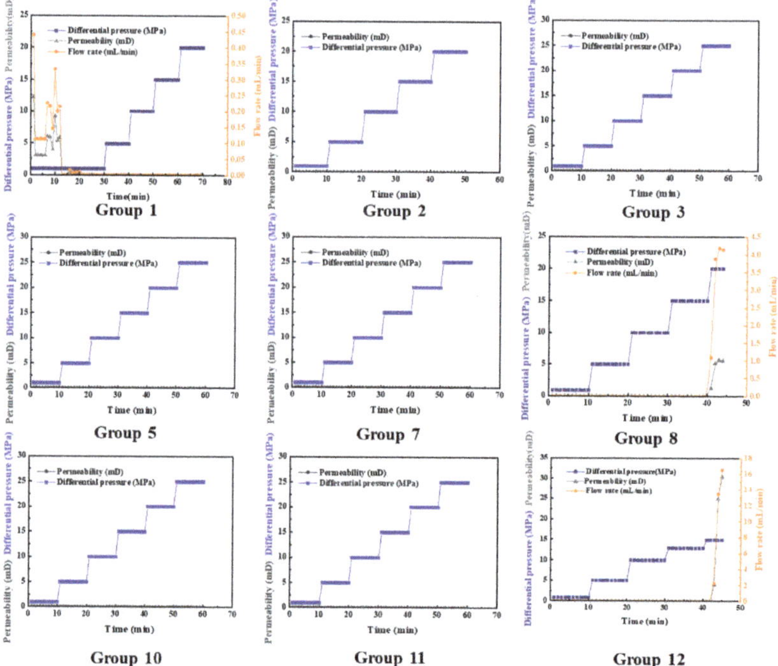

**Figure 6.** Plugging curves of different groups.

### 3.2.2. Solubility Test

The gas solubility test is a key experiment to evaluate whether the plugging agent can automatically degrade in contact with gas. Considering the dissolution of plugging agents at different temperatures and densities over time at corresponding temperatures after curing, different sets of experiments were conducted. The experimental parameters are shown in Table 3.

**Table 3.** Parameters in the gas solubility test.

| Group | Density | Test Temperature (°C) | Solvent |
|---|---|---|---|
| 1 | Low density, 1.15 g/cm$^3$ | 90 | Methane |
| 2 | Low density, 1.15 g/cm$^3$ | 60 | Methane |
| 3 | High density, 1.74 g/cm$^3$ | 60 | Methane |
| 4 | High density, 1.90 g/cm$^3$ | 60 | Methane |
| 5 | High density, 2.00 g/cm$^3$ | 60 | Methane |
| 6 | High density, 2.00 g/cm$^3$ | 90 | Methane |

**Figure 7.** Optical photos of the plugging agent and the proppant after the test.

The six groups of plugging agents in the above experimental program were weighed into a paper cup, and fracturing fluid was added and mixed thoroughly. They were then put in an oven and continuously heated at the corresponding experimental temperature until they were cured and taken out. The photos of the cured plugging agent are shown in Figure 8. Table 4 shows the experimental results of the dissolution capacity of the obtained plugging agent.

Group 1      Group 2      Group 3

Group 4      Group 5      Group 6

**Figure 8.** Optical photographs of the plugging agent after solidification in an oven.

**Table 4.** Solubility of the plugging agent.

| Group | Heat for 0 h | | Heat for 1 h | | Heat for 2 h | | Heat for 3 h | | Heat for 4 h | | Heat for 5 h | | Heat for 6 h | |
|---|---|---|---|---|---|---|---|---|---|---|---|---|---|---|
| | Weight (g) | DR (%) | Weight (g) | DR (%) | Weight (g) | DR (%) | Weight (g) | DR (%) | Weight (g) | DR (%) | Weight (g) | DR (%) | Weight (g) | DR (%) |
| 1 | 3.32 | 0 | 0.12 | 96.4 | 0.11 | 96.7 | 0.11 | 96.7 | 0.11 | 96.7 | | | | |
| 2 | 2.96 | 0 | 1.11 | 62.5 | 0.81 | 72.6 | 0.53 | 82.1 | 0.33 | 88.9 | 0.15 | 94.9 | 0.07 | 97.6 |
| 3 | 3.21 | 0 | 1.80 | 43.9 | 1.48 | 53.9 | 1.24 | 61.4 | 1.10 | 65.7 | 1.00 | 68.8 | 0.99 | 69.2 |
| 4 | 3.48 | 0 | 1.78 | 48.9 | 1.55 | 55.5 | 1.39 | 60.1 | 1.30 | 62.6 | 1.23 | 64.7 | 1.20 | 65.5 |
| 5 | 3.17 | 0 | 2.04 | 35.8 | 1.64 | 48.2 | 1.46 | 53.8 | 1.38 | 56.5 | 1.31 | 58.6 | 1.28 | 59.6 |
| 6 | 3.26 | 0 | 1.92 | 41.2 | 1.65 | 49.3 | 1.52 | 53.5 | 1.43 | 56.1 | 1.36 | 58.3 | 1.33 | 59.2 |

Note: DR—dissolution rate.

It is clear from Table 4 and Figure 8 that the dissolution rate of the plugging agent increases over time. The solubility of the low-density (1.15 g/cm$^3$) plugging agent reached 96.7% and 94.9% after heating for 5 h, respectively. The plugging agent in the beaker was reduced from almost full to only a little residue at the bottom of the beaker. After heating for 6 h, the solubility at 60 °C reached 97.6%, and the plugging agent in the beaker was significantly reduced again, showing good solubility. For the high-density plugging agents, the solubility was relatively low (59% to 69%) due to the addition of insoluble weighting agents. The higher the density of a plugging agent, the more insoluble material is left. Therefore, when the dissolution rate of plugging agent is calculated by mass difference method, the higher the density, the lower the dissolution rate of plugging fracturing technology controlled by plugging agent with small displacement and low displacement fracture. The optical photographs show the plugging agent residue after each hour. After 6 h, there is still a thick layer of solid at the bottom of the beaker. The dissolution of different groups of plugging agents at different times is shown in Figure 9. Generally, the plugging agent with different densities exhibited good gas solubility, with high dissolution rates achieved within a few hours.

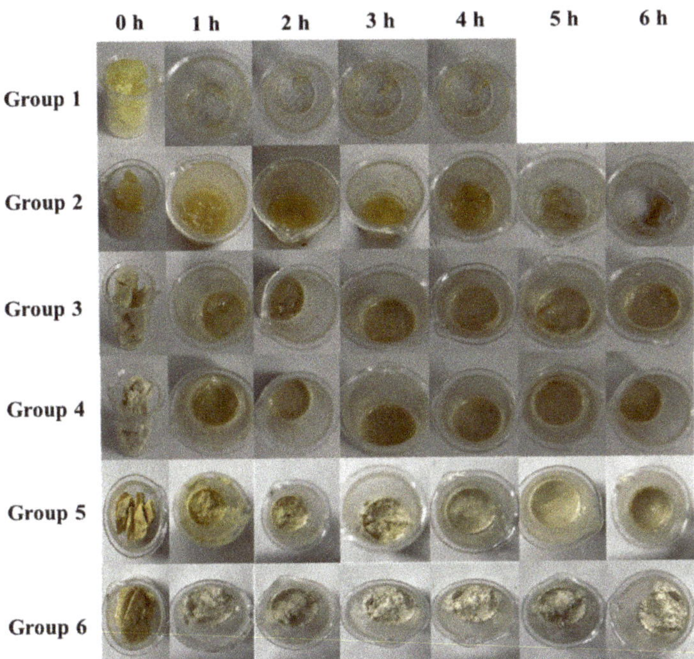



**Figure 9.** Dissolution of the plugging agent at different times.

### 3.2.3. Settlement Rule of Plugging Agent

Using proppant sedimentation law experiments for reference, we simulated underground fractures, measured the sedimentation law and plugging performance of the water shut-off material alone and mixed with proppant in the fracturing fluid, and clarified the effect of plugging the bottom water and preventing the blockage of the gas layer.

- Rheological properties of fracturing fluid.

The fracturing fluid system was prepared according to the method in Section 2.6, and its rheological properties were tested (Figure 10).

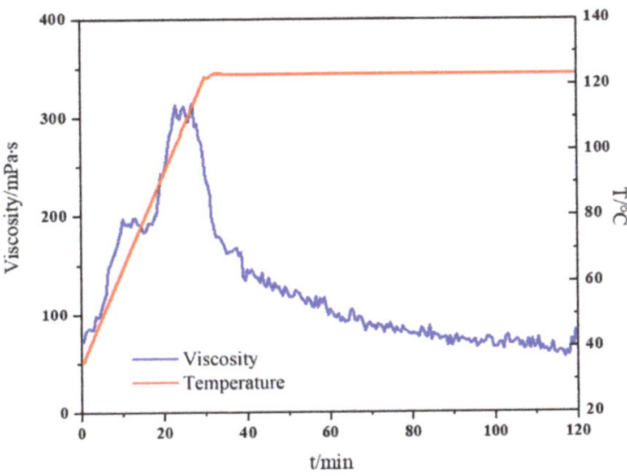



**Figure 10.** High temperature rheological properties of fracturing fluid.

According to the rheological properties, the viscosity of fracturing fluid at 120 °C remains above 60 mPa·s after 120 min, which meets the requirements of fracturing (The experiment method is carried out according to Chinese oil and gas standard SY/T 5107-2016 (water-based fracturing fluid performance evaluation method)).

- Test in water.

The settlement of the plugging agent and proppant after testing the density of 1.15 g/cm$^3$ and increasing it to 2.0 g/cm$^3$ in clear water is shown in Figures 11 and 12, and the settlement rates are shown in Tables 5 and 6.

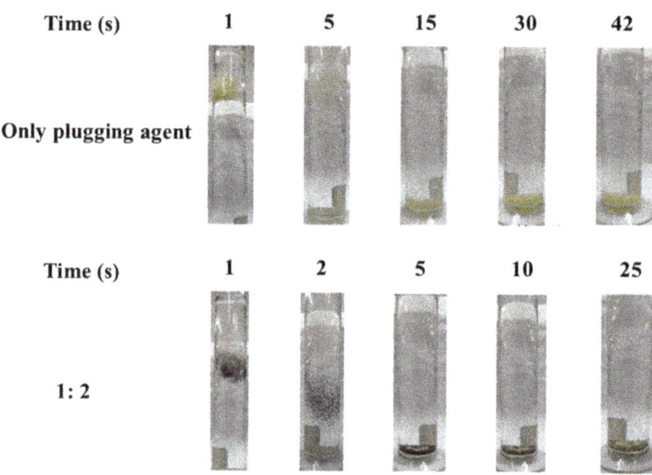

**Figure 11.** Settlement of the plugging agent and proppant (density = 1.15 g/cm$^3$) in water.

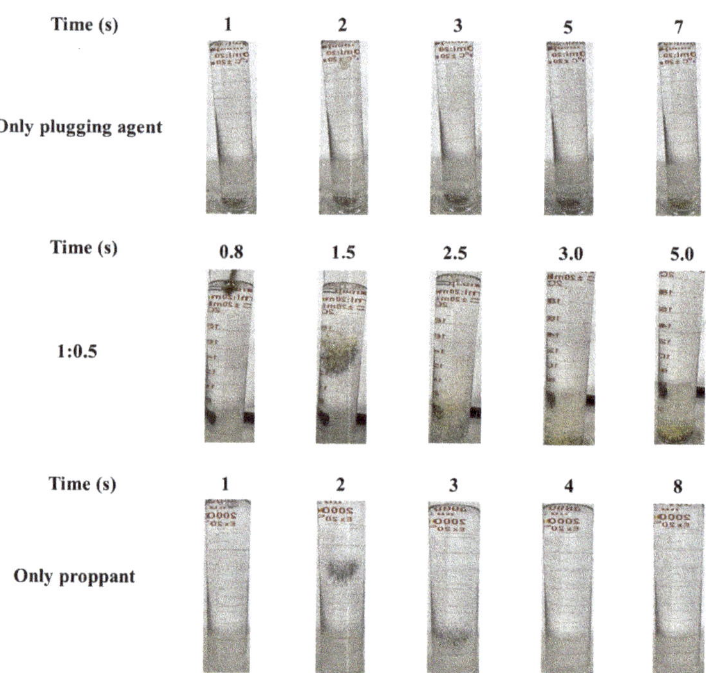

**Figure 12.** Settlement of the plugging agent and proppant (density = 2.00 g/cm$^3$) in water.

**Table 5.** Sedimentation rates (cm/s) of the plugging agent and proppant (density = 1.15 g/cm$^3$) in water.

| Volume Ratio of the Plugging Agent to the Proppant | First Test | Second Test | Average |
| --- | --- | --- | --- |
| Only the plugging agent | 0.71 | 0.67 | 0.69 |
| 1:2 | 1.18 | 1.21 | 1.20 |

**Table 6.** Sedimentation rate (cm/s) of the plugging agent and proppant (density = 2.00 g/cm$^3$) in water.

| Volume Ratio of the Plugging Agent to the Proppant | Settling Time (s) | Settling Velocity (cm/s) |
| --- | --- | --- |
| Only the plugging agent | 5.997 | 6.67 |
| 1:0.5 | 4.201 | 9.52 |
| Only the proppant | 3.799 | 10.53 |

The results show that the settling velocity of the proppant in water is much higher than that of the plugging agent at different densities. As shown in Figure 11 and Table 5, when the plugging agent density is 1.15 g/cm$^3$ and the volume ratio of plugging agent to proppant is 1:2, the settling velocity of the plugging agent/proppant mixture is about twice that of the pure plugging agent. The pure plugging agent was added into the clean water, and the settlement did not occur at 1 s. At 5 s, the plugging agent began to disperse and settle. At 15 s, the settlement was almost complete, and, 30 s later, the settlement was complete. However, when the 1:2 volume ratio of plugging agent and proppant was added into the water, the plugging agent and proppant settled in the first and second seconds, the proppant almost settled in the 5 s, the plugging agent almost settled in the 10 s, and settlement was completed in 25 s. Table 5 shows the sedimentation rates of the two systems after two repeated experiments. The average settlement rate of pure plugging agent is 0.69 cm/s, and that of proppant to plugging agent is 1.20 cm/s. At a proppant density of 2.00 g/cm$^3$, similarly, the higher the proppant ratio, the faster the settling time and rate. Figure 12 and Table 6 show that the settlement rate of the plugging agent is slow, and the settlement rate is 6.67 cm/s when the plugging agent settles completely at 5.997 s. When the volume ratio of plugging agent and proppant is 1:0.5, the solid settlement rate is faster than that of pure plugging agent, and the settlement rate is 9.52 cm/s when the plugging agent settles completely at 4.201 s. The pure proppant has the highest settlement rate, with a settlement time of 3.799 s and a settlement rate of 10.53 cm/s. The results show that the presence of proppant in water is beneficial to the rapid settlement of a plugging agent.

Test in guanidine gum solution. Sedimentation at densities of 1.15 g/cm$^3$ up to 2.0 g/cm$^3$ was tested in a guanidine gum solution. In actual pumping, the plugging agent and proppant will disperse in the guanidine gum solution under agitation. Therefore, the influence of dispersion on sedimentation in the guanidine gum was discussed, as shown in Figures 13 and 14 and Tables 7 and 8, and the sedimentation and the sedimentation rates in the guanidine gum were given.

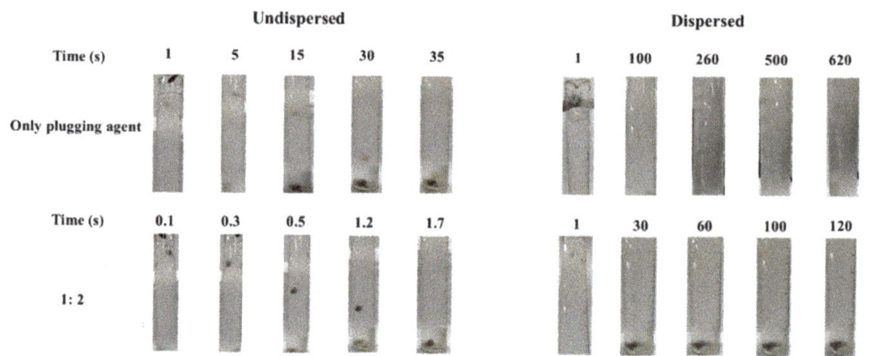

**Figure 13.** Settlement of the plugging agent and proppant (density = 1.15 g/cm$^3$) in the guanidine gum solution.

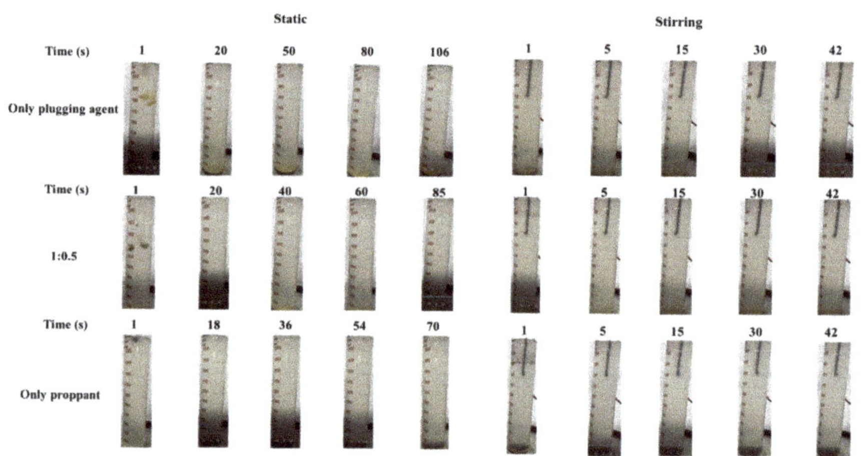

**Figure 14.** Settlement of the plugging agent and proppant (density = 2.00 g/cm$^3$) in the guanidine gum solution.

**Table 7.** Sedimentation rates of the plugging agent and proppant (density = 1.15 g/cm$^3$) in water.

| Settlement Form | Undispersed | | Dispersed | |
|---|---|---|---|---|
| Volume ratio of the plugging agent to the proppant | Only the plugging agent | 1:2 | Only the plugging agent | 1:2 |
| Settling time (s) | 35.8 | 1.7 | 684 | 120 |
| Settling velocity (cm/s) | 0.84 | 17.65 | 0.04 | 0.25 |

**Table 8.** Sedimentation rates of the plugging agent and proppant (density = 2.00 g/cm$^3$) in water.

| Settlement Form | Static | | | Stirring | | |
|---|---|---|---|---|---|---|
| Volume ratio of the plugging agent to the proppant | Only the plugging agent | 1:0.5 | Only the proppant | Only the plugging agent | 1:0.5 | Only the proppant |
| Settling time (s) | 106 | 85 | 70 | 180 | 145 | 117 |
| Settling velocity (cm/s) | 0.38 | 0.47 | 0.57 | 0.22 | 0.28 | 0.34 |

The experimental test results in Figure 13 show that, when the density is 1.15 g/cm$^3$, the plugging agent and proppant in guanidine gum solution completely settle and bond together, making it difficult to separate without dispersing. With the increased volume ratio of proppant, the settling velocity of the plugging agent increases significantly. Table 7 shows that, when the ratio is 1:2, the settling velocity of the plugging agent is about 21 times that of pure plugging agent. The settling time of the pure plugging agent is 35.8 s and the settlement rate is 0.84. When the volume ratio of the propped collector is 1:2, the settling time is reduced to 1.7 s and the settlement rate is 17.65 cm/s. However, in the actual pumping, there is a stirring effect, and the plugging agent and proppant will disperse in guanidine gum base liquid. Therefore, the dispersion settlement experiment is carried out and the plugging agent and proppant are dispersed into guanidine gum artificially, and the settlement process is observed. Figure 14 shows a dispersion settlement experiment where the plugging agent and proppant settle together. With the increased volume ratio of the proppant, the settling velocity of the plugging agent increases significantly. When the ratio is 1:2, the settling velocity of the plugging agent is about six times that of the pure plugging agent. Table 7 shows that the presence of the proppant facilitates rapid settlement of the plugging agent. The settling time of the pure plugging agent is 684 s and the settlement rate is 0.04 cm/s. When the volume ratio is 1:2, the settlement time is reduced to 120 s and the settlement rate is 0.25 cm/s. At a density of 2.00 g/cm$^3$, Figure 15 shows that, as the proppant ratio increases, the faster the settling time increases, the faster the settlement rate increases. Again, the presence of the proppant is conducive to rapid settlement of the plugging agent. Table 8 shows that, when the settling time of the pure plugging agent, the volume ratio of the plugging agent and proppant is 1:0.5, the pure proppant in guanidine gum solution is 106 s, 85 s, and 70 s, respectively, and the corresponding settlement rates are 0.38 cm/s, 0.47 cm/s, and 0.57 cm/s, respectively. In addition, it is proved that stirring can reduce the settling velocity. In the stirring experiment in Figure 14, the overall settlement rate slows down. In Table 8, when the settlement time of the volume ratio of pure plugging agent, plugging agent, and proppant is 1:0.5, pure proppant in guanidine gum solution is 180 s, 145 s, and 117 s, respectively, and the corresponding settlement rate is 0.22 cm/s, 0.28 cm/s, and 0.34 cm/s, respectively.

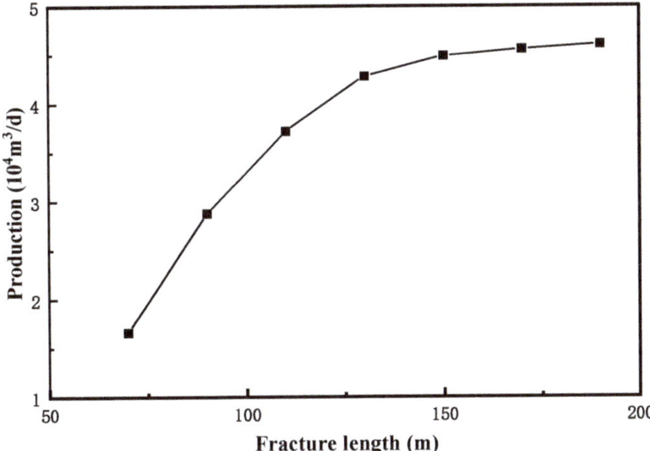

**Figure 15.** Production curve after fracturing at different fracture lengths.

### 3.3. Field Test and Post-Fracturing Evaluation

The preliminary field tests of water plugging and fracturing were conducted in a typical gas well with bottom water, and the post-fracturing effects were evaluated and analyzed. Then, improvement measures were put forward to provide bases for further experiments.

### 3.3.1. Basic Information on the Reservoir and Well

A typical bottom water gas well (J66-8-3) in Shanxi Formation was selected to perform a preliminary field test of water plugging and fracturing. The well was drilled to a depth of 2611.85 m. The well was completed with a 41/2″ casing and perforated at 2570–2572 m and 2574–2576 m. The formation closure pressure gradient was about 0.0158 MPa/m, and the formation closure pressure was 40.3 MPa. Other specific data for well J66-8-3 are shown in Table 9.

**Table 9.** Data for well J66-8-3.

| Name | Sounding (m) | Apparent Thickness (m) | Lithology (%) | Porosity (%) | Permeability (mD) | Gas Saturation (%) |
|---|---|---|---|---|---|---|
| He 2[1] | 2546.3–2550.6 | 4.3 | Grayish pebbly coarse sandstone | 16.00 | 3.68 | 58.37 |
| He 1[3] | 2558.8–2566.0 | 7.2 | Light gray coarse sandstone | 9.53 | 0.41 | 41.40 |
| | 2566.0–2576.6 | 10.6 | | 15.77 | 3.87 | 55.17 |

### 3.3.2. Treatment Design

The following treatment processes and parameters were designed to ensure a suitable fracturing and water shut-off process:

(1) Through logging interpretation, it is found that the gas layer He 1[3] (2558.8–2576.6 m) of well J66-8-3 is adjacent to the lower gas and water layer (2579.2–2598.0 m). The well layer was suitable for the sealing bottom water fracturing technology. As the distance between He 2[1] and He 1[3] was relatively long and the intermediate layer was better, the sealing method was used for fracturing separately. Therefore, He 2[1] was not considered during water plugging and fracturing.

(2) Because of the large amount of water produced by well J66-8-3 in the early stage, fracturing was often performed by controlling the scale and displacement, resulting in relatively short fracture lengths (118 m) in the early-stage fracturing wells and insufficient reservoirs. The main idea of blocking bottom water fracturing involves uncontrollable scales, uncontrollable fracture heights, and long fractures. Therefore, the length of the fracture must be optimized from a full transformation perspective, and the amount of water plugging material must be optimized according to the expansion of the height of the fracture.

(3) Another parameter is the selection of the temperature of the water-blocking material. According to the geothermal gradient of the block and the vertical depth of the well, the temperature of the target layer was calculated to be 67 °C. According to the research results of the project, a water shut-off material with a temperature of 60 °C can be selected.

(4) The density of the water-blocking material was then selected. According to the average pressure coefficient of the target layer (0.83), the formation pressure of the target layer must be lower than 20.8 MPa. Because of the large thickness of the gas and water layer in this well, from a cost-saving perspective, a water shut-off method combining a water shut-off material and a proppant should be considered. At the same time, high-density water shut-off materials should be used to reduce the shut-in time of the pump. In consideration of the plugging experiment of the water shut-off material, using a 1.9 g/cm$^3$ water shut-off material is recommended, and the water shutoff should be based on a 1:0.5 volume ratio of the water shutting material and proppant. The production pressure difference should not be greater than 15 MPa. We also used the sinking time to determine the experiment and put forward reasonable suggestions.

- Fracture length optimization.

The fracture length was optimized with production as the goal to fully transform the reservoir. The basic parameters used for calculation were as follows: reservoir thickness,

15 m; porosity, 15%; permeability, 2.5 mD; gas saturation, 38%; and diversion capacity, 20 D·cm. Figure 15 shows the relationship between the fracture length and the output after pressing. Production first increases with the increase in seam length. When the fracture length reaches about 150 m, production hardly increases and the curve appears to plateau. Finding a 150 m fracture length was advisable, so a 150 m fracture length was taken as the target layer.

- Optimization of the plugging agent dosage.

According to the simulation, the hydraulic fractures in the well entered the water layer 13.7 m after water plugging and fracturing. According to the calculation using the fracture half-length of 150 m, the well should pump 10.8 $m^3$ of the plugging agent and proppant mixture during the pre-fluid-plugging stage (where the plugging agent composes 7.2 $m^3$ of the total volume, whereas the proppant composes 3.6 $m^3$). We calculated the amount of plugging agent and proppant (1:0.5 volume ratio of the plugging agent to the proppant) required to plug the gas and water layers of the well under different fracture lengths, and the results are shown in Figure 16. As the required length increases, the amount of proppant and plugging agent increase. When the length is about 80 m, 3.8 $m^3$ of plugging agent and 1.9 $m^3$ of proppant are required. At 160 m, 7.7 $m^3$ plugging agent and 3.9 $m^3$ proppant were required. The longer the fracture, the more proppant and plugging agents are needed to create the fracture and plug the water.

- Determination of sinking time.

According to the basic characteristics of the well layers of J66-8-3, the fracture morphology after water plugging and fracturing was simulated (Figure 17). The length of the support fracture was 151.5 m (13.7 m into the water layer), and the height of the support fracture was 40.6 m (2552.3–2592.9 m).

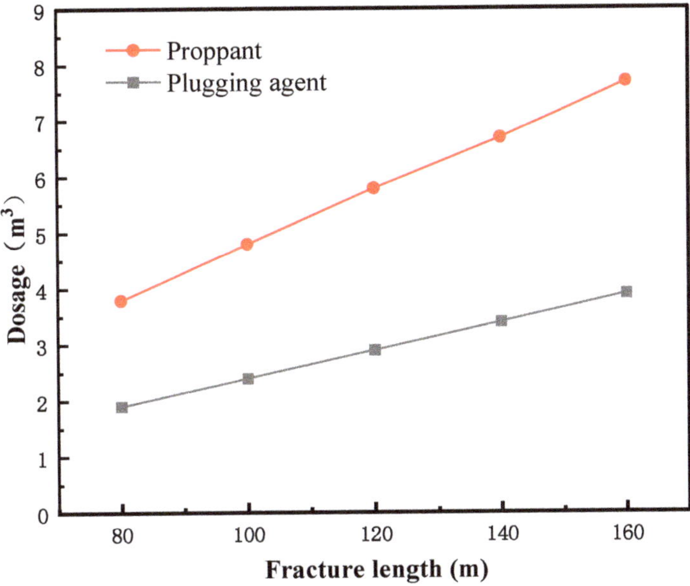


**Figure 16.** Plugging agent and proppant dosages for different fracture lengths.

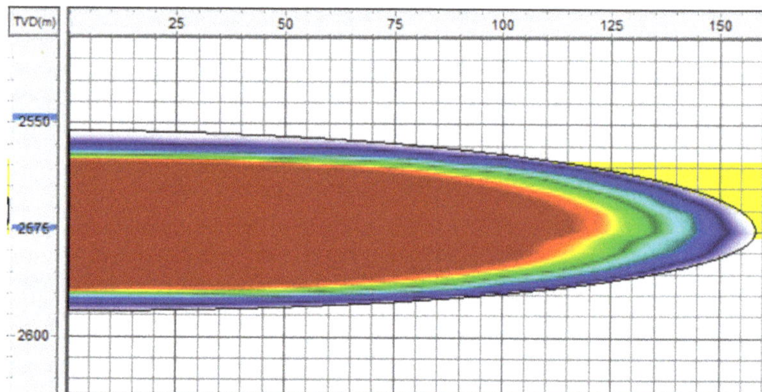

**Figure 17.** Fracture morphology after fracturing.

According to the research results, the sinking timing should be calculated according to the fracture height and the settlement speed of the plugging agent. When the volume ratio of the plugging agent to the proppant was 1:0.5, the settlement time of the plugging agent with a density of 1.9 g/cm$^3$ in the gel breaker was calculated as shown in Figure 18. It was found that the settling time of the plugging agent and proppant mixture increased linearly as the height of the fracture increased. The settling time of the plugging agent was 2.8 h in the case of a fracture height of 40.6 m. After the well pump was injected with the plugging agent, the settling time of the plugging agent was 2.8 h. We recommend stopping the pump for 3.0 h to allow the plugging agent to fully settle.

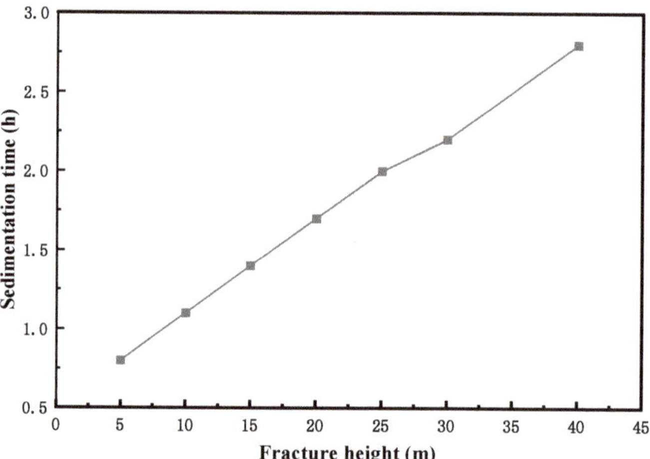

**Figure 18.** Settlement time at different fracture heights.

The water plugging and fracturing construction procedures for well J66-8-3 were adjusted considering the requirements of block reconstruction and the above water plugging and fracturing optimization results, and the results are shown in Table 10.

**Table 10.** Fracturing treatment pump injection procedure.

| Construction Phase | Pump Injection Stage | Stage | Stage Liquid Volume (m³) | Displacement (m³/min) | Proppant Type |
|---|---|---|---|---|---|
| Water plugging stage | Preflush | Base fluid | 10 | 4 | |
| | Cementing fluid | Crosslinked fluid | 40 | 4 | A 20/40-mesh proppant mixed with the plugging agent |
| | | Crosslinked fluid | 40 | 4 | A 20/40-mesh proppant mixed with the plugging agent |
| | Spacer fluid | Crosslinked fluid | 3 | 4 | |
| | Displacing liquid | Base fluid | 12 | 4 | |
| | | After a 3-h shut-in, the plugging agent settled and set | | | |
| Fracture | Preflush | Base fluid | 15 | 4 | |
| | Sand-carrying fluid | Crosslinked fluid | 25 | 4 | 20/40-mesh ceramsite |
| | | Crosslinked fluid | 30 | 4 | 20/40-mesh ceramsite |
| | | Crosslinked fluid | 30 | 4 | 20/40-mesh ceramsite |
| | | Crosslinked fluid | 40 | 4 | 20/40-mesh ceramsite |
| | | Crosslinked fluid | 30 | 4 | 20/40-mesh ceramsite |
| | Displacing liquid | Base fluid | 12 | 4 | |
| Total | | | 287 | | |

### 3.3.3. Evaluation Methods for the Post-Fracturing Wells

- Fracture height assessment.

The propagation of the fracture height after water plugging and fracturing can be obtained through temperature logging and net pressure fitting, and the relationship between the fracture height obtained and the gas–water layer interpreted by logging was determined to distinguish situations where the fracture height enters the water layer.

(1) If the fracture height extends but does not enter the lower water layer, it indicates that the scale of water plugging fracturing is small or the isolation layer condition is good (i.e., the water plugging fracturing technology is unnecessary).

(2) If the fracture height extends into the lower water layer, the bottom of the fracture height and the position of the water layer are further compared to obtain the depth of the fracture height into the water layer, which is compared with the water plugging fracturing design to determine whether the amount of the designed plugging agent is enough to block the water layer.

- Production analysis.

Water production is the direct display of the water plugging effect, and gas production is the ultimate goal of water plugging fracturing. Therefore, evaluating the effect of water plugging and fracturing from the aspects of water production and gas production is necessary. This can be conducted by comparing local wells and adjacent wells.

(1) Comparison in a local well. If a well is produced before fracturing, we can directly compare its daily water and gas production before and after plugging and fracturing under the same production system. We can then calculate the increase in gas production and decrease in water production of the well to quantify the effect of plugging and fracturing.

(2) Comparison of adjacent wells. Adjacent wells with similar formation conditions and without water plugging and fracturing technology were compared with the example

well. Under the same production system, we compared the daily gas production and daily water production of the adjacent wells and the example well. Then, we calculated the increase in the daily gas production of the example well and the decrease in the daily water production compared with that of the adjacent wells.

The rationality of a water plugging fracturing design and the effectiveness of water plugging fracturing can be effectively evaluated by evaluating the fracture height and analyzing the production situation. On this basis, improvement measures are put forward to constantly improve the water plugging fracturing technology and improve the pertinence and effectiveness of the water plugging fracturing technology.

### 4. Conclusions

This study includes laboratory study and field applications of a plugging agent, and the experimental results obtained led us to the following conclusions:

(1) Using pyrophosphoryl chloride, p-hydroxybenzaldehyde, and lactide as raw materials, a polylactide polymer plugging agent was successfully synthesized by a multi-step method.

(2) The plugging agent has excellent plugging performance. When the ratio of plugging agent to proppant is 1:1 and 1:2, the plugging rate of $1.15 \text{ g/cm}^3$ water plugging material can achieve 100%. The plugging rate of $1.56 \text{ g/cm}^3$ and $1.74 \text{ g/cm}^3$ density plugging materials is 100% at 1:0.5, and the plugging rate of $1.90 \text{ g/cm}^3$ and $2.0 \text{ g/cm}^3$ density plugging materials is 100% at 1:0.5.

(3) The solubility of the low-density ($1.15 \text{ g/cm}^3$) plugging agent reached 96.7% and 94.9% after heating for 4 h, and 97.6% after heating for 6 h, showing excellent solubility. For high-density plugging agents, the solubility is relatively low (59 to 69%) due to the addition of insoluble weighting agents.

(4) The settlement rate of a pure plugging agent in water and guar solution is slower than that of a proppant. As the proppant volume ratio increases, the sedimentation rate increases. In general, the presence of a proppant accelerates the settlement of a plugging agent.

(5) According to the actual well conditions of J66-8-3, the fracture length, plugging agent addition, and plugging agent sinking time were optimized and the single-well fracturing scheme was improved, providing reference for the realization of bottom water plugging fracturing technology in similar wells.

**Author Contributions:** Conceptualization, A.H.; methodology, A.H. and K.L.; software, H.F.; validation, H.F. and Y.Z.; formal analysis, Y.F.; investigation, Y.Z.; resources, A.H.; data curation, Y.F.; writing—original draft preparation, A.H.; writing—review and editing, Y.F. and Y.Z.; visualization, K.L.; supervision, H.F.; project administration, Y.F.; funding acquisition, Y.Z. All authors have read and agreed to the published version of the manuscript.

**Funding:** This research was funded by State Key Laboratory of Oil and Gas Reservoir Geology and Exploitation (Chengdu University of Technology), grant number PLC20210306.

**Institutional Review Board Statement:** Not applicable.

**Informed Consent Statement:** Not applicable.

**Data Availability Statement:** Not applicable.

**Conflicts of Interest:** The authors declare no conflict of interest.

### References

1. Wang, H.; Ma, F.; Tong, X.; Liu, Z.; Zhang, X.; Wu, Z.; Li, D.; Wang, B.; Xie, Y.; Yang, L. Assessment of global unconventional oil and gas resources. *Pet. Explor. Dev.* **2016**, *43*, 925–940. [CrossRef]
2. Pang, X.Q.; Jia, C.Z.; Wang, W.Y.; Chen, Z.X.; Li, M.W.; Jiang, F.J.; Hu, T.; Wang, K.; Wang, Y.X. Buoyance-driven hydrocarbon accumulation depth and its implication for unconventional resource prediction. *Geosci. Front.* **2021**, *12*, 101133. [CrossRef]

3.   Tang, X.; Zhang, J.; Shan, Y.; Xiong, J. Upper Paleozoic coal measures and unconventional natural gas systems of the Ordos Basin, China. *Geosci. Front.* **2012**, *3*, 863–873. [CrossRef]
4.   King, G.E.; Durham, D. Environmental aspects of hydraulic fracturing: What are the facts? In *Hydraulic Fracturing: Environmental Issues*; American Chemical Society: Washington, DC, USA, 2015; Volume 1216, pp. 1–44.
5.   Osborn, S.G.; Vengosh, A.; Warner, N.R.; Jackson, R.B. Methane contamination of drinking water accompanying gas-well drilling and hydraulic fracturing. *Proc. Natl. Acad. Sci. USA* **2011**, *108*, 8172–8176. [CrossRef]
6.   Shaffer, D.L.; Chavez, L.H.A.; Ben-Sasson, M.; Castrillon, S.R.V.; Yip, N.Y.; Elimelech, M. Desalination and Reuse of High-Salinity Shale Gas Produced Water: Drivers, Technologies, and Future Directions. *Environ. Sci. Technol.* **2013**, *47*, 9569–9583. [CrossRef]
7.   Arthur, M.A.; Cole, D.R. Unconventional Hydrocarbon Resources: Prospects and Problems. *Elements* **2014**, *10*, 257–264. [CrossRef]
8.   Song, Y.; Li, Z.; Jiang, L.; Hong, F. The concept and the accumulation characteristics of unconventional hydrocarbon resources. *Pet. Sci.* **2015**, *12*, 563–572. [CrossRef]
9.   Xie, Z.; Yang, C.; Li, J.; Zhang, L.; Guo, J.; Jin, H.; Hao, C. Accumulation characteristics and large-medium gas reservoir-forming mechanism of tight sandstone gas reservoir in Sichuan Basin, China: Case study of the Upper Triassic Xujiahe Formation gas reservoir. *J. Nat. Gas Geosci.* **2021**, *6*, 269–278. [CrossRef]
10.  Qin, Y.; Yao, S.; Xiao, H.; Cao, J.; Hu, W.; Sun, L.; Tao, K.; Liu, X. Pore structure and connectivity of tight sandstone reservoirs in petroleum basins: A review and application of new methodologies to the Late Triassic Ordos Basin, China. *Mar. Pet. Geol.* **2021**, *129*, 105084. [CrossRef]
11.  Zhong, Y.; Zhang, H.; Kuru, E.; Kuang, J.; She, J. Mechanisms of how surfactants mitigate formation damage due to aqueous phase trapping in tight gas sandstone formations. *Colloids Surf. A Physicochem. Eng. Asp.* **2019**, *573*, 179–187. [CrossRef]
12.  Huang, L.; Dontsov, E.; Fu, H.; Lei, Y.; Weng, D.; Zhang, F. Hydraulic fracture height growth in layered rocks: Perspective from DEM simulation of different propagation regimes. *Int. J. Solids Struct.* **2022**, *238*, 111395. [CrossRef]
13.  Fu, S.; Hou, B.; Xia, Y.; Chen, M.; Wang, S.; Tan, P. The study of hydraulic fracture height growth in coal measure shale strata with complex geologic characteristics. *J. Pet. Sci. Eng.* **2022**, *211*, 110164. [CrossRef]
14.  Pandey, V.J.; Rasouli, V. A semi-analytical model for estimation of hydraulic fracture height growth calibrated with field data. *J. Pet. Sci. Eng.* **2021**, *202*, 108503. [CrossRef]
15.  Zhang, Y.; Yu, R.; Yang, W.; Tian, Y.; Shi, Z.; Sheng, C.; Song, Y.; Du, H. Effect of temporary plugging agent concentration and fracturing fluid infiltration on initiation and propagation of hydraulic fractures in analogue tight sandstones. *J. Pet. Sci. Eng.* **2022**, *210*, 110060. [CrossRef]
16.  Wang, T.; Chen, M.; Wu, J.; Lu, J.; Luo, C.; Chang, Z. Making complex fractures by re-fracturing with different plugging types in large stress difference reservoirs. *J. Pet. Sci. Eng.* **2021**, *201*, 108413. [CrossRef]
17.  Wang, D.; Dong, Y.; Sun, D.; Yu, B. A three-dimensional numerical study of hydraulic fracturing with degradable diverting materials via CZM-based FEM. *Eng. Fract. Mech.* **2020**, *237*, 107251. [CrossRef]
18.  Xu, L. In Study on the shut-off mechanism of volumetric swell agent and productivity simulation for bottom water reservoir. In Proceedings of the 2016 International Conference on Progress in Informatics and Computing (PIC), Shanghai, China, 23–25 December 2016; pp. 449–454.
19.  Li, L.; Wu, Y.; Guo, N.; Zhou, J.; Liu, T.; Zhang, Z.; Li, Y.; Cai, G. Application of selective water plugging technology for slotted pipe horizontal well. *IOP Conf. Ser. Earth Environ. Sci.* **2019**, *252*, 052109. [CrossRef]
20.  Zhang, L.; Xiong, Y.M.; Xu, J.N.; Cheng, Y.; Liu, L.M.; Shan, Y.K. Delaying bottom water coming intelligent well completion technology research. In Proceedings of the Asia-Pacific Power and Energy Engineering Conference (APPEEC), Chengdu, China, 28–31 March 2010.
21.  Zhao, J.S.; Li, T.T.; Shi, Y.; An, C.Q.; Liu, Y.L. A new sinking agent used in water-control fracturing for low-permeability/bottom-water reservoirs: Experimental study and field application. *J. Pet. Sci. Eng.* **2019**, *177*, 215–223. [CrossRef]
22.  Qi, N.; Li, B.; Chen, G.; Liang, C.; Ren, X.; Gao, M. Heat-Generating Expandable Foamed Gel Used for Water Plugging in Low-Temperature Oil Reservoirs. *Energy Fuels* **2018**, *32*, 1126–1131. [CrossRef]
23.  Liu, J.; Zhang, Y.; Li, X.; Dai, L.; Li, H.; Xue, B.; He, X.; Cao, W.; Lu, X. Experimental Study on Injection and Plugging Effect of Core–Shell Polymer Microspheres: Take Bohai Oil Reservoir as an Example. *ACS Omega* **2020**, *5*, 32112–32122. [CrossRef]
24.  Li, J.; Niu, L.; Lu, X. Migration characteristics and deep profile control mechanism of polymer microspheres in porous media. *Energy Sci. Eng.* **2019**, *7*, 2026–2045. [CrossRef]
25.  Guo, Y.; Pu, W.; Zhao, J.; Guo, Y.; Lian, P.; Liu, Y.; Wang, L.; Yu, Y. Effect of the foam of sodium dodecyl sulfate on the methane hydrate formation induction time. *Int. J. Hydrog. Energy* **2017**, *42*, 20473–20479. [CrossRef]
26.  Zhang, Y.; Cai, H.; Li, J.; Cheng, R.; Wang, M.; Bai, X.; Liu, Y.; Sun, Y.; Dai, C. Experimental study of acrylamide monomer polymer gel for water plugging in low temperature and high salinity reservoir. *Energy Sources Part A Recovery Util. Environ. Eff.* **2018**, *40*, 2948–2959. [CrossRef]
27.  Liu, E.H. Research and application of high temperature gel plugging agent. *Appl. Mech. Mater.* **2014**, *675–677*, 1565–1568. [CrossRef]
28.  Xie, K.; Su, C.; Liu, C.; Cao, W.; He, X.; Ding, H.; Mei, J.; Yan, K.; Cheng, Q.; Lu, X. Synthesis and Performance Evaluation of an Organic/Inorganic Composite Gel Plugging System for Offshore Oilfields. *ACS Omega* **2022**, *7*, 12870–12878. [CrossRef]
29.  Zhang, X.; Liu, W.; Yang, L.; Zhou, X.; Yang, P. Experimental Study on Water Shutoff Technology Using In-Situ Ion Precipitation for Gas Reservoirs. *Energies* **2019**, *12*, 3881. [CrossRef]

30. Gao, Y.; Chen, S.; Huang, F.; Gao, Y.; Song, T.; Wang, S.; Jia, P.; Liu, C. Micro-occurrence of formation water in tight sandstone gas reservoirs of low hydrocarbon generating intensity: Case study of northern Tianhuan Depression in the Ordos Basin, NW China. *J. Nat. Gas Geosci.* **2021**, *6*, 215–229. [CrossRef]
31. Zhao, X.; Zhu, S. Prediction of water breakthrough time for oil wells in low-permeability bottom water reservoirs with barrier. *Pet. Explor. Dev.* **2012**, *39*, 504–507. [CrossRef]
32. He, Y.; Qin, J.; Cheng, S.; Chen, J. Estimation of fracture production and water breakthrough locations of multi-stage fractured horizontal wells combining pressure-transient analysis and electrical resistance tomography. *J. Pet. Sci. Eng.* **2020**, *194*, 107479. [CrossRef]
33. Zhang, Y.; Zhang, D.; Wen, Q.; Zhang, W.; Zheng, S. Development and evaluation of a novel fracture diverting agent for high temperature reservoirs. *J. Nat. Gas Sci. Eng.* **2021**, *93*, 104074. [CrossRef]

*Article*

# Characterization of Microstructures in Lacustrine Organic-Rich Shale Using Micro-CT Images: Qingshankou Formation in Songliao Basin

Yan Cao [1,2], Qi Wu [3], Zhijun Jin [1,2,*] and Rukai Zhu [4]

1 Institute of Energy, Peking University, Beijing 100871, China
2 School of Earth and Space Sciences, Peking University, Beijing 100871, China
3 State Key Laboratory for Turbulence and Complex Systems, College of Engineering, Peking University, Beijing 100871, China
4 Research Institute of Petroleum Exploration & Development, PetroChina, Beijing 100083, China
* Correspondence: jinzj1957@pku.edu.cn

**Abstract:** In order to explore the development characteristics and influencing factors of microscale pores in lacustrine organic-rich muddy shale, this study selected five shale samples with different mineral compositions from the Qingshankou Formation in the Songliao Basin. The oil content and mineralogy of the shale samples were obtained by pyrolysis and X-ray diffraction analysis, respectively, while the porosity of the samples was computed by micro-CT imaging. Next, based on the CT images, the permeability of each sample was calculated by the Avizo software. Results showed that the continuous porosity of Qingshankou shale in the Songliao Basin was found between 0.84 and 7.79% (average 4.76%), the total porosity between 1.87 and 12.03% (average 8.28%), and the absolute permeability was calculated between 0.061 and $2.284 \times 10^{-3}$ $\mu m^2$. The total porosity of the samples has a good positive correlation with the continuous porosity and permeability. This means higher values of total porosity suggested better continuous porosity and permeability. Both total porosity and continuous porosity are positively correlated with the content of clay minerals. Moreover, the oil content of the samples (the $S_1$ peak from programmed pyrolysis) exhibits a good positive correlation with the total porosity, continuous porosity, permeability, and clay mineral content. Therefore, pores that are developed by clay minerals are the main storage space for oil and flow conduits as well. Clay minerals were found to be the main controlling factor in the porosity, permeability, and the amount of oil content in the pores in the study area.

**Keywords:** organic-rich mud shale; Songliao Basin; Qingshankou Formation; micro-CT; simulation; porosity and permeability

**Citation:** Cao, Y.; Wu, Q.; Jin, Z.; Zhu, R. Characterization of Microstructures in Lacustrine Organic-Rich Shale Using Micro-CT Images: Qingshankou Formation in Songliao Basin. *Energies* **2022**, *15*, 6712. https://doi.org/10.3390/en15186712

Academic Editors: Xingguang Xu, Kun Xie and Yang Yang

Received: 23 July 2022
Accepted: 7 September 2022
Published: 14 September 2022

**Publisher's Note:** MDPI stays neutral with regard to jurisdictional claims in published maps and institutional affiliations.

## 1. Introduction

Shale oil and gas resources are abundant around the globe, and drilling horizontally and hydraulic fracturing have been applied successfully in North America, enabling large-scale commercial development of shale oil and gas [1,2], and accounting for a significant growth in hydrocarbon production. In 2018, the global crude oil production was $44.5 \times 10^8$ t, of which 14% was from unconventional shale plays. Additionally, natural gas production was $3.97 \times 10^{12}$ $m^3$, with 25% from unconventional plays [3]. In addition to the US, China is also rich in plays with a huge exploration potential, heading towards large-scale production [1,2].

The shale oil–producing layers in North America are mainly distributed in marine or foreland basins, in a large depositional area, with good continuity, mostly over pressured and high thermal maturity. Conversely, in China, such plays are mostly distributed in depressions and rift basins of continental deposition, which are characterized by strong heterogeneity, low overall pore pressure, and lower thermal maturity [4]. Generally speaking, regardless of the basin in China, shale reservoirs have ultra-low permeability ($1 \times 10^{-9}$~$1 \times 10^{-4}$ µm$^2$), nanoscale pore diameter (1~200 nm), and complex pore structures [5]. The pore structure and permeability in shale are often considered key indicators of the storage space and flow capacity of shale oil and gas [1,2,5]. Therefore, the pore structure and permeability of Chinese terrestrial shales have been a hot topic of research [1,2,5].

A large number of research tools have been used to reveal the pores of shale, including scanning electric microscopes (SEM) [5], atomic force microscopes (AFM) [6], mercury intrusion porosimetry (MIP) [7], low-pressure gas adsorption [8], micro-CT and nano-CT [9,10], small-angle X-ray or neutron scattering [11], and so on. Loucks et al. (2009) [5] reported that the pores of Mississippian Barnett shale are mainly nanoscale (nanopores), and the pore size can be as low as 5 nm. Organic matter pore space is the dominant pore type in the shale and is strongly controlled by maturity, and pyrite and shale matrix can also provide some pore space for shale [5]. There are several factors that control the enrichment of gas in the marine shales, including well-developed micropores, a large surface area, and a high gas adsorption capacity [8]. Inter-particle pores between organic matter and clay minerals may be responsible for those three controls [8]. Sun et al. (2018) [11] reported that the storage and flow mechanisms of gas in shale reservoirs can be greatly affected by pore structure. There were rarely closed pores in illite but mainly in organic matter by means of small-angle neutron scattering (SANS) [11]. Moreover, geometrical tortuosity and matrix permeability are negatively correlated with the fraction of closed pores [11]. However, most of the current methods only reveal shale pore space but ignore the pore connectivity of shale; furthermore, shale oil and gas development depends heavily on pore connectivity [12]. Among those means, micro-CT and 3D reconstruction of pore spaces via digital rock physics (DRP) methods can effectively distinguish various type of pores, including the continuous pores and isolated ones [13].

In confined nanopores of shale, the solid-liquid intermolecular forces result in complex fluid properties which impact fluid flow. This means the traditional macroscopic Darcy flow equation is no longer applicable for accurately characterizing the fluid flow in such confined very fine spaces [13,14], which has encouraged a large amount of research [15–19]. Currently, the permeability of shales can be determined in the laboratory using three methods, including (i) gas measurements of plunger cores, (ii) gas analysis of particle samples, and (iii) the use of mercury (Hg) intrusion curves [20]. However, it is not feasible to measure shale permeability using steady-flow methods because of measurement of extremely small pressure drops or flow rates requires highly complex instrumentation. The development of pulse decay techniques followed, which could measure pressure decay on an upstream end of a confined core as well as pressure increase on the downstream end. As little as $10^{-9}$ millidarcies ($1 \times 10^{-15}$ mD $\approx 1$ m$^2$) of permeability can be measured with pulse decay techniques in minutes to hours or even days, depending on the application [21]. Helium (He) is utilized for permeability measurements of granular samples of shale, and pressure decay is measured and quantified as permeability [22]. The particle density or skeletal density of the rock is obtained by He porosimetry using crushed samples, and the porosity of shale can also be calculated by combining the capacity of the impregnated mercury [23]. Numerous studies have examined the relationship between permeability and Hg injection curves [24,25]. Permeability is calculated based on Hg saturation and capillary pressure at the apex of the hyperbolic log-log Hg injection plot [26]. Although the above experimental methods can characterize the permeability of shale to some extent, there are still problems such as large errors and low reproducibility. Considering low saturation and mobility of oil in shale, it would be difficult to intuitively capture oil flow in shale samples in experimental studies. Therefore, theoretical methods via simulation of the flow process [19] were used to investigate permeability in our selected shales.

The shale of the Qingshankou Formation in the Gulong Sag is a key target for shale oil development in China. The relationship between total porosity, continuous porosity, and permeability in studied shale has not been previously revealed by the combination of micro-CT experiments and simulations. Therefore, it is necessary to reveal the characteristics of porosity and permeability of shale in the region and its influence factors through a combination of experiments and simulations.

In this study, five shale samples from the Well Songyeyou 1HF, which was drilled through the Qing 1–3 Member in the Gulong Sag of the Songliao Basin, were selected for analysis and testing. Through micro-CT imaging combined with modeling, the difference in porosity, permeability, and oil content of the shale samples with different mineral assemblages is recognized first, then flow behavior in each is studied by simulation method to enable us to judge the possibility of commercial development of shale layers in the block.

## 2. Geological Setting and Sampling

The Songliao Basin is a large-scale lacustrine facies basin located in northeastern China (Figure 1A). There were three major tectonic stages in the formation of the basin: fault subsidence, thermal subsidence, and inversion [27]. The basin can be divided into 6 tectonic units: the western slope, northern steep slope, central depression, northeast uplift belt, southeast uplift belt, and southwest uplift belt (Figure 1B). The Gulong Sag is located in the western side of the Central Sag, which was formed during the depositional period of the Qingshankou Formation, the main source rock in the basin (Figure 1C). The point B' (i.e., red star), the well location of the studied samples, is located at the junction between profile AA' and BB' (Figure 1C).

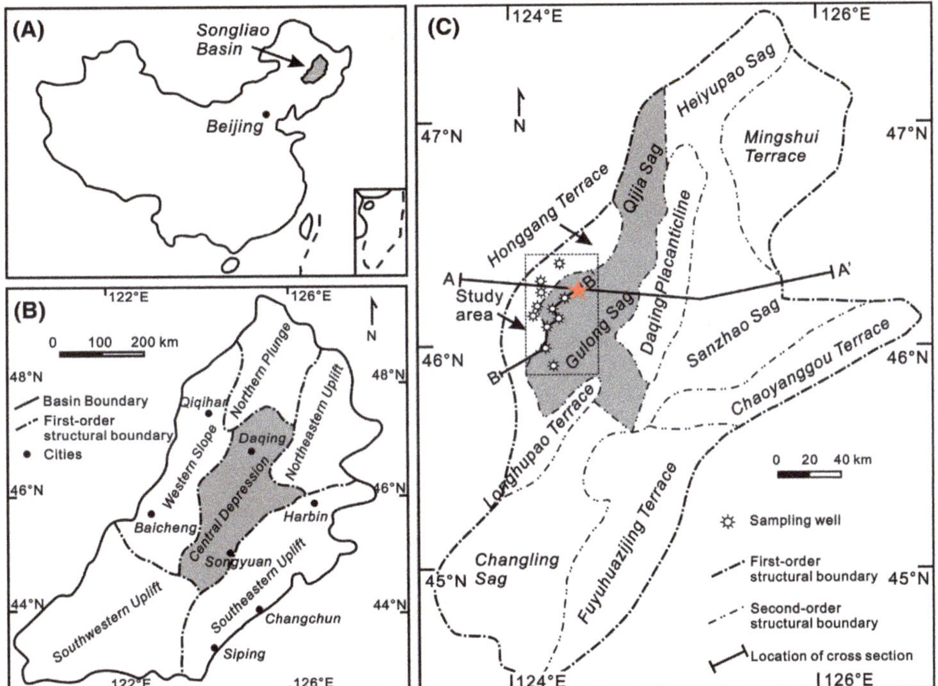



**Figure 1.** Location of the Songliao Basin on a generalized map (**A**). Central Depression (**B**). Well locations in the study area (the Gulong Sag) (**C**). (Modified from Liu et al., 2019 [27]).

The Qingshankou Formation ($K_2qn$), a moderately deep lacustrine environment, was influenced by periodic marine intrusions. The lithology divides the Qingshankou Formation ($K_2qn$) into three subsections ($K_2qn^1$, $K_2qn^2$, and $K_2qn^3$) based on lithology (Figure 2). The first member of the Qingshankou Formation (section $K_2qn^1$) is widely distributed throughout the basin and is one of the most favorable hydrocarbon source rocks in the Songliao Basin [27]. The first member of the Qingshankou Formation (section $K_2qn^1$)

is up to 500 m thick, and formed during rapid, large-scale lake transgression. Lake levels rose several times due to stepwise subsidence of the basement during the emplacement of $K_2qn^1$, leading to the interbedded accumulation of dark shale and siltstone [27]. The first member of the Qingshankou Formation (section $K_2qn^1$) was selected as the target layer for this study, and five typical dark shale samples in this member were selected for our study.

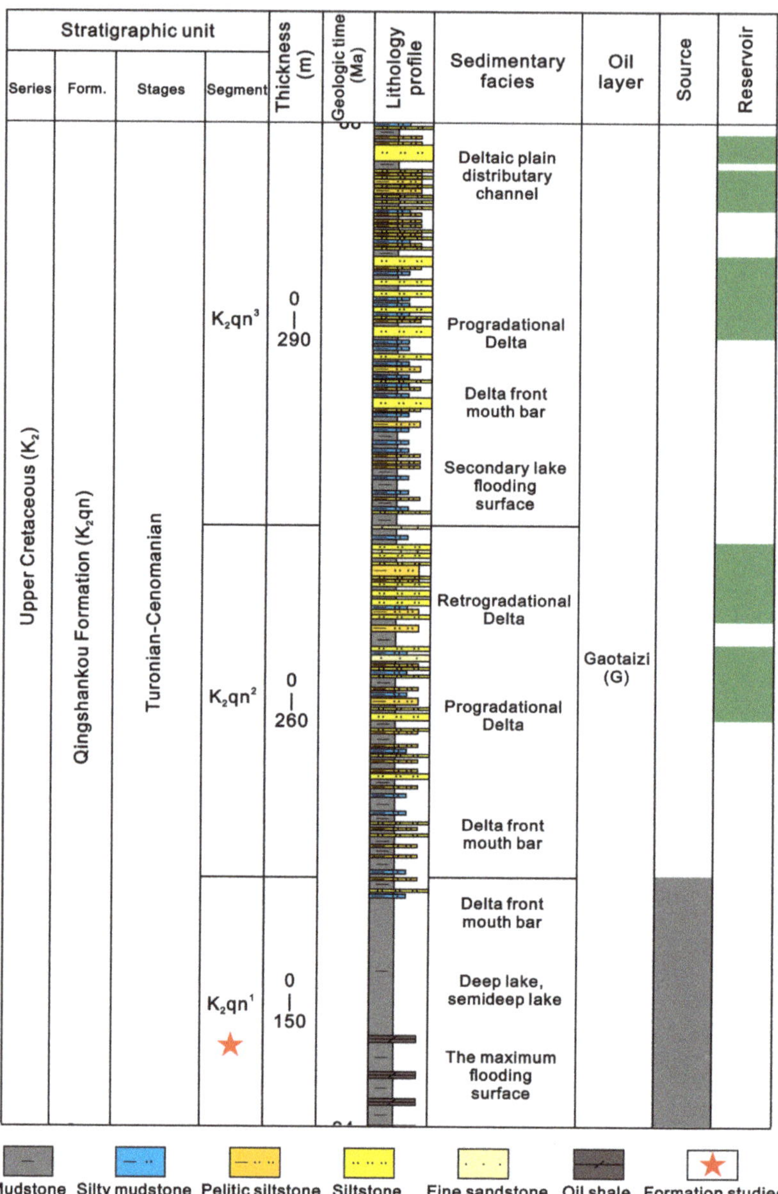


**Figure 2.** The strata and sedimentary characteristics of the Qingshankou Formation in western Songliao Basin (modified from Liu et al., 2019 [27]).

## 3. Experiments

### 3.1. Analysis of Oil Content and Mineral Content

Oiliness analysis is performed following the ASTM standards [28]. Next, samples were powdered to the mesh size of 100 and were analyzed for thermal maturity with programmed pyrolysis. This procedure provided us the $S_1$ peak (mg HC/g rock), which is the quantity of free hydrocarbons volatized at 300 °C. This peak can represent the oiliness of the sample. Furthermore, powder finer than 200 mesh (i.e., <0.075 mm) was analyzed by quantitative X-ray diffraction (XRD) to determine the mineral content of the studied samples. The D/max-2500 diffractometer was used for the measurements, following two separate CPSC procedures [29].

### 3.2. Micro CT Experiment

First, a cylindrical core plug with a diameter of 1 mm and a length of 1 mm parallel to the bedding from the main core was retrieved. The micro-CT imaging was completed in the China Petroleum Exploration and Development Research Institute using the Nano-CTX Radia scanning equipment (Model Ultra XRM-L200) from ZEISS, with a maximum resolution of 1 μm. During the scanning process, the sample is rotated from −90 to 90° and the X-ray information is continuously acquired [30].

In the micro-CT experiment, the scanning voltage was set to 8 kV, at 20 °C, and the exposure time was 90 s. A total of 901 two-dimensional plane images along the Z-axis were obtained, which can be stacked to form a three-dimensional data volume with a diameter of 65 μm and a height of 60 μm [30]. Using the Avizo software, the 3D model of the sample can be reconstructed which is shown in Figure 3 for the S41 sample. In this study, we analyzed five shale samples from the Qingshankou Formation in the Gulong Sag labeled as: S41, S189, S201, S317, and S353.

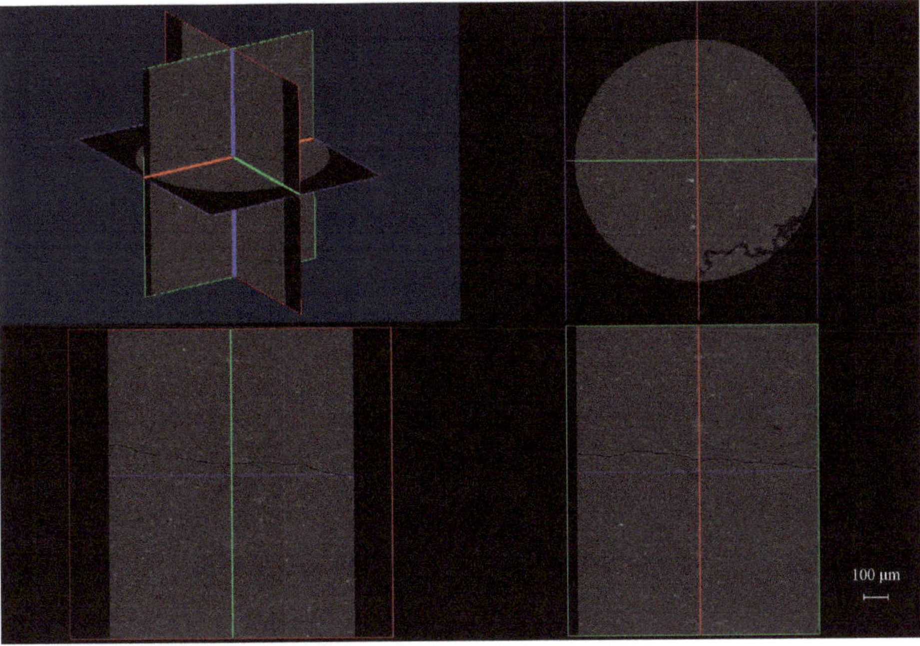


**Figure 3.** 3D reconstruction of the S41 shale sample.

### 3.3. Avizo Simulation Computing

Avizo software was used to reconstruct the 3D shale models from cross-sectional images from the micro-CT data. To separate pores from the matrix, the threshold segmentation method based on the gray scale was used to select using the Avizo software. To be more specific, the Gaussian deconvolution threshold segmentation method for identifying different phases in the CT images was employed which converts each image into binary mode and further will be used to reconstruct the pore structure network. Figure 4 represents the process that was performed for the S41 sample as an example, where the blue part is the pore distribution extracted from the area with higher values on the gray scale.

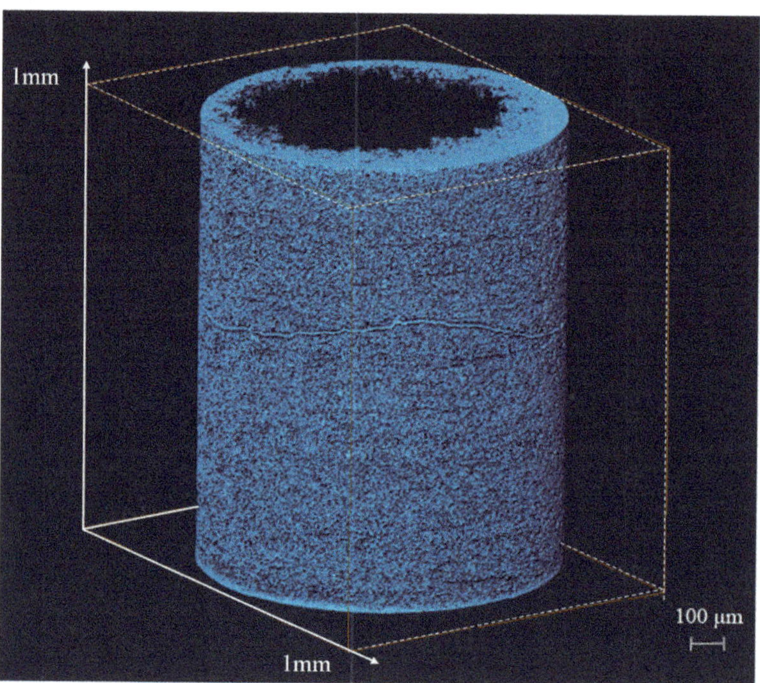


**Figure 4.** 3D reconstruction of S41 sample after threshold segmentation.

Th permeability of the reservoir refers to the property of the rock that allows fluid to pass through its continuous pores under a certain pressure difference. In other words, permeability refers to the conductivity of the rock to fluids. The permeability of the reservoir determines the ease of hydrocarbon penetration, which is one of the main parameters for evaluating reservoir quality. In the petroleum industry, absolute permeability is a common parameter that has been used frequently as a measure of the reservoir productivity. This parameter can be calculated from the CT-based 3D models which can be verified with experimental analysis as well. Flow experiments on core samples were conducted under steady state, and the following equation was used to calculate permeability using Darcy's law:

$$k_g = \frac{2p_a\mu q_g L}{P_1{}^2 - P_2{}^2 A} \tag{1}$$

$$v_g = \frac{q_g}{A} = \frac{P_1{}^2 - P_2{}^2}{L} \times \frac{k_g}{2p_a\mu} \tag{2}$$

In order to make sure that the flow conditions satisfy Darcy's Law, tests were carried out under different flow rates. In practice, permeability is calculated from the slope of the curve of flow velocity, $v_g$ vs $(P_1{}^2 - P_2{}^2)/L$. A similar approach can be followed in numerical simulation as well. In the simulation models, air density in the flow process was ignored and the flow was considered incompressible viscous, which satisfies the three laws of mass, momentum, and energy conservation. Therefore, the flow can be described by the Navier–Stokes equation, which is defined by the following formula:

$$\rho\left[\frac{\partial v}{\partial t} + (v\cdot\nabla)v\right] = \rho f - \nabla p + \mu\nabla^2 v \tag{3}$$

Using Equation (2), if $q_g$ is known, permeability $k_g$ can be computed. Hence, porous media parameters such as the pore structure and permeability can be easily estimated from the digital shale model.

*3.4. The Porosity and Permeability of the Sample by Conventional Method*

Based on a rock sample's bulk volume, grain volume, and pore volume, the porosity is calculated. Porosity studies were per-formed using typical equipment and conventional methods (i.e., GRI [31,32]).

Using virtually the same method as GRI, samples were analyzed in this laboratory [33]. Samples were weighed to a precision of 0.001 g and their bulk volumes measured to a precision of 0.001 cm$^3$. In the next step, a core plug was drilled perpendicular to the lamination. After crushing the remaining sample with a mechanical rock crusher, the 20/35 US mesh fraction was sieved. In order to limit the evaporation of fluids from the sample, these steps were performed as quickly as possible. The 20/35 fraction was then divided into two subsamples and sealed in airtight vials. Using the GRI method, one subsample was measured for porosity and permeability. Afterwards, a second subsample was refluxed for 7 days in toluene in a Dean Stark apparatus. Water extraction was verified twice a day by checking fluid volumes. After being dried for 2 weeks at 110 °C, the samples were weighed until weight stabilization (0.001 g) was achieved. After that, the samples were kept in a desiccator. Helium gas at approximately 200 psig was used for measuring permeability. We measured pressure at 0.25 s intervals for a maximum of 2000 s. In the end, we measured the permeability of core plugs with the PDP technique described by Jones [34] at a helium pressure of 1000 psi and confining pressure of 5000 psi [33].

*3.5. Experimental Procedure for Studying Samples*

Combining above mentioned experiments, we summarize the schematic diagram of the experimental procedure of the studied sample in Figure 5. The porosity and permeability of the studied samples were obtained by means of micro-CT and simulation calculations, respectively. The reliability of micro-CT and simulation is verified by comparing the reservoir characteristics obtained above with the porosity and permeability measured by typical experimental equipment. Correlation of reservoir characteristics with XRD results and free hydrocarbon S$_1$ to obtain control factors of shale reservoir.

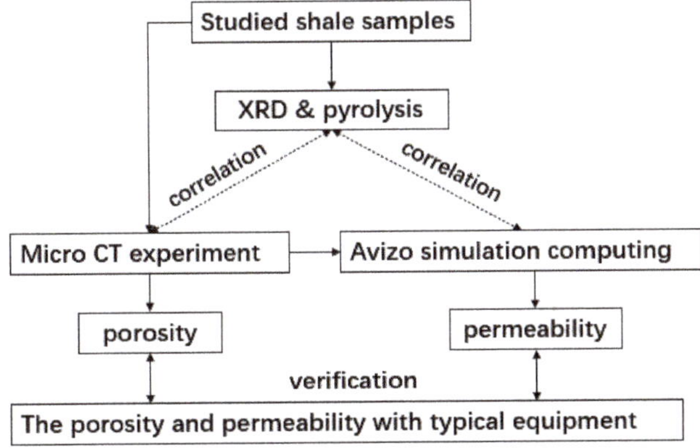


**Figure 5.** The schematic diagram of the experimental procedure of the studied sample.

## 4. Results

### 4.1. Mineral Composition and Oil Content

XRD results confirm all five samples from the Qingshankou Formation in the Gulong Sag consisted of large amounts of clay, varying between 32.6 wt. and 42.3 wt.%, with an average value of 36.46 wt.%. In addition, quartz is also abundant in the samples ranging between 29.9 and 34 wt.%, with an average of 32.4 wt.%, and plagioclase (18.1–23.4 wt.%, with and average of 20.16 wt.%) was also detected in the samples. Other minerals, including calcite, siderite, and pyrite, were detected with different amounts in the samples. Movable hydrocarbon ($S_1$) content was found relatively high, mainly at 1.36–8.54 mg/g, with an average of 5.67 mg/g. The detailed mineralogy and oil-bearing characteristics of all samples are summarized in Table 1.

**Table 1.** Mineralogy and oil-bearing characteristics of the studied samples.

| Sample | Depth | Member | Lithological Description | Quartz (wt.%) | Potassium Feldspar (wt.%) | Plagioclase (wt.%) | Calcite (wt.%) | Iron Dolomite (wt.%) | Siderite (wt.%) | Pyrite (wt.%) | Clay Mineral (wt.%) | $S_1$ (mg/g) |
|---|---|---|---|---|---|---|---|---|---|---|---|---|
| S41 | 2335.8 | $K_1qn^1$ | Dark gray lamellar shale | 34 | 0.9 | 19.5 | 1.3 | 3.9 | 0.6 | 3.8 | 35.9 | 6.73 |
| S189 | 2409.8 | $K_1qn^1$ | Dark gray lamellar shale | 32.2 | 0.6 | 18.1 | 1.7 | 2.1 | 1.3 | 5.4 | 38.6 | 8.54 |
| S201 | 2415.1 | $K_1qn^1$ | Dark gray lamellar shale | 29.9 | 1.2 | 23.4 | 5.2 | 0.0 | 1.3 | 6.0 | 32.9 | 1.36 |
| S317 | 2473.93 | $K_1qn^1$ | Dark gray lamellar shale | 32.9 | 2.6 | 21.6 | 8.6 | 0.0 | 0.0 | 1.9 | 32.6 | 3.48 |
| S353 | 2491.8 | $K_1qn^1$ | Dark gray lamellar shale | 33.0 | 2.0 | 18.2 | 0.3 | 0.0 | 0.0 | 4.2 | 42.3 | 8.26 |

### 4.2. Porosity and Permeability

Using DRP methods and the 3D model that was obtained by Avizo software, the total and effective porosity of five samples is calculated and reported in Table 2. Furthermore, the flow simulation module provided by the Avizo software can compute the permeability of the samples which is also listed in Table 2. Results showed that these values vary notably among the samples. The continuous porosity of these five samples was found between 0.84 and 7.79% (average 4.76%), the total porosity between 1.87 and 12.03% (average 8.28%), and the absolute permeability was calculated between 0.061 and $2.284 \times 10^{-3}$ $\mu m^2$. Moreover, average open pores account for 55.79% of the total pores (Table 2), which means the average porosity in this area is relatively low, while the proportion of continuous pores is high, and the permeability is low. Based on the results we can categorize the shale samples in the study area as tight reservoir with low porosity and permeability.

**Table 2.** Total porosity, continuous porosity and permeability of the studied samples.

| Sample | S41 | S189 | S201 | S317 | S353 | Average |
|---|---|---|---|---|---|---|
| Continuous porosity % | 4.83 | 7.79 | 0.84 | 4.14 | 6.20 | 4.76 |
| Total porosity % | 10.16 | 12.03 | 1.87 | 6.31 | 11.05 | 8.28 |
| Continuous pore/Total pore % | 47.54 | 64.75 | 44.92 | 65.61 | 56.11 | 55.79 |
| Permeability ($10^{-3}$ $\mu m^2$) | 1.666 | 2.284 | 0.061 | 1.524 | 1.496 | 1.137 |

The measured results of porosity and permeability of the samples using the typical equipment are shown in Table 3, and the above test results are similar to those obtained by micro-CT and simulation; the error produced by comparing the results of the two methods is acceptable (Tables 2 and 3). Therefore, the combination of micro-CT and simulation is a reliable method to reveal shale porosity and permeability.

**Table 3.** The porosity and permeability of the sample with typical equipment.

| Sample | S41 | S189 | S201 | S317 | S353 | Average |
|---|---|---|---|---|---|---|
| Porosity % | 9.39 | 11.26 | 1.22 | 5.11 | 10.48 | 7.492 |
| Permeability ($10^{-3}$ $\mu m^2$) | 1.563 | 2.211 | 0.054 | 1.231 | 1.321 | 1.276 |

Based on the above summary, the low saturation and mobility of oil and gas in shale makes it difficult to obtain plausible permeability in shale samples in conventional experimental studies (i.e., gas measurements of plunger core, gas analysis of particle sample, and the use of mercury (Hg) intrusion curves). However, the combination of micro-CT and simulation calculation could obtain credible total porosity, continuous porosity, and permeability of shale samples, while avoiding the consumption caused by permeability experiments.

## 5. Discussions

### 5.1. Relationship between Porosity and Permeability

Previous studies have shown that porosity characterizes the storage capacity of the reservoir rock [35]. However, if the porosity of the tight reservoirs is too large, the pore-throat ratio increases, which makes it difficult to establish an effective driving pressure difference during production. Therefore, it is necessary to comprehensively evaluate the porosity of the rock samples from the perspective of reservoir property and flow behavior during reservoir quality evaluation [35]. In this study, we found that the total porosity and the continuous porosity have a linear relationship (Figure 6a), which means higher total porosity leads to more developed continuous pores. Total porosity and continuous porosity were both significantly positively correlated with permeability (Figure 6b,c), which indicates higher values of total porosity and continuous porosity suggested better permeability. Findings from this study are also consistent with the results from Burnham (2017) [36], which studies shale samples from the Green River Formation. In addition, our study found no positive correlation between the total porosity of the studies samples and the burial depth (Figure 6d), which indicates the effect of formation compaction on the porosity of shale in the study area is not obvious. The possible reason is that the hydrocarbon generation of organic matter in the studied strata generates strong pressure, which could counteract the compaction of the overlying strata [37].

### 5.2. Relationship between Porosity and Mineral Content

Previous studies have explained that clay minerals are the main constituent component of shale and closely control the occurrence and enrichment of shale plays [38,39]. Considering clay minerals, their special crystal structure causes different types of pores to form between their crystal layers, also creating inter and intraparticle pore spaces. Furthermore, size, morphology and specific surface area of these pores determine the shale oil storage capacity. Previously it was documented that clay minerals are mostly composed of different types of porous structures, i.e., montmorillonite mostly develops in a circular and slit-like meso/micropores and has the largest total specific surface area. As the burial temperature increases, it transforms into mixed layers of other clay minerals, specifically illite, and the number of pores in the corresponding clay minerals gradually decreases [39]. Kaolinite is dominated by medium and large primary pores of 20–100 nm in size, which can be altered into illite in an alkaline environment [39]. Mesopores and macropores are mostly developed in illite and kaolinite. Additionally, it was found that the content of clay minerals in the same TOC range has a positive correlation with the pore volume and pore specific surface area of the organic-rich shale [40], and the pores between clay minerals can be filled with organic matter that migrates over a short distance. Therefore, pores of organo–clay mineral nanocomponents and organo–clay mineral complexes [41,42] can be considered as the main contributors to shale pore structure. Therefore, in general, pores of clay minerals are in general very well developed [43,44], which is the main controlling factor for the development of various types of pores in shale. In addition, previous studies found that illite and kaolinite are the main clay types in the shales of the Qingshankou Formation of the Gulong Sag [45]. Comparing our results with previous studies confirm that the total and continuous porosity of the samples in this study have a linear relationship with the content of clay minerals (Figure 7a,b), which means clay minerals are providing adequate pore space to the for the fluid to flow as well as storing the generated hydrocarbons. Our findings echo, to some extent, those of previous studies that suggested the variation of shale permeability depends on the nature of the clay mineral surface [46], the presence of a large number of pore structures in typical clay minerals (i.e., illite and chlorite) in shale has a positive effect on permeability [47]. Notably, our porosity values are lower than the typical porosity range of shale in the depth, according to porosity data compilation in Kim et al. [48], which could be attributed to a sequential microquartz cementation process and the quartz cement preferentially blocked the small mudstone pores during diagenesis [49].

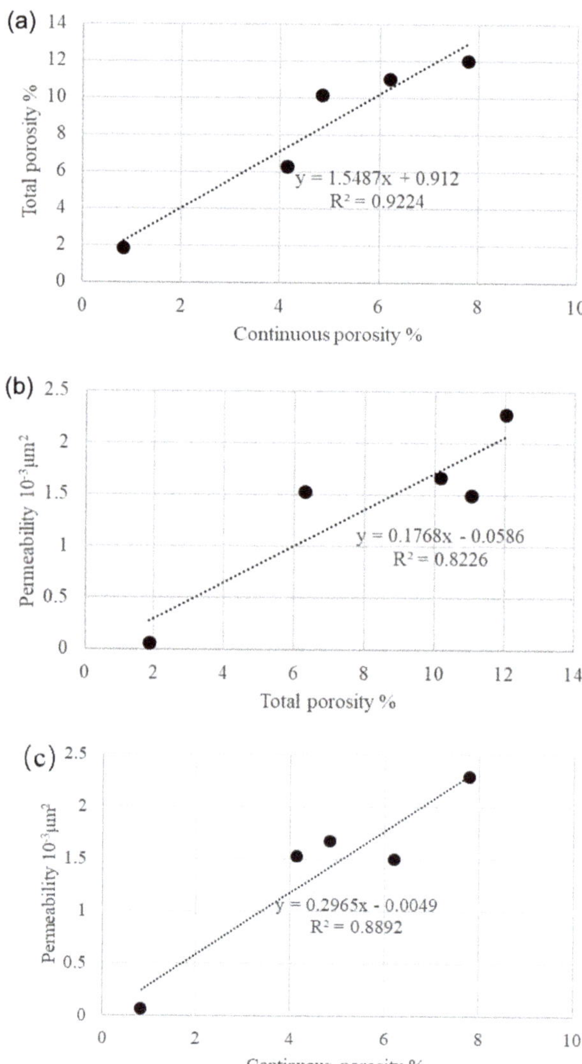

**Figure 6.** *Cont.*

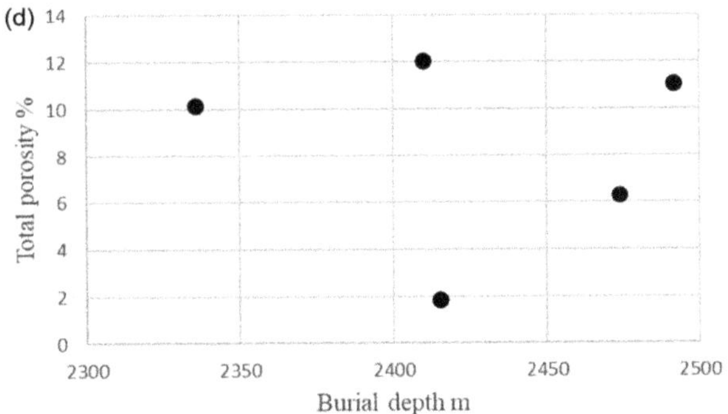

**Figure 6.** Relationship between porosity and permeability in studied samples. (**a**) Relationship between total porosity and continuous porosity of the studied samples. (**b**) The crossplot of total porosity versus permeability for the studied samples. (**c**) The relationship between continuous porosity and permeability of the studied samples. (**d**) The relationship between total porosity and burial depth of the studied samples.

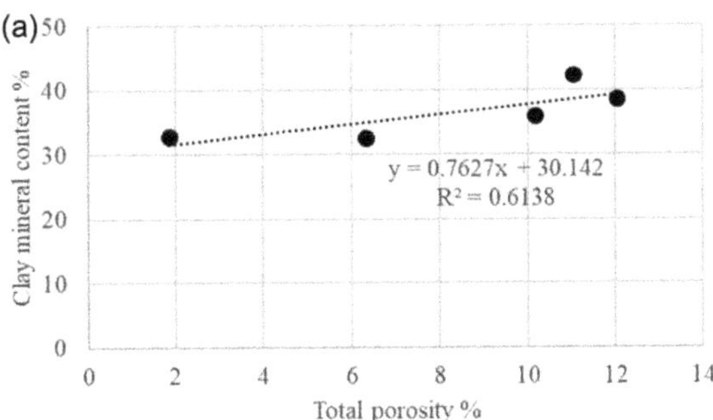

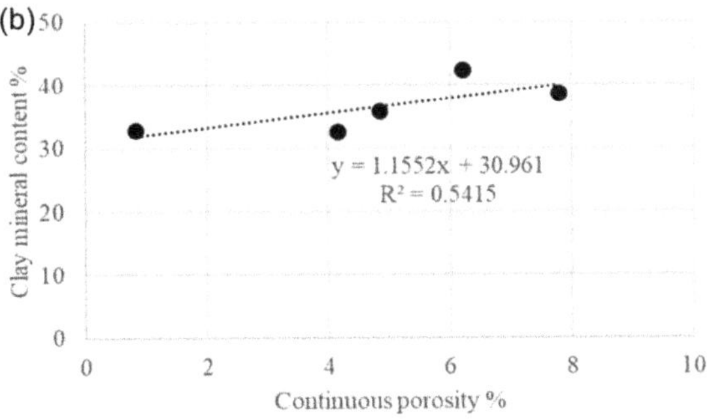

**Figure 7.** (**a**) Crossplot of total porosity versus clay content for the studied samples. (**b**) Continuous porosity versus clay content of the studied samples.

### 5.3. Relationship between Porosity and Oil-Bearing $S_1$

High-yield shale oil formations are often accompanied by a variety of minerals such as kaolinite and $Fe_2O_3$ that can promote hydrocarbon generation and transformation [50]. Natural attapulgite, kaolinite, clinoptilolite, and other minerals can catalyze the in situ upgrading of oil shale. Moreover, clay minerals are naturally micro-mesoporous materials with good thermochemical stability and are widely used as catalyst carriers and adsorbents [51,52], which not only improves the hydrocarbon production from shale oil but also reduces the activation energy [53–56], promoting the thermal maturity of oil shale kerogen and increasing the oil generation from the source rock. In this regard, simulation studies argue that montmorillonite has a catalytic effect on hydrocarbon generation from the organic matter [57], while the catalytic ability of illites is relatively small. These clay-based catalysts have good thermal stability and a simple development process and show significant potential in improving oil shale conversion efficiency and hydrocarbon yield [58] which makes us to conclude that the value of $S_1$ with the amount of clay minerals in the samples should be closely related.

Shale oil mainly accumulates in matrix pores, and there is a positive correlation between pore development and the S1 peak [40,43]. Clay minerals have been shown to have a high adsorption capacity for shale oil [59]. In this regard, hydrocarbons show different adsorption properties in the pore structure of clay minerals where the adsorption capacity of montmorillonite for hydrocarbons is stronger than that of illite and kaolinite. In addition to the pore size of the adsorbent, the structure and morphology of hydrocarbon molecules also strongly affects the adsorption behavior [56]. However, clay mineral content and pore volume are negatively correlated, according to some studies, which indicates developing large pores or enriching movable oil in clay is not possible [60].

Here, we found that the oil content of the samples ($S_1$) has a good linear relationship with the total and continuous porosity as well as the permeability and the clay content of the samples (Figure 8). Therefore, we can conclude the pores that are developed by clay minerals are the main storage space for oil in these samples.

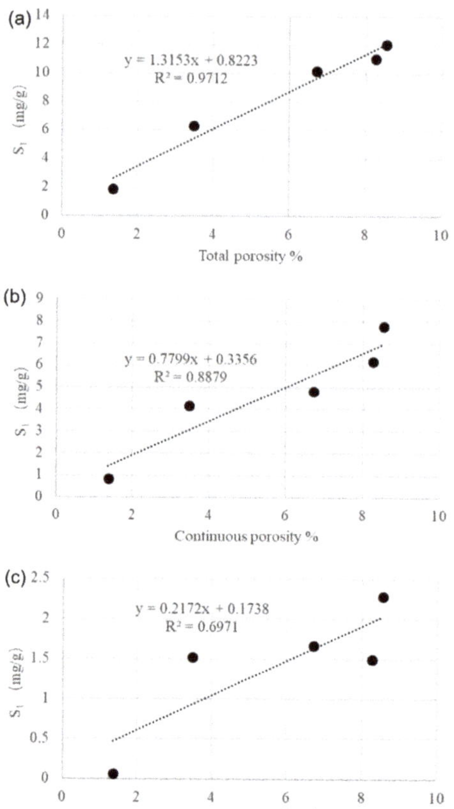


**Figure 8.** *Cont.*

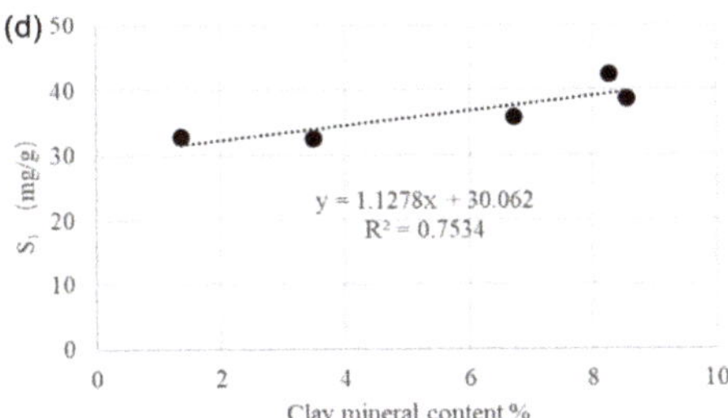

Figure 8. (a) The relationship between total porosity and oil content, $S_1$ peak of the studied samples. (b) The relationship between continuous porosity and the oil-bearing capacity, $S_1$, of the studied samples. (c) The relationship between permeability and the oil content ($S_1$) of the studied samples. (d) The relationship between clay mineral content and the oil content ($S_1$) of the studied samples.

## 6. Conclusions

In order to obtain the pore and permeability characteristics of the shales in the Qingshankou Formation and their relationship with oil-bearing, micro-CT and simulation calculations, instead of traditional experiments, were performed on five typical samples. Based on the results the following conclusions can be made:

(1)  The permeability obtained using micro-CT and simulation approximate those obtained with typical equipment. The combination of micro-CT and simulation in this study is reliable for revealing shale reservoir characteristics.

(2)  The continuous porosity of Qingshankou shale in the Songliao Basin was found between 0.84 and 7.79% (average 4.76%), the total porosity between 1.87 and 12.03% (average 8.28%), and the absolute permeability was calculated between 0.061 and $2.284 \times 10^{-3}$ $\mu m^2$. The total porosity of the Qingshankou shale in the Songliao Basin has a good positive correlation with the continuous porosity and permeability. Higher the total porosity in the study area denotes better development of continuous pores, thus higher permeability.

(3)  Both total and continuous porosity of the samples are positively correlated with the content of clay minerals. The pores developed by clay minerals are the main space for oil storage in these shale samples, and the amount of clay minerals would be the main controlling parameter that impacts samples porosity, permeability, and oil content in the study area.

**Author Contributions:** Conceptualization, Y.C.; Data curation, Y.C. and R.Z.; Formal analysis, Y.C.; Funding acquisition, Z.J.; Investigation, Z.J. and R.Z.; Methodology, Q.W. and R.Z.; Resources, Z.J. and R.Z.; Software, Q.W.; Supervision, Z.J.; Writing—original draft, Y.C.; Writing—review & editing, Y.C. All authors have read and agreed to the published version of the manuscript.

**Funding:** This study was jointly funded by the National Science Foundation of China (42090020/42090025) and the 2022 American Association of Petroleum Geologists Foundation Grants-in-Aid Program (Grants-in-Aid General Fund Grant).

**Conflicts of Interest:** The authors declare no conflict of interest.

## References

1.  Cao, Y.; Han, H.; Liu, H.-W.; Jia, J.-C.; Zhang, W.; Liu, P.-W.; Ding, Z.-G.; Chen, S.-J.; Lu, J.-G.; Gao, Y. Influence of solvents on pore structure and methane adsorption capacity of lacustrine shales: An example from a Chang 7 shale sample in the Ordos Basin, China. *J. Pet. Sci. Eng.* **2019**, *178*, 419–428. [CrossRef]

2.  Loucks, R.G.; Reed, R.M.; Ruppel, S.C.; Hammes, U. Spectrum of pore types and networks in mudrocks and a descriptive classification for matrix-related mudrock pores. *AAPG Bull.* **2012**, *96*, 1071–1098. [CrossRef]

3.  Zou, C.N.; Pan, S.Q.; Jing, Z.H.; Gao, J.L.; Yang, Z.; Wu, S.T.; Zhao, Q. Shale oil and gas revolution and its impact (in Chinese with English abstract). *Acta Petrol. Sin.* **2020**, *41*, 1–12.

4.   Jin, X.; Li, G.X.; Meng, S.W.; Wang, X.Q.; Liu, C.; Tao, J.P.; Liu, H. Microscale comprehensive evaluation of continental shale oil recoverability (in Chinese with English abstract). *Pet. Explor. Dev.* **2021**, *48*, 256–268. [CrossRef]

5.   Loucks, R.G.; Reed, R.M.; Ruppel, S.C.; Jarvie, D.M. Morphology, Genesis, and Distribution of Nanometer-Scale Pores in Siliceous Mudstones of the Mississippian Barnett Shale. *J. Sediment. Res.* **2009**, *79*, 848–861. [CrossRef]

6.   Li, C.X.; Ostadhassan, M.; Guo, S.; Gentzis, T.; Kong, L.Y. Application of PeakForce tapping mode of atomic force microscope to characterize nanomechanical properties of organic matter of the Bakken Shale. *Fuel* **2018**, *233*, 894–910. [CrossRef]

7.   Clarkson, C.R.; Solano, N.; Bustin, R.M.; Bustin, A.M.M.; Chalmers, G.R.L.; He, L.; Melnichenko, Y.B.; Radliński, A.P.; Blach, T.P. Pore structure characterization of North American shale gas reservoirs using USANS/SANS, gas adsorption, and mer-cury intrusion. *Fuel* **2013**, *103*, 606–616. [CrossRef]

8.   Han, H.; Zhong, N.N.; Ma, Y.; Huang, C.X.; Wang, Q.; Chen, S.J.; Lu, J.G. Gas storage and controlling factors in an over-mature marine shale: A case study of the Lower Cambrian Lujiaping shale in the Dabashan arc-like thrust–fold belt, southwestern China. *J. Nat. Gas Sci. Eng.* **2016**, *33*, 839–853. [CrossRef]

9.   Sisk, C.; Diaz, E.; Walls, J.; Grader, A.; Suhrer, M. 3D Visualization and Classification of Pore Structure and Pore Filling in Gas Shales, SPE-paper 134582. In Proceedings of the SPE Annual Technical Conference and Exhibition, Florence, Italy, 19–22 September 2010.

10.   Walls, J.D.; Diaz, E.; Cavanaugh, T. Shale reservoir properties from digital rock physics, SPE-Paper 152752. In Proceedings of the SPE/EAGE European Unconventional Resources Conference and Exhibition, Vienna, Austria, 20–22 March 2012.

11.   Sun, M.; Yu, B.; Hu, Q.; Yang, R.; Zhang, Y.; Li, B.; Melnichenko, Y.B.; Cheng, G. Pore structure characterization of organ-ic-rich Niutitang shale from China: Small angle neutron scattering (SANS) study. *Int. J. Coal Geol.* **2018**, *186*, 115–125. [CrossRef]

12.   Wang, Y.; Pu, J.; Wang, L.; Wang, J.; Jiang, Z.; Song, Y.-F.; Wang, C.-C.; Wang, Y.F.; Jin, C. Characterization of typical 3D pore networks of Jiulaodong formation shale using nano-transmission X-ray microscopy. *Fuel* **2016**, *170*, 84–91. [CrossRef]

13.   Wu, K.L.; Chen, Z.X.; Li, J.; Li, X.F.; Xu, J.Z.; Dong, X.H. Wettability effect on nanoconfined water flow. *Proc. Natl. Acad. Sci. USA* **2017**, *114*, 3358–3363. [CrossRef] [PubMed]

14.   Wang, H.; Su, Y.L.; Wang, W.D.; Sheng, G.L.; Li, H.; Zafar, A. Enhanced water flow and apparent viscosity model consid-ering wettability and shape effects. *Fuel* **2019**, *253*, 1351–1360. [CrossRef]

15.   Wang, H.; Su, Y.L.; Zhao, Z.F.; Wang, W.D.; Sheng, G.L.; Zhan, S.Y. Apparent permeability model for shale oil transport through elliptic nanopores considering wall-oil interaction. *J. Pet. Sci. Eng.* **2019**, *176*, 1041–1052. [CrossRef]

16.   Wang, H.; Su, Y.L.; Wang, W.D.; Li, L.; Sheng, G.L.; Zhan, S.Y. Relative permeability model of oil-water flow in nanoporous media considering multi-mechanisms. *J. Pet. Sci. Eng.* **2019**, *183*, 106361. [CrossRef]

17.   Zhang, T.; Li, X.F.; Yin, Y.; He, M.X.; Liu, Q.; Huang, L.; Shi, J.T. The transport behaviors of oil in nanopores and nanoporous media of shale. *Fuel* **2019**, *242*, 305–315. [CrossRef]

18.   Cui, J.F.; Wu, K.L. Equivalent permeability of shale rocks: Simple and accurate empirical coupling of organic and inorganic matter. *Chem. Eng. Sci.* **2020**, *216*, 115491. [CrossRef]

19.   Javadpour, F.; Singh, H.; Rabbani, A.; Babaei, M.; Enayati, S. Gas Flow Models of Shale: A Review. *Energy Fuels* **2021**, *35*, 2999–3010. [CrossRef]

20.   Cui, X.; Bustin, A.M.M.; Bustin, R.M. Measurements of gas permeability and diffusivity of tight reservoir rocks: Different ap-proaches and their applications. *Geofluids* **2009**, *9*, 208–223. [CrossRef]

21.   Brace, W.F.; Walsh, J.B.; Frangos, W.T. Permeability of granite under high pressure. *J. Geophys. Res.* **1968**, *73*, 2225–2236. [CrossRef]

22.   Egermann, P.; Lenormand, R.; Longeron, D.; Zarcone, C. A fast and direct method of permeability measurements on drill cuttings. *Soc. Petrol. Eng. Reserv. Eval. Eng.* **2005**, *4*, 269–275. [CrossRef]

23.   American Petroleum Institute (API). *Recommended Practices for Core Analysis—Recommended Practice 40*, 2nd ed.; API: Dallas, TX, USA, 1998.

24.   Kamath, J. Evaluation of Accuracy of Estimating Air Permeability From Mercury- Injection Data. *SPE Form. Eval.* **1992**, *7*, 304–310. [CrossRef]

25.   Carles, P.; Egermann, P.; Lenormand, R.; Lombard, J.M. Low permeability measurements using steady-state and transient methods, Paper SCA2007-07. In Proceedings of the 2007 International Symposium of the Society of Core Analysis, Calgary, AB, Canada, 10–12 September 2007.

26.   Swanson, B.F. A Simple Correlation between Permeabilities and Mercury Capillary Pressures. *J. Pet. Technol.* **1981**, *33*, 2498–2504. [CrossRef]

27.   Liu, B.; Wang, H.L.; Fu, X.F.; Bai, Y.F.; Bai, L.H.; Jia, M.C.; He, B. Lithofacies and depositional setting of a highly prospective lacustrine shale oil succession from the Upper Cretaceous Qingshankou Formation in the Gulong sag, northern Songliao Basin, northeast China. *AAPG Bull.* **2019**, *103*, 405–432. [CrossRef]

28.   ASTM. *Annual Book of ASTM Standards, 2010—Gaseous Fuel, Coal and Coke, 5.06*; ASTM International: West Conshohocken, PA, USA, 2010.

29.   Guo, H.J.; Jia, W.L.; Peng, P.A.; Lei, Y.H.; Luo, X.R.; Cheng, M.; Wang, X.Z.; Zhang, L.X.; Jiang, C.F. The composition and its impact on the methane sorption of lacustrine shales from the Upper Triassic Yanchang Formation, Ordos Basin, China. *Mar. Pet. Geol.* **2014**, *57*, 509–520. [CrossRef]

30.   Jiang, C.B.; Niu, B.W.; Yin, G.Z.; Zhang, D.M.; Yu, T.; Wang, P. CT-based 3D reconstruction of the geometry and propagation of hydraulic fracturing in shale. *J. Pet. Sci. Eng.* **2019**, *179*, 899–911. [CrossRef]

31. Sun, J.M.; Dong, X.; Wang, J.J.; Schmitt, D.R.; Xu, C.L.; Mohammed, T.; Chen, D.W. Measurement of total porosity for gas shales by gas injection porosimetry (GIP) method. *Fuel* **2016**, *186*, 694–707. [CrossRef]

32. Zhou, S.; Dong, D.; Zhang, J.; Zou, C.; Tian, C.; Rui, Y.; Liu, D.X.; Jiao, P.F. Optimization of key parameters for po-rosity measurement of shale gas reservoirs. *Nat. Gas Ind. B* **2021**, *8*, 455–463. [CrossRef]

33. Fisher, Q.; Lorinczi, P.; Grattoni, C.; Rybalcenko, K.; Crook, A.J.; Allshorn, S.; Burns, A.D.; Shafagh, I. Laboratory charac-ter-ization of the porosity and permeability of gas shales using the crushed shale method: Insights from experiments and numerical modelling. *Mar. Petrol. Geol.* **2017**, *86*, 95–110. [CrossRef]

34. Jones, S.C. A Technique for Faster Pulse-Decay Permeability Measurements in Tight Rocks. *SPE Form. Eval.* **1997**, *12*, 19–25. [CrossRef]

35. Pang, Y.; Soliman, M.Y.; Deng, H.; Emadi, H.C. Analysis of Effective Porosity and Effective Permeability in Shale-Gas Res-ervoirs With Consideration of Gas Adsorption and Stress Effects. *SPE J.* **2017**, *22*, 1739–1759. [CrossRef]

36. Burnham, A.K. Porosity and permeability of Green River oil shale and their changes during retorting. *Fuel* **2017**, *203*, 208–213. [CrossRef]

37. Price, L.C.; Wenger, L.M. The influence of pressure on petroleum generation and maturation as suggested by aqueous py-rolysis. *Org. Geochem.* **1992**, *19*, 141–159. [CrossRef]

38. Cao, Z.; Jiang, H.; Zeng, J.H.; Saibi, H.; Lu, T.Z.; Xie, X.M.; Zhang, Y.C.; Zhou, G.G.; Wu, K.Y.; Guo, J.R. Nanoscale liquid hydrocarbon adsorption on clay minerals: A molecular dynamics simulation of shale oils. *Chem. Eng. J.* **2021**, *420*, 127578. [CrossRef]

39. Liu, B.; Shi, J.X.; Fu, X.F.; Lyu, Y.F.; Sun, X.D.; Gong, L.; Bai, Y.F. Petrological characteristics and shale oil enrichment of la-custrine fine-grained sedimentary system: A case study of organic-rich shale in first member of Cretaceous Qingshankou Formation in Gulong Sag, Songliao Basin, NE China. *Pet. Explor. Dev.* **2018**, *45*, 884–894. [CrossRef]

40. Xu, L.; Chang, Q.S.; Feng, L.L.; Zhang, N.; Liu, H. The reservoir characteristics and control factors of shale oil in Permian 183 Fengcheng Formation of Mahu Sag, Junggar Basin. *China Pet. Explor.* **2019**, *24*, 649.

41. Ji, Y.Q.; Qiu, L.M.; Xia, J.L.; Zhang, T.W. Micro-pore characteristics and methane adsorption properties of common clay minerals by electron microscope scanning. *Acta Petrol. Sin.* **2012**, *33*, 249.

42. Wu, C.J.; Tuo, J.C.; Zhang, L.F.; Zhang, M.F.; Li, J.; Liu, Y.; Qian, Y. Pore characteristics differences between clay-rich and clay-poor shales of the Lower Cambrian Niutitang Formation in the Northern Guizhou area, and insights into shale gas storage mechanisms. *Int. J. Coal Geol.* **2017**, *178*, 13–25. [CrossRef]

43. Gao, B.; He, W.Y.; Feng, Z.H.; Shao, H.M.; Zhang, A.D.; Pan, H.F.; Chen, G.L. Lithology, physical property, oil-bearing property and their controlling factors of Gulong shale in Songliao Basin (in Chinese with English abstract). *Pet. Geol. Oilfield Dev. Daqing* **2022**, *41*, 68–79.

44. Cao, Y.; Han, H.; Guo, C.; Pang, P.; Ding, Z.-G.; Gao, Y. Influence of extractable organic matters on pore structure and its evolution of Chang 7 member shales in the Ordos Basin, China: Implications from extractions using various solvents. *J. Nat. Gas Sci. Eng.* **2020**, *79*, 103370. [CrossRef]

45. Hou, L.-H.; Wu, S.-T.; Jing, Z.-H.; Jiang, X.-H.; Yu, Z.-C.; Hua, G.L.; Su, L.; Yu, C.; Liao, F.-R.; Tian, H. Effects of types and content of clay minerals on reservoir effectiveness for lacustrine organic matter rich shale. *Fuel* **2022**, *327*, 125043. [CrossRef]

46. Kwon, O.; Herbert, B.E.; Kronenberg, A.K. Permeability of illite-bearing shale: 2. Influence of fluid chemistry on flow and functionally connected pores. *J. Geophys. Res. Earth Surf.* **2004**, *109*. [CrossRef]

47. Meng, Z.; Sun, W.; Liu, Y.; Luo, B.; Zhao, M. Effect of pore networks on the properties of movable fluids in tight sandstones from the perspective of multi-techniques. *J. Pet. Sci. Eng.* **2021**, *201*, 108449. [CrossRef]

48. Kim, Y.; Lee, C.; Lee, E.Y. Numerical analysis of sedimentary compaction: Implications for porosity and layer thickness variation. *J. Geol. Soc. Korea* **2018**, *54*, 631–640. [CrossRef]

49. Thyberg, B.; Jahren, J. Quartz cementation in mudstones: Sheet-like quartz cement from clay mineral reactions during burial. *Pet. Geosci.* **2011**, *17*, 53–63. [CrossRef]

50. Lai, D.G.; Chen, Z.H.; Lin, L.X.; Zhang, Y.M.; Gao, S.Q.; Xu, G.W. Secondary cracking and upgrading of shale oil from py-ro-lyzing oil shale over shale ash. *Energy Fuels* **2015**, *29*, 2219–2226. [CrossRef]

51. Ariskina, K.A.; Yuan, C.D.; Abaas, M.; Emelianov, D.A.; Rodionov, N.; Varfolomeev, M.A. Catalytic effect of clay rocks as natural catalysts on the combustion of heavy oil. *Appl. Clay Sci.* **2020**, *193*, 105662. [CrossRef]

52. Zhou, C.H. An overview on strategies towards clay-based designer catalysts for green and sustainable catalysis. *Appl. Clay Sci.* **2011**, *53*, 87–96. [CrossRef]

53. Rahman, H.M.; Kennedy, M.; Löhr, S.; Dewhurst, D.N.; Sherwood, N.; Yang, S.Y.; Horsfield, B. The influence of shale depo-sitional fabric on the kinetics of hydrocarbon generation through control of mineral surface contact area on clay catalysis. *Geochim. Cosmochim. Acta* **2018**, *220*, 429–448. [CrossRef]

54. Zhu, X.J.; Cai, J.G.; Wang, G.L.; Song, M.S. Role of organo-clay composites in hydrocarbon generation of shale. *Int. J. Coal Geol.* **2018**, *192*, 83–90. [CrossRef]

55. Wei, H.B.; Zhang, Y.P.; Cui, J.H.; Han, L.L.; Li, Z.Q. Engineering and environmental evaluation of silty clay modified by waste fly ash and oil shale ash as a road subgrade material. *Constr. Build. Mater.* **2019**, *196*, 204–213. [CrossRef]

56. Williams, P.F. Thermogravimetry and decomposition kinetics of British Kimmeridge Clay oil shale. *Fuel* **1985**, *64*, 540–545. [CrossRef]

57. Jurg, J.W.; Eisma, E. Petroleum Hydrocarbons: Generation from Fatty Acid. *Science* **1964**, *144*, 1451–1452. [CrossRef] [PubMed]
58. Song, R.R.; Meng, X.L.; Yu, C.; Bian, J.J.; Su, J.Z. Oil shale in-situ upgrading with natural clay-based catalysts: Enhancement of oil yield and quality. *Fuel* **2022**, *314*, 123076. [CrossRef]
59. Wu, T.; Pan, Z.; Liu, B.; Connell, L.D.; Sander, R.; Fu, X. Laboratory characterization of shale oil storage behavior: A comprehensive review. *Energy Fuels* **2021**, *35*, 7305–7318. [CrossRef]
60. Su, S.; Jiang, Z.; Xuanlong, S.; Zhang, C.; Zou, Q.; Li, Z.; Zhu, R. The effects of shale pore structure and mineral components on shale oil accumulation in the Zhanhua Sag, Jiyang Depression, Bohai Bay Basin, China. *J. Pet. Sci. Eng.* **2018**, *165*, 365–374. [CrossRef]

*Article*

# FTCN: A Reservoir Parameter Prediction Method Based on a Fusional Temporal Convolutional Network

**Hongxia Zhang [1,*], Kaijie Fu [1], Zhihao Lv [1], Zhe Wang [2], Jiqiang Shi [3], Huawei Yu [2,4] and Xinmin Ge [2,4]**

[1] Qingdao Institute of Software, College of Computer Science and Technology, Qingdao 266580, China
[2] School of Geosciences, China University of Petroleum (East China), Qingdao 266580, China
[3] Geophysical Research Institute of Shengli Oilfeld Branch, Sinopec, Dongying 257022, China
[4] Shandong Provincial Key Laboratory of Deep Oil & Gas, China University of Petroleum (East China), Qingdao 266580, China
* Correspondence: zhanghx@upc.edu.cn

**Abstract:** Predicting reservoir parameters accurately is of great significance in petroleum exploration and development. In this paper, we propose a reservoir parameter prediction method named a fusional temporal convolutional network (FTCN). Specifically, we first analyze the relationship between logging curves and reservoir parameters. Then, we build a temporal convolutional network and design a fusion module to improve the prediction results in curve inflection points, which integrates characteristics of the shallow convolution layer and the deep temporal convolution network. Finally, we conduct experiments on real logging datasets. The results indicate that compared with the baseline method, the mean square errors of FTCN are reduced by 0.23, 0.24 and 0.25 in predicting porosity, permeability, and water saturation, respectively, which shows that our method is more consistent with the actual reservoir geological conditions. Our innovation is that we propose a new reservoir parameter prediction method and introduce the fusion module in the model innovatively. Our main contribution is that this method can well predict reservoir parameters even when there are great changes in formation properties. Our research work can provide a reference for reservoir analysis, which is conducive to logging interpreters' efforts to analyze rock strata and identify oil and gas resources.

**Keywords:** reservoir parameter prediction; temporal convolutional network; porosity; permeability; water saturation

**Citation:** Zhang, H.; Fu, K.; Lv, Z.; Wang, Z.; Shi, J.; Yu, H.; Ge, X. FTCN: A Reservoir Parameter Prediction Method Based on a Fusional Temporal Convolutional Network. *Energies* **2022**, *15*, 5680. https://doi.org/10.3390/en15155680

Academic Editors: Xingguang Xu, Kun Xie and Yang Yang

Received: 14 July 2022
Accepted: 2 August 2022
Published: 5 August 2022

**Publisher's Note:** MDPI stays neutral with regard to jurisdictional claims in published maps and institutional affiliations.



## 1. Introduction

Reservoir parameters are very important in petroleum exploration and development and also a significant reference foundation to analyze reservoir geology and evaluate oil and gas reservoirs accurately. In the actual exploitation process, obtaining reservoir parameters is expensive from core data, and the amount of data obtained is limited. At the same time, the actual development environment is changeable, and the underground geological situation is complex and diverse. Affected by the original data, logging cost, the level of explorers, the empirical coefficients, response logging curve selection, heterogeneous formation, depositional environment and tectonic location, it becomes extremely complex to obtain accurate reservoir parameters.

In recent years, artificial intelligence technology has provided a possibility for intelligent exploration [1]. Deep learning methods such as BP (back propagation) network [2], recurrent neural network (RNN) [3], long short-term memory (LSTM) [4] and gated recurrent unit (GRU) [5] have been applied to petroleum field by many researchers. Dos [6] proposed a computational system based on deep recurrent neural networks (RNNs) as an effective method to automatically identify lithofacies patterns from well logs. For forecasting petroleum production, a novel method based on a gated recurrent neural network has been proposed. It has multiple hidden layers, and each layer has a number of nodes.

The robustness of this model is very good [7]. Heghedus [8] concentrated on pressure-rate datasets accumulated with massive installation of permanent downhole gauge production and injection wells based on an LSTM network. The research results provide a basis for filling gaps in well monitoring data. Meanwhile, it also becomes a research hotspot for intelligent reservoir parameter prediction.

For the sake of enhancing the prediction results of reservoirs in intricate geology effectively, a new parameter prediction method named the fusional temporal convolutional network (FTCN) has been proposed. Firstly, the relationship between logging curves and reservoir parameters is analyzed deeply, and the logging curves, which are closely related to reservoir parameters, are obtained. We use the selected correlation logging curves to predict the reservoir parameters. Secondly, we present a fusion module on TCN to improve the affection of reservoir parameter prediction innovatively. We weigh the output of different network layers and combine the output to enrich the data characteristics obtained by the predictive network model. A fusion module is raised to utilize the information from different network layers adequately, which decreases the large deviation in the fluctuation of local peak values. Finally, experiments are set up on the actual logging data. Our method provides better support to understand and analyze reservoir conditions technically and thus provides a novel reference to the exact interpretation of the logging data.

There are a lot of measurement data in the petroleum industry. These data usually contain momentous information describing the characteristics of strata and reservoir properties and play an important role in production and reservoir management. Logging records provide a data source for logging interpretation experts to analyze reservoir properties. Logging experts can obtain the information of reservoir characteristics by analyzing logging records. Reservoir parameters are exceedingly significant for logging interpreters to analyze formation properties and reservoir capacity and are also the foundation for fine logging reservoir evaluation and analysis of oil-gas. Reservoir parameters, such as porosity, permeability and water saturation, reflect the reservoir's storing ability. Generally, the greater the porosity is, the more likely it is to store oil and natural gas in the pores. The higher the permeability is, the stronger the fluidity of oil, and the easier it is to be exploited. Predicting reservoir parameters effectively can provide a reference for analyzing reservoir properties, characterizing oil reservoirs and providing accurate interpretation of well logging data, which can assist logging interpreters in judging formation conditions and evaluating oil-gas potential. Thus, it provides a support for reserve calculation, flow unit identification and reservoir evaluation. Using the effective porosity parameter of the fracture and cavity reservoirs in combination with the effective thickness, oil-bearing area and other reserve calculation parameters, Kuanzhi [9] formulated a reserve estimation scheme for fractured vuggy carbonate reservoirs so as to guide the exploration and development of oil and gas reservoirs. In the research of reservoir flow unit identification in the North Rumaila Oilfield, according to the logging curve similarity, the porosity and permeability crossplot, the capillary pressure data, the porosity and water saturation and depth relationship and the flow zone indication method, Al-Jawad [10] subdivided the primary reservoir units in the oilfield, interpreted and classified the sub units and thus identified the good reservoirs. Therefore, reservoir parameters play an important role in the exploration and development of the petroleum industry.

The main contributions of this paper are as follows:

(1)     Our method can predict reservoir porosity, permeability and water saturation effectively. It is beneficial for logging interpreters to analyze rock strata and identify oil and gas resources;

(2)     A new reservoir parameter prediction method named FTCN is proposed. We introduce a fusion module based on TCN to utilize the information from different network layers adequately, which improves the effect of curve prediction;

(3)     We have conducted various experiments on the real logging dataset to illustrate the effectiveness of this method. FTCN can predict well where the log curve changes abruptly and is effective when the properties of underground strata change greatly.

## 2. Previous Work

Complex reservoir analysis is a significant field in oil reservoir description. Well logging data involves abundant geological information. By analyzing logging data, logging interpreters can judge stratum properties and identify oil and gas reservoirs. In recent years, with the incessant development of machine learning, there have been many outstanding works in the field of engineering applications of deep learning or convolutional networks. For example, a novel method based on deep convolutional neural networks to identify and localise damages to building structures equipped with smart control devices has been proposed [11]. In addition, Yu [12] developed a vision-based crack diagnosis method using a deep convolutional neural network (DCNN) and enhanced chicken swarm algorithm (ECSA). It has also been applied in the domain of petroleum logging successfully [13]. Scholars have done a great deal of research and achieved many achievements in the integration of oil exploration and development and artificial intelligence [14]. For example, the porosity classification and quantification scheme [15] mainly introduces a thorough understanding of the carbonate pore system, which is essential to hydrocarbon prospecting and the prediction of petroleum reservoir properties [16]. Other examples include sedimentary facies classification, reservoir evaluation [17], and so on.

A variety of methods have been proposed to solve the problem of reservoir parameter prediction, which continuously promotes the development of reservoir analysis and logging interpretation technology and provides an important basis for geological experts to analyze reservoirs. In the early stage, based on years of experience in the analysis of reservoir parameters and geological conditions, researchers [18–20] established many empirical formulas to determine the reservoir parameters in the research field. However, because the geological conditions of the new and old exploration areas are different, there will be great differences. A new, undeveloped study area is likely to have rich potential oil and gas resources and good reservoir physical properties. However, in an old study area, due to long-term development, the physical properties and lithology of the reservoir will have changed, and the geological conditions become very poor, which brings challenges to the development of the remaining oil and gas resources. Some empirical coefficient values cannot be generalized, and the formula is also affected by the subjective factors of logging interpreters. Thus, the results are uncertain. Empirical formulas can only be used as a reference. In addition, cross plots can also be used to analyze reservoir properties in exploration and development [21]. The cross plot draws a two-factor or multi-factor rendezvous map using logging curve readings or calculation parameters. Geological experts interpret geological models and analyze and evaluate strata according to the observation of cross plots, which is also uncertain. Therefore, the above methods are influenced by the subjective factors of logging interpreters and the great changes in geological conditions, so the accuracy of the reservoir parameter prediction needs to be improved.

As mature oilfields turn into a later exploitation period of the ultra-high water cut stage, the geological situation becomes complex and changeable, and the quality of oil resources gradually deteriorates [22]. The search for oil and gas fields with complex reservoirs has become difficult, and traditional methods have been unable to meet the demand. The development of machine learning technology makes it possible to improve the effect of reservoir parameter prediction [23].

Through the analysis of tight sandstone reservoirs, Zhu [24] considered that the clay content, the irreducible water saturation, the porosity and the diagenetic coefficient were important factors affecting the reservoir parameter permeability. The studied samples of the model were selected based on the representative core analysis data. According to these influencing factors and samples, permeability was predicted based on an improved BP neural network. Mahdaviara [25] pointed out that the prediction of permeability was a challenge in carbonate heterogeneous rock and built a model to predict reservoir permeability based on Gaussian process regression. The evaluation of permeability in the southern Yellow Sea basin showed that it can be used as a supplement to the neural network prediction methods. In addition, researchers [26,27] have used machine learning methods, such as support-vector machines, particle swarm optimization algorithm [28], and artificial neural networks [29–31] to study and analyze reservoir parameters, achieving good results.

As a research hotspot in the domain of machine learning, deep learning has achieved fruitful results in many fields, such as agriculture [32], ultrasound imaging [33], smart cities [34] and so on. Many researchers have applied deep learning to the field of petroleum exploration and development [35,36]. In the study of predicting reservoir parameters, deep learning technology has been combined to improve the prediction accuracy. As two significant parameters of the oil and gas storage, porosity and shale content express the sedimentary characteristics of various historical stages and have an intense nonlinear mapping with logging parameters. Deep learning has a powerful data mining capacity. Therefore, AN [37] applied an LSTM network to predict the shale content and porosity of a reservoir. The prediction accuracy of this network was more superior than the conventional deep neural network. The hardship of gaining porosity increases gradually with increasing drilling depth, and the cost for gaining intact porosity by the conventional coring method is large comparatively. Thus, for the sake of achieving low-cost and high-efficiency porosity prediction, Chen [38] proposed a logging method found on a multi-layer LSTM network, which performs well for logging at different depths and predicts the changing trend of porosity in strata effectively. The logging curves gained from deep to shallow stratum indicate the sedimentary features of distinct geological stages. The porosity, as a vital reservoir parameter, reflects the capacity of the oil and gas storage. It is very meaningful for the exact description of a reservoir to use logging parameters to acquire reservoir porosity [39]. The application effect in a certain research region of the Ordos basin showed that a gated recurrent unit (GRU) network combined with various logging curves was more effective in predicting reservoir porosity than multiple regression analysis as well as RNN. In addition, convolution structures [40,41] also have certain advantages in predicting sequence tasks. However, although the above methods solve the problem of reservoir parameter prediction in some practical areas effectively, the effect on geological complex reservoir prediction is general, such as in an old oilfield with serious water flooding and intense inhomogeneity. The structure of the reservoir sand body is loose, and the lithology is complex. Development is difficult, and the effect needs to be further improved. The generalization of the model is limited to a certain extent, so these methods have some limitations in practical application and cannot meet the requirements of all kinds of fine reservoir prediction.

## 3. Methodology

LSTM [4] was first proposed by Hochreiter and Schmidhuber to solve the long-term dependence problem of general RNN, which can avoid gradient vanishing and gradient exploding. GRU [5] is an important variant of LSTM. It improves the design of gates in LSTM and optimizes the forgetting gate, input gate and output gate in LSTM into two gates called the reset gate and update gate, respectively. A temporal convolutional network (TCN) [40] is a special convolution network that has the advantages of a flexible receptive field and stable gradient.

### 3.1. FTCN Network Model

For the purpose of resolving problems of the resulting uncertainty, reservoir area limitation and low prediction accuracy, a fusional temporal convolutional network (FTCN) based on TCN is proposed in this paper by digging into the nonlinear relationship between logging curves and reservoir parameters in complex reservoirs. Firstly, the input curves are optimized by selecting the logging curves, which are sensitive to porosity, permeability and water saturation, and excluding the non-correlation curves. Then, a fusion module is designed to improve the prediction results of the inflection point of the curve, which can reduce the local deviation of reservoir parameter prediction parameters efficiently.

The framework of FTCN is shown in Figure 1. A variety of logging curves are preprocessed. The middle part is the main structure of the prediction network. The right part shows the optimization of the network. Specifically, the original logging data mainly include acoustic travel time, density, compensated neutron logging, natural gamma ray, spontaneous potential, micro-potential resistivity, micro-gradient resistivity, and so on. We use the numerical values of these logging curves as the input data of the network model.

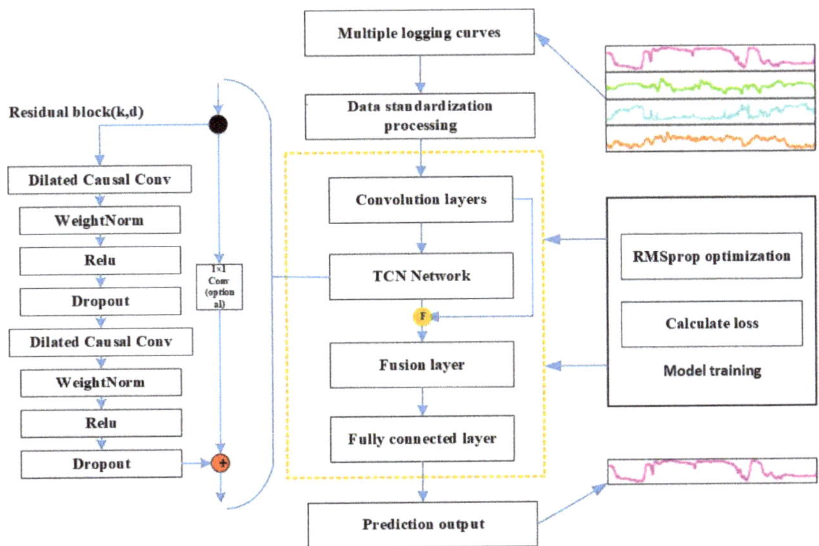


**Figure 1.** Framework of reservoir parameter prediction model named FTCN.

There are great differences in the value range of different logging curve data. For the sake of reducing the effect of different dimensions of the original logging data, the data are standardized and preprocessed as

$$y_{ij} = \frac{x_{ij} - \mu_i}{s_i},\tag{1}$$

where $i$ is the $i$th kind of curve parameter, $j$ is the $j$th sample of the curve parameter, $x_{ij}$ denotes the original data, $y_{ij}$ denotes the standardized data, and $\mu_i$ and $s_i$ are the mean and standard deviation of data, respectively.

The standardized data are used as the input data of the predictive network model and first enter the convolution layers, where convolution and pooling operations are carried out, which are mainly used to obtain the low-level features of the network. Then, they enter the TCN network, including dilated causal convolution and residual connection blocks, and the deep-level features of the data are obtained through the TCN network. Then, they enter the fusional module. After passing the network fusional module and then going through the full connection layer, the predicted reservoir parameters are output. In the process of adjusting the training network, we chose the RMSProp algorithm [42] with adaptive learning rate. Through continuous iterative training, the reservoir parameter prediction model is established.

### 3.2. Fusional Module

The variation of well logging curves reflects the change in reservoir parameters in some ways, and it has a good correlation among adjacent wells in the same horizon. Convolution networks have strong data mining abilities. Dilated convolution expands the receptive field by introducing a dilation factor to the convolution. This dilation factor defines the distance between values when the network processes data. The dilated convolution obtains the information farther from the current input by skipping part of the input value. In order to make the network remember more effective information, we introduce dilated convolution that can expand the receptive field into our network. Thus, the network can pay more attention to global information and capture more feature information. However, with the increase in network depth, dilated convolution will lose the continuity of data information and weaken the attention to local information. To get the most out of the non-linear mapping relation between logging curves and reservoir parameters, we combine the output weighting of shallow convolution layer and deep TCN network to enrich the data information obtained by prediction network model, and a network fusional module is designed as

$$F(x) = \frac{(1 + \alpha^2) * T(x) * C(x)}{(\alpha^2 * T(x)) + C(x)},\tag{2}$$

where $F(x)$ is the output of the fusion module, $C(x)$ and $T(x)$ are the output of the convolution layer and after the TCN, and $\alpha$ is a balance factor, which is used to weigh the integration with the network layer.

This design considers the information of the deep and shallow network to enhance the prediction performance of the network comprehensively and maximize the characteristic information of the logging curves.

### 3.3. TCN Network

The network uses causal convolution to handle time series data. There is a causal relationship between different layers of the network, so that it does not have information leakage from the future into the past, as shown in Figure 2. At time t, the output is only convolved with t and earlier elements in an anterior layer. In order to make the network produce the same output as the input length, a one-dimensional fully convolutional network structure is used for TCN, in which the length of the hidden layer is the same as the input layer. In addition, it adds zero padding of kernel size 1 length to ensure that the length of the subsequent layer is equal to the preceding layers [40].

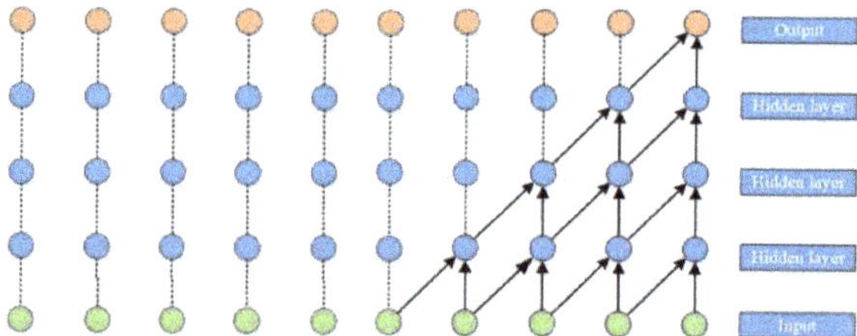


**Figure 2.** Causal convolution.

An ordinary causal convolution can merely review a history of linear size in the depth of the network, which results in difficulties when using causal convolution in assignments that require longer history. In order to remember long effective historical information, dilated convolution [43] is introduced into the network, which increases the receptive field of the kernel and maintains the parameters unchanged. Specifically, for a filter f: $\{0, \ldots, k-1\} \rightarrow R$, a 1-D sequence input $X \in R^n$, and sequence element $e$, the dilated convolution operational express $F$ is denoted as

$$F(e) = (X *_d f)(e) = \sum_{i=0}^{k=1} f(i) \cdot X_{e-d \cdot i},\tag{3}$$

where $d$ denotes the dilation factor. $k$ denotes the filter size, and $e - d \cdot i$ indicates the direction of the past. Therefore, the dilation corresponds to bringing in a fixed step in the middle of every two contiguous filter taps. When the $d$ value is 1, dilated convolution turns into regular convolution. This utilization of greater dilation makes the top-level output express a broader input scope. Therefore, the receptive field in ConvNet is expanded effectively.

Dilated convolution obtains the information farther from the current input by skipping part of the input values. The dilation factor $d(d = O(2^i))$ increases along with the depth of network exponentially. This is shown in Figure 3, which represents a dilated convolution when the filter size $k = 3$ and dilated factors $d = 1, 2, 4$. Therefore, with the increase in the network layer, the receptive field of the network continues to increase, thus ensuring that the network can remember more historical information while avoiding an excessively deep network.

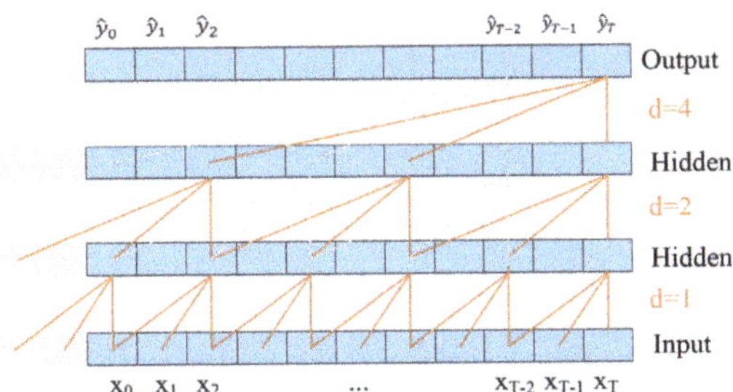


**Figure 3.** Dilated convolution.

In general, the expression ability of neural networks increases with the increase in network depth, but a deeper and larger network can easily produce exploding gradients, vanishing gradients and so on. Residual connections are used in this network [44].

$$o = \text{Activation}\ (x + \mathcal{F}(x)), \tag{4}$$

The residual block consists of a branch that leads to a train of transformations $\mathcal{F}(x)$, and its output is appended to input $x$ of this block, which avoids the degradation of very deep networks. The left part of the FTCN framework shows the residual block, which contains two dilated causal convolution layers, weight normalization layers and rectified linear unit layers. The dropout is used to discard some neurons to prevent overfitting. In addition, $1 \times 1$ convolution is applied to assure the element-wise addition $\oplus$ takes over tensors which have an identical shape. This design can retain information as much as possible and improve the performance of the network model.

### 3.4. FTCN Network Flow

The RMSProp algorithm is an effective and practical depth network optimization algorithm. It adjusts changes in the learning rate by combining an exponential moving average of the square of the gradient and can converge well in the case of the unstable objective function. According to the mean absolute error loss calculated by the prediction network model, the algorithm updates the model parameters by computing the gradient of each weight to optimize the network. The pseudo-code of the FTCN algorithm flow is shown in Algorithm 1, in which AC, CNL, DEN, GR and SP represent acoustic travel time, compensated neutron logging, density, natural gamma ray and spontaneous potential, respectively, and POR, PERM and SW represent porosity, permeability and water saturation, respectively.

---

**Algorithm 1:** FTCN RP = FTCN(LCS).

---

**Input:** Logging Curves(AC, CNL, DEN, GR, SP $\cdots$).
**Output:** Reservoir Parameter(For example: POR, PERM, SW).

1   Data standardization(Logging curves);
2   Create Network layers to build a FTCN prediction framework;
3   Training Network Model;
4   Initialize the weight and the threshold, accumulation variables $\gamma = 0$, decay rate $\rho$;
5   **while** stopping criterion not met **do**
6     Sample a mini-batch of m examples from the training set$\left\{ x^{(1)}, \cdots, x^{(m)} \right\}$,
      corresponding targets $y^{(i)}$;
7     Forward propagation layer by layer;
8     Compute gradient, $g \leftarrow \frac{1}{m} \nabla_\theta \sum_i L\left( f\left( x^{(i)}; \theta \right), y^{(i)} \right)$;
9     Accumulate squared gradient, $\gamma \leftarrow \rho\gamma + (1-\rho)g \odot g$;
10     Compute parameter update,
      $\Delta\theta = -\frac{\epsilon}{\sqrt{\delta+\gamma}} \odot g \cdot \left( \frac{1}{\sqrt{\delta+\gamma}} \text{ applied element} - \text{wise} \right)$;
11     Apply update, $\theta \leftarrow \theta + \Delta\theta$;
12   **end**
13   Calculate the MAE, MSE and RMSE on the test to measure the effect of the model;
14   Improve the network (step 2) or exit according to the prediction effect of the model.

---

## 4. Results and Discussion

### 4.1. Geological Setting and Data Source

As shown in Figure 4, the experiment was set up on a real oil field reservoir, which is located in the east China. The Figure shows the relative position of the well. In this area, the braided river deposit is composed of a mid-channel bar and watercourse. The existence of different configuration units leads to the diversity and complexity of reservoir properties and development characteristics. In turn, the differences in reservoir properties and development characteristics also reflect the differences in sedimentary facies distribution or geological flow unit distribution. The oil field has entered the mid-to-late stage of development, and it is in the stage of high and ultra-high water cut. After long-term water flooding, the heterogeneity has become stronger, and the physical properties, electrical properties and oil-bearing properties of the reservoir have also changed.

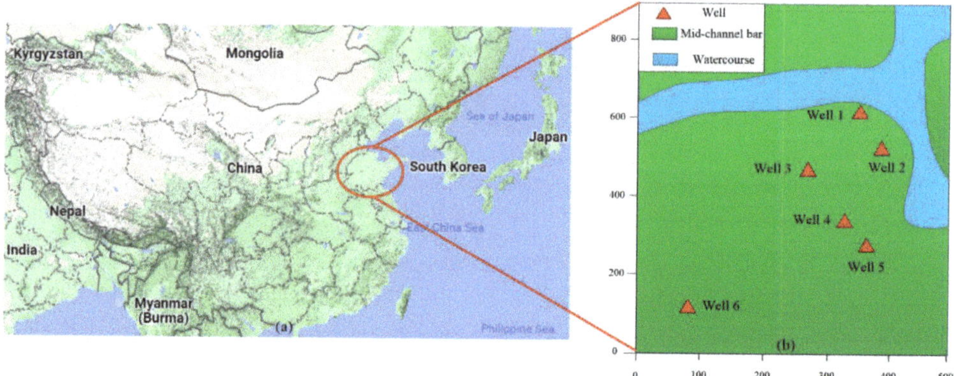


**Figure 4.** Situation in the study area. (**a**) the map of the location of the study area; (**b**) the relative position of the well and architecture analysis.

Figure 5 shows different sedimentary facies in this study area. The thick oil layer in this area is a braided river reservoir, which mainly develops parallel-bedding sandstone facies, trough and plate cross-bedding sandstone facies and conglomerate, with a flushing surface at the bottom. Parallel bedding sandstone facies generally form the top of the braided river channel and the center beach of the braided river. In most cases, due to erosion, the preservation is incomplete, and the thickness is thin, so it is easy for a high-permeability layer to form. The study area is seriously flooded, and it is difficult to stabilize production. Affected by the development and distribution of sand bodies, the reservoir has serious heterogeneity, loose structure and easy sand production, so it is difficult to evaluate accurately. Therefore, the fine reservoir parameter prediction, such as porosity, permeability and water saturation, is important for the analysis of reservoir properties in the research region particularly.

**Figure 5.** Different sedimentary facies in the study area.

The actual exploration logging and core data of 6 wells in this area were used to study the reservoir parameter prediction in this paper. There are many kinds of logging curves. In our scenario and actual logging, based on actual engineering experience, we obtained acoustic travel time, density, compensated neutron logging, natural gamma ray, spontaneous potential, micro-potential resistivity, micro-gradient resistivity, deep investigation induction log, medium investigation induction log, induced conductivity, microspherically focused logging, high-resolution array-induced resistivity (M2R1, M2R2, M2R3, M2R6, M2R9, M3RX), and 4 m bottom gradient resistivity curves, which are considered to be important in the region. Our experiments are based on the field data. The above logging curves were used for predicting reservoir porosity, permeability and water saturation, and the true values of reservoir parameters were determined by core data in this area. Because different well logs may respond to each one of the predicted parameters differently, we used the logs showing direct responses to each one of the reservoir parameters. Specifically, we analyzed the correlation between logging curves and reservoir parameters and optimized the input logging curves during data preprocessing. Taking the analysis of porosity correlation as an example, as shown in Figure 6, the cross-plots show the relationship between input well logs and porosity. It can be seen that the tri-porosity logging curves are closely related to porosity. As the logging curves assist in calculating porosity, the natural gamma ray and the spontaneous potential are also related to porosity. Therefore, acoustic travel time, density, compensated neutron logging, natural gamma ray, and spontaneous potential were selected as the input data for predicting porosity. Similarly, micro-potential resistivity, micro-gradient resistivity, acoustic travel time, density, compensated neutron logging, natural gamma ray, spontaneous potential, deep investigation induction log, medium investigation induction log, and high-resolution array-induced resistivity were selected to predict permeability. Additionally, deep investigation induction log, medium investigation induction log, induced conductivity, high-resolution array-induced resistivity and 4 m bottom gradient resistivity were selected to predict water saturation. Among them, the porosity is about 18% to 46%, the permeability varies widely, from about

50 md to 18,000 md, and the reservoir water saturation is about 10% to 100%, which has the characteristics of strong watered-out layers such as high permeability and low water saturation. The depth of the well section in this area is 1240 m to 1350 m, and the lithology is complex, mainly sand-mudstone, sometimes bottom conglomerate or gravel-bearing sandstone, with vertical accretive sedimentary corrugated siltstone and silty mudstone interbedded at the top. As shown in Figure 7, the core images of well2 show its internal structure clearly.

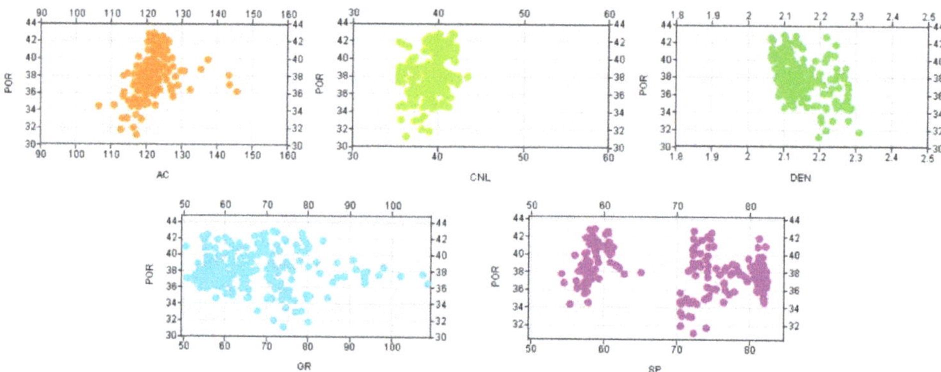

**Figure 6.** The cross-plots of logging curve correlations. Shown are the AC-POR cross-plot, CNL-POR cross-plot, DEN-POR cross-plot, GR-POR cross-plot and SP-POR cross-plot, respectively. These show the correlation between different logging curves and POR.

**Figure 7.** The core images of well2. (**a**) trough cross-bedding gravelly sandstone facies; (**b**) massive bedding gravelly sandstone facies; (**c**) wavy cross bedding siltstone facies; (**d**) horizontal-bedding siltstone facies; (**e**) trough cross-bedding sandstone facies; (**f**) planar cross-bedding sandstone facies; (**g**) parallel-bedding sandstone facies; (**h**) massive mudstone facies.

The lithofacies types are trough cross-bedding gravelly sandstone facies (1343.40–1343.50 m), massive bedding gravelly sandstone facies (1334.81–1334.88 m), wavy cross-bedding siltstone facies (1330.31–1330.40 m), horizontal-bedding siltstone facies (1329.27–1329.36 m), trough cross-bedding sandstone facies(1343.20–1343.30 m), planar cross-bedding sandstone facies (1341.14–1341.29 m), parallel-bedding sandstone facies (1339.79–1339.91 m) and massive mudstone facies (1343.53–1343.70 m), respectively.

Additionally, as shown in Figure 8, we can analyse the sedimentary structures characteristics from the cored well5 further. According to the observation and analysis of well5, we know that the lithofacies types are retained conglomerate facies (1330.44–1330.53 m), trough cross-bedded sandstone facies (1329.96–1330.07 m), tabular cross-bedded sandstone facies (1329.26–1329.37 m), parallel-bedding sandstone facies (1331.37–1331.48 m), wavy-bedding siltstone facies (1327.03–1327.15 m) and massive mudstone facies (1340.35–1340.46 m). During the experiment, the logging data have been relocated deeply. Because there are some missing values in the original core data, and the data values of similar depth are very close, we complement the missing data with its nearest value so as to improve the missing data. The experimental dataset consists of 6734 groups of logging data samples. In this paper, we separated the dataset for network training and testing by the size of the data set and general experience. Then, 80% of logging data samples were randomly selected as the training set of FTCN prediction model, and 20% of the logging data samples were selected as the test set.

**Figure 8.** Sedimentary structure characteristics of cored well5. (**a**) retained conglomerate facies; (**b**) trough cross-bedded sandstone facies; (**c**) tabular cross-bedded sandstone facies; (**d**) parallel-bedding sandstone facies; (**e**) wavy-bedding siltstone facies; (**f**) massive mudstone facies.

*4.2. Evaluation Metrics*

For estimating the prediction effect of FTCN model, mean absolute error (MAE), mean square error (MSE) and root-mean-square error (RMSE) are used as evaluating indicators. The calculation equations are as follows:

$$MAE = \frac{1}{n}\sum_{i=1}^{n}|\hat{y}_i - y_i|, \tag{5}$$

$$MSE = \frac{1}{n}\sum_{i=1}^{n}(\hat{y}_i - y_i)^2, \tag{6}$$

$$RMSE = \sqrt{\frac{1}{n}\sum_{i=1}^{n}(\hat{y}_i - y_i)^2}, \tag{7}$$

where $\hat{y}_i$ and $y_i$ are the predicted and actual values of the model, respectively, and $n$ is the number of samples.

*4.3. FTCN Parameter Setting*

4.3.1. Setting $\alpha$

We define $\alpha$ in Equation (2). The $\alpha$ is a balance factor of the fusional module in the FTCN model, and it is used to weigh the integration with the network layers. Its value represents the integration degree of the network layers. By adjusting $\alpha$, we can better realize the integration of the network layers. In order to study the effect of various $\alpha$ for the experimental prediction results, the experimental values of $\alpha$ were 0.2, 0.4, 0.6, 0.8, 1, 1.2, 1.4, 1.6, 1.8 respectively.

As shown in Figure 9, the MSE, MAE and RMSE of the FTCN model in the prediction of three kinds of reservoir parameters can reach lower values when $\alpha = 1$. In the experiment of porosity prediction (shown as the blue line), changes in $\alpha$ have little influence on the prediction result except $\alpha = 1.8$ and $\alpha = 1.6$. The FTCN model is relatively stable when $\alpha <= 1.4$. When $\alpha = 1.8$, the errors of porosity, permeability and water saturation become larger in varying degrees, indicating that $\alpha = 1.8$ is not conducive to the prediction of the FTCN model. This shows that the larger the $\alpha$ value is, the less the effectiveness of the fusion of different network layers will be. When predicting water saturation, the difference between $\alpha = 0.2$ and $\alpha = 1$ is very small, and when $\alpha = 1$, the FTCN model performs best in permeability prediction. Considering three reservoir parameters, $\alpha = 1$ is the optimal parameter for the FTCN model. In the follow-up experiments, $\alpha$ was set to 1.

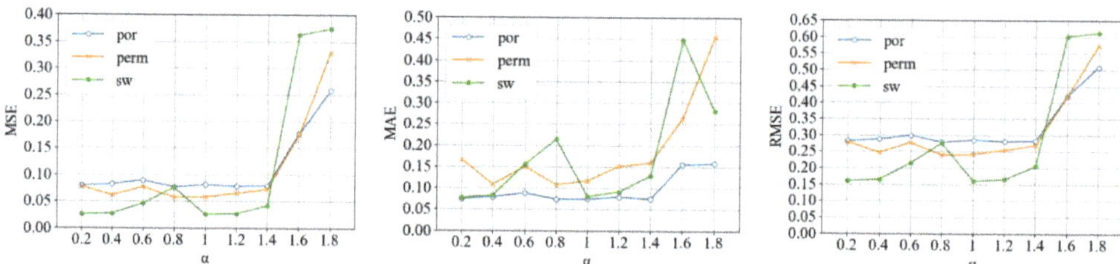

**Figure 9.** The influence of different $\alpha$ on FTCN model.

### 4.3.2. Verification of Fusional Module

For the sake of verifying the effect of the proposed fusional module, we compared it with the unimproved TCN and AddTCN, ConTCN and AveTCN combined with Add, Concat and Average fusion methods. Among them, Add, Concat and Average are all commonly used methods of network multi-layer feature fusion. In various network models, such as ResNet [44] and FPN [45], add is used to fuse features, while in DenseNet [46], concat is used to fuse features. Experiments were carried out in the prediction of three reservoir parameters that include porosity, permeability and water saturation. Among them,

(1)  AddTCN refers to the method of the add operation, which is to carry out a sum operation on the output of the merged layer to fuse the information;

(2)  ConTCN refers to the method of the concat operation, which splices the output of the layer to be merged along the last dimension and then fuses the information;

(3)  AveTCN refers to the method of the average operation, which fuses the output of the layer to be merged by the element mean.

As shown in Figure 10, the FTCN model is superior to the TCN model in predicting porosity (por), permeability (perm) and water saturation (sw) in MAE, MSE and RMSE. In the experiment of predicting porosity, the predictive result of the FTCN model is obviously better than AddTCN and is very similar to that of ConTCN and AveTCN. When predicting permeability, ConTCN has the best performance, and the evaluation metrics obtained by FTCN are slightly higher than ConTCN but also significantly lower than the TCN and AddTCN models. In the experiment of predicting water saturation, the performance of the FTCN model is superior to other models and achieves the best prediction effect. This may be because the fusional module is more suitable for the information fusion of different layers in the reservoir parameter prediction network compared with other fusional methods.

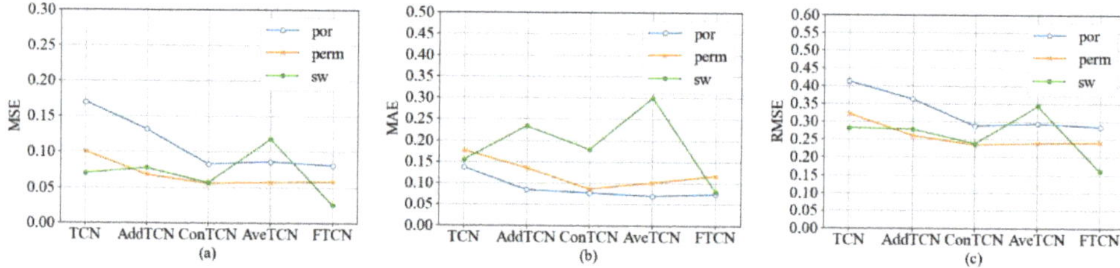

**Figure 10.** Influence of different fusion methods on the TCN model. (**a**) the MSE; (**b**) the MAE; (**c**) the RMSE.

### 4.3.3. Influence of Filter Size k and Residual Block

We explore the influence of the filter size (k) and residual block in the FTCN model based on experiments. The effect of porosity predicted by the FTCN model is demonstrated in Figure 11.

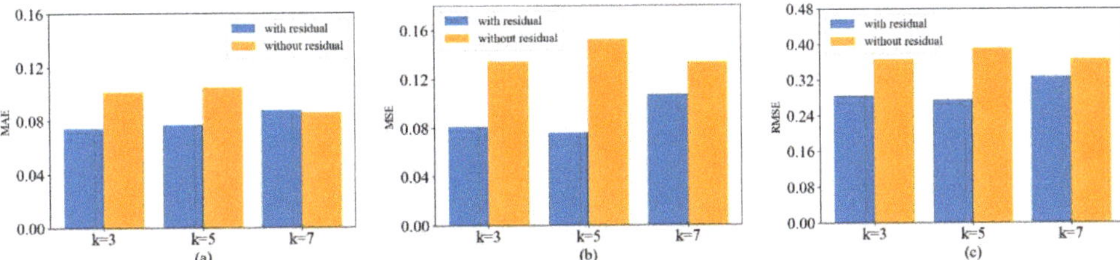

**Figure 11.** Effect of different k and residual blocks on FTCN-predicted porosity. (**a**) the MAE; (**b**) the MSE; (**c**) the RMSE.

The MAE reaches the lowest level when $k = 3$ and the residual block exists. The MSE of the model with the residual block is significantly better than that of the model without the residual block when $k = 3$, $k = 5$ and $k = 7$. When looking at the RMSE, the performance is relatively better when $k = 5$, and the model with the residual block also has a better performance.

We also explore the performance of the FTCN model for predicting permeability. As illustrated in Figure 12, this model has the best effect when $k = 3$, and the performance of this model with the residual block is better as a whole.

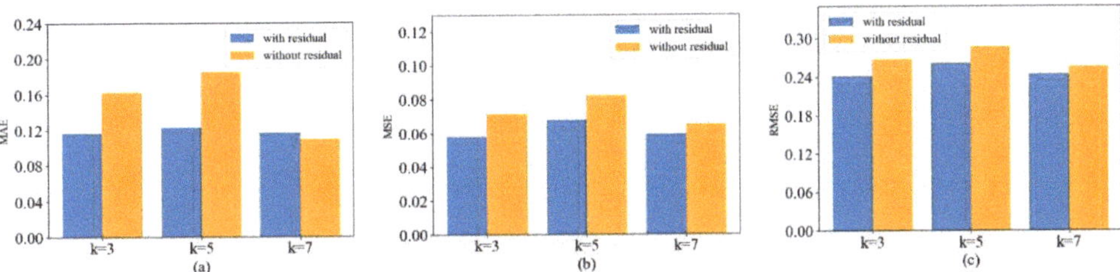

**Figure 12.** Effect of different k and residual blocks on FTCN predicted permeability. (**a**) the MAE; (**b**) the MSE; (**c**) the RMSE.

We also explored the experiment of the FTCN model for predicting water saturation. As shown in Figure 13, when $k = 3$, the model performs better than $k = 5$ and $k = 7$ on MAE, MSE and RMSE, and the model with the residual block is better than the FTCN model without the residual block.

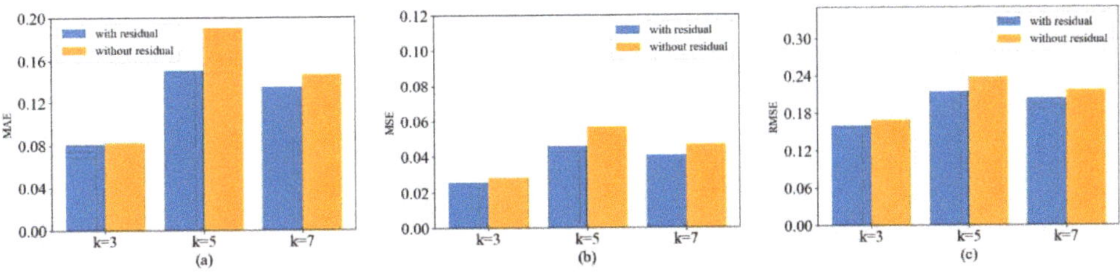

**Figure 13.** Influence of different k and residual blocks on FTCN-predicted water saturation. (**a**) the MAE; (**b**) the MSE; (**c**) the RMSE.

From the above experiments on the FTCN model predicting porosity, permeability and water saturation, it can be seen that the prediction effect of the model with $k = 3$ and the residual block is better. That is because it avoids the degradation of the very deep network and advances the effect of the network.

*4.4. Influence of Input Logging Curves*

For the purpose of estimating the validity of the optimized logging curves used for the reservoir parameter prediction, the effects of different input logging curves on the experimental prediction results are explored.

In logging interpretation, permeability is related to porosity, and the commonly used three-porosity logging includes acoustic travel time, density and compensated neutron logging curves. Acoustic logging mainly measures the time difference of formation sliding waves. Using the interaction between gamma ray and formation, density logging can reflect the formation porosity by measuring the gamma count of gamma rays emitted by the source and arriving at the detector after one or more scatterings through the formation. The compensated neutron logging mainly reflects the deceleration ability of the formation to fast neutrons and shows the change of hydrogen content in the formation. They have different responses in different formations, are closely related to the determination of porosity, and have great advantages in calculating porosity, permeability and fluid properties. In addition, natural gamma ray and spontaneous potential curves are also often used to assist calculation. Natural gamma logging measures natural radioactivity in strata. Spontaneous potential logging is used to measure the variation of the potential naturally generated on the shaft with depth in an open hole so as to study the stratigraphic properties of the well profile. In the permeability prediction experiment, we split all the input logs into three different input log sets. Among them,

(1) Curve_Set1 contains the acoustic travel time, density, compensated neutron logging, natural gamma ray, and spontaneous potential, which are closely related to porosity.
(2) Curve_Set2 adds micro potential resistivity, micro-gradient resistivity, deep and medium investigation induction log, and high-definition induction logging curve M2R10, which are closely related to permeability calculation to the Curve_Set1.
(3) Curve_Set3 contains all the input logs used for permeability prediction in this paper.

As shown in Figure 14, compared with Curve_ Set1, the prediction error of the Curve_Set2 experiment is significantly lower in MSE and RMSE. The results of MAE, MSE and RMSE of Curve_Set3 are all optimal, but it performs significantly better than Curve_Set1 and slightly lower than Curve_Set2. The reason may be that Curve_Set3 is very similar to the Curve_Set2, except that Curve_Set3 has several additional high-definition induction logging curves of different feet, which are very similar to the M2R10 logging curve in Curve_Set2. In practical applications, it can be considered to reduce the cost of logging by removing the high-resolution array induction logs of different feet and retain only the M2R10.

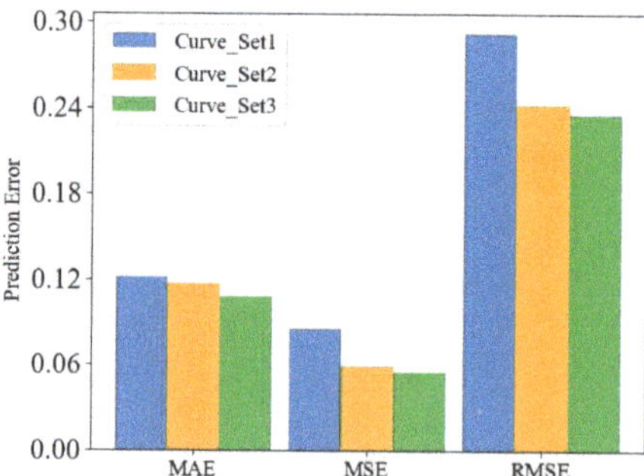


**Figure 14.** Comparison of different logging input sets.

### 4.5. Comparison of Methods

For the purpose of estimating the effectiveness of the FTCN prediction method proposed in this paper, a series of comparative experiments are carried out to compare the proposed FTCN with LSTM, GRU and unimproved TCN. In our experiments, we employed the adaptive learning rate RMSProp algorithm to optimize the network. The initial learning rate was set to 0.001, and batch size the batch size was set to 32 to predict reservoir parameters such as porosity (POR), permeability (PERM) and water saturation (SW).

We used 20% of the logging data samples to test, and the results are shown in Table 1. The RMSE of the FTCN model in porosity prediction is 0.23, 0.19 and 0.13 lower than the LSTM, GRU and TCN models, respectively. For the purpose of estimating the effect of the FTCN prediction model on different reservoir parameters, the FTCN model is used to predict permeability and water saturation. The MAE, MSE and RMSE of the FTCN model reach 0.12, 0.06 and 0.24, respectively. The MAE, MSE and RMSE predicted by the FTCN model are 0.08, 0.03 and 0.16, respectively, and the RMSE is 0.25 and 0.2 lower than the LSTM and GRU model, respectively. The possible reason is that the FTCN model's fusional module considers the effects of different network layers and learns more accurate response relationships comprehensively. It can make better use of logging curves; thus, it achieves better performance than other methods. Compared with LSTM, GRU and TCN models, the FTCN model has a more accurate prediction effect and stable performance in the prediction of different reservoir parameters.

**Table 1.** Prediction performance of four models in three reservoir parameters.

| Method | POR | | | PERM | | | SW | | |
|---|---|---|---|---|---|---|---|---|---|
| | MAE | MSE | RMSE | MAE | MSE | RMSE | MAE | MSE | RMSE |
| LSTM | 0.23 | 0.26 | 0.51 | 0.28 | 0.23 | 0.48 | 0.22 | 0.17 | 0.41 |
| GRU | 0.17 | 0.22 | 0.47 | 0.25 | 0.19 | 0.44 | 0.20 | 0.13 | 0.36 |
| TCN | 0.13 | 0.17 | 0.41 | 0.17 | 0.10 | 0.32 | 0.15 | 0.07 | 0.28 |
| FTCN | 0.07 | 0.08 | 0.28 | 0.12 | 0.06 | 0.24 | 0.08 | 0.03 | 0.16 |

Figures 15–17 show the experimental results of reservoir parameters predicted by the four models, respectively. It can be seen that the LSTM and GRU models can predict the parameter values at different depths of the reservoir roughly, but there is a prediction deviation in the detailed value of the curves. The prediction performance of TCN is better than the former, but the problem is still not solved. The FTCN model more accurately reflects the slight fluctuations in the curve, and its prediction results are more consistent with real reservoir conditions. This may be because of the design of the unique fusional module in FTCN, which makes it achieve better results than other methods. In addition, as shown in Figure 15, the lower parts of the well section are water-flooded layers, and the upper part is a mudstone section, which reflects the characteristics of this area.

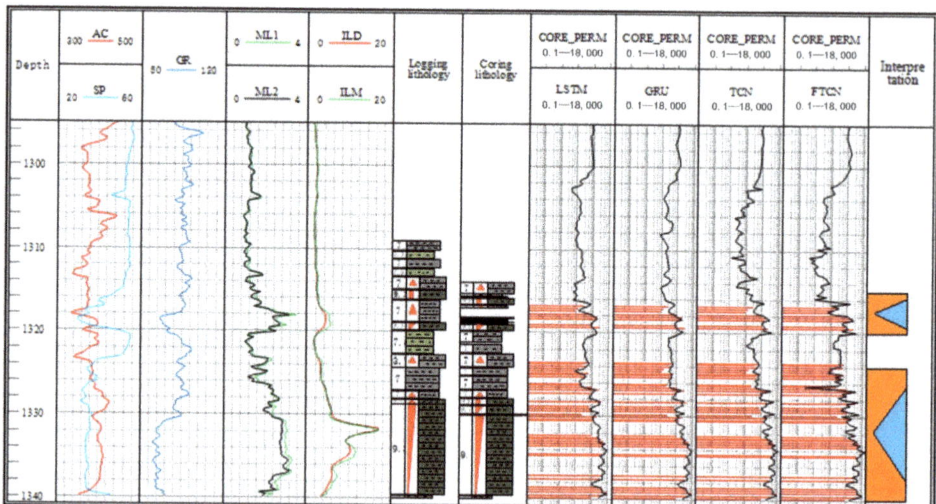

**Figure 15.** Contrast of permeability prediction of distinct models.

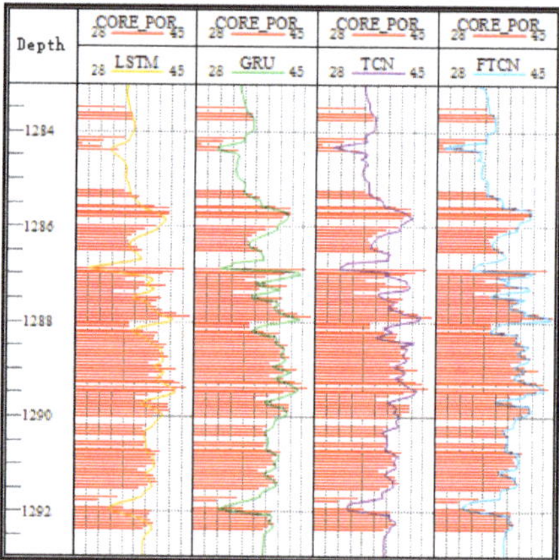

**Figure 16.** Contrast of porosity prediction of distinct models.

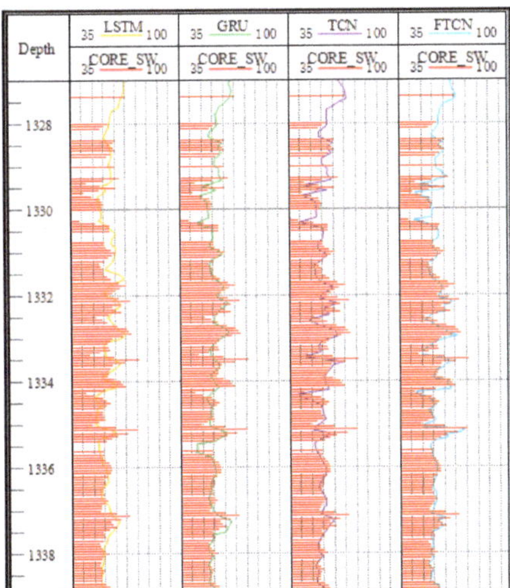

**Figure 17.** Contrast of water saturation prediction of distinct models.

## 5. Conclusions and Future Works

Reservoir parameter prediction is exceedingly significant in petroleum exploration and development. Predicting reservoir parameters effectively can provide a reference for analyzing reservoir properties and assist interpreters in evaluating oil and gas reservoirs. In this paper, a reservoir parameter prediction method named FTCN is proposed. Firstly, a fusion module is designed to fully exploit the nonlinear mapping on curves by using data information of different network layers, which makes our method more sensitive to the relationship between logging curves and reservoir parameters. Secondly, we design the structure of the FTCN. Dilated causal convolution and residual connection are taken, which expands the receptive field of the network so that the effective information obtained is richer and the model is more stable. At last, experiments on real logging datasets show that the prediction results of FTCN are more consistent with the real formation conditions in reservoir parameter prediction, even if there are great changes in stratus. Therefore, our work can provide a reference for well interpreters to analyze reservoirs.

The reservoir parameter prediction effect may be improved by selecting representative and sensitive curves. In the future, we will conduct research in many different exploration areas, and further study the improved reservoir parameter prediction method by considering the curve quality improvement and the response curve selection strategy to improve the prediction effect under poor geological environments and imperfect well data. Additionally, we will explore whether considering the mixed application of multiple models for different strata can further enhance the anti-interference ability to improve the prediction effect.

**Author Contributions:** Conceptualization, H.Z. and K.F.; methodology, H.Z.; software, K.F.; validation, Z.L., Z.W. and J.S.; formal analysis, Z.L.; investigation, K.F.; resources, H.Z.; data curation, H.Y.; writing—original draft preparation, K.F.; writing—review and editing, H.Z., H.Y. and X.G.; visualization, Z.L.; supervision, Z.W.; project administration, J.S.; funding acquisition, H.Z. All authors have read and agreed to the published version of the manuscript.

**Funding:** This research was funded by The Major Scientific and Technological Projects of CNPC (No. ZD2019-183-004), The Fundamental Research Funds for the Central Universities (No. 20CX05019A) and Sponsored by CNPC Innovation Found (2021DQ02-0402).

**Institutional Review Board Statement:** Not applicable.

**Informed Consent Statement:** Not applicable.

**Data Availability Statement:** Not applicable.

**Conflicts of Interest:** The authors declare no conflict of interest.

## References


1. Xu, B.; Wu, H.; Feng, Z.; Li, Y.; Wang, K.; Liu, P. Application status and prospects of artificial intelligence in well logging and formation evaluation. *Acta Pet. Sin.* **2021**, *42*, 508–522.
2. Rumelhart, D.E.; Hinton, G.E.; Williams, R.J. Learning representations by back-propagating errors. *Nature* **1986**, *323*, 533–536. [CrossRef]
3. Hopfield, J.J. Neural networks and physical systems with emergent collective computational abilities. *Proc. Natl. Acad. Sci. USA* **1982**, *79*, 2554–2558. [CrossRef]
4. Hochreiter, S.; Schmidhuber, J. Long short-term memory. *Neural Comput.* **1997**, *9*, 1735–1780. [CrossRef] [PubMed]
5. Cho, K.; van Merrienboer, B.; Gulcehre, C.; Bougares, F.; Schwenk, H.; Bengio, Y. Learning phrase representations using RNN encoder-decoder for statistical machine translation. In Proceedings of the Conference on Empirical Methods in Natural Language Processing (EMNLP 2014), Doha, Qatar, 25–29 October 2014.
6. dos Santos, D.T.; Roisenberg, M.; dos Santos Nascimento, M. Deep Recurrent Neural Networks Approach to Sedimentary Facies Classification Using Well Logs. *IEEE Geosci. Remote Sens. Lett.* **2021**, *19*, 1–5. [CrossRef]
7. Al-Shabandar, R.; Jaddoa, A.; Liatsis, P.; Hussain, A.J. A deep gated recurrent neural network for petroleum production forecasting. *Mach. Learn. Appl.* **2021**, *3*, 100013. [CrossRef]
8. Heghedus, C.; Shchipanov, A.; Rong, C. Advancing deep learning to improve upstream petroleum monitoring. *IEEE Access* **2019**, *7*, 106248–106259. [CrossRef]
9. Kuanzhi, Z.; Zhang, L.; Zheng, D.; Chonghao, S.; Qingning, D. A reserve calculation method for fracture-cavity carbonate reservoirs in Tarim Basin, NW China. *Pet. Explor. Dev.* **2015**, *42*, 277–282.
10. Al-Jawad, S.N.; Saleh, A.H.; Al-Dabaj, A.A.A.; Al-Rawi, Y.T. Reservoir flow unit identification of the Mishrif formation in North Rumaila Field. *Arab. J. Geosci.* **2014**, *7*, 2711–2728. [CrossRef]
11. Yu, Y.; Wang, C.; Gu, X.; Li, J. A novel deep learning-based method for damage identification of smart building structures. *Struct. Health Monit.* **2019**, *18*, 143–163. [CrossRef]
12. Yu, Y.; Rashidi, M.; Samali, B.; Mohammadi, M.; Nguyen, T.N.; Zhou, X. Crack detection of concrete structures using deep convolutional neural networks optimized by enhanced chicken swarm algorithm. *Struct. Health Monit.* **2022**, 14759217211053546. [CrossRef]
13. Sircar, A.; Yadav, K.; Rayavarapu, K.; Bist, N.; Oza, H. Application of machine learning and artificial intelligence in oil and gas industry. *Pet. Res.* **2021**, *6*, 379–391. [CrossRef]
14. Choubey, S.; Karmakar, G. Artificial intelligence techniques and their application in oil and gas industry. *Artif. Intell. Rev.* **2021**, *54*, 3665–3683. [CrossRef]
15. Janjuhah, H.T.; Kontakiotis, G.; Wahid, A.; Khan, D.M.; Zarkogiannis, S.D.; Antonarakou, A. Integrated Porosity Classification and Quantification Scheme for Enhanced Carbonate Reservoir Quality: Implications from the Miocene Malaysian Carbonates. *J. Mar. Sci. Eng.* **2021**, *9*, 1410. [CrossRef]
16. Adeniran, A.A.; Adebayo, A.R.; Salami, H.O.; Yahaya, M.O.; Abdulraheem, A. A competitive ensemble model for permeability prediction in heterogeneous oil and gas reservoirs. *Appl. Comput. Geosci.* **2019**, *1*, 100004. [CrossRef]
17. Liu, J.J.; Liu, J.C. An intelligent approach for reservoir quality evaluation in tight sandstone reservoir using gradient boosting decision tree algorithm-A case study of the Yanchang Formation, mid-eastern Ordos Basin, China. *Mar. Pet. Geol.* **2021**, *126*, 104939. [CrossRef]
18. Yong Hong, W.; Zhengnan, Z.; Shufa, Z. Estimation of porosity using seismic and logging data. *Oil Geophys. Prospect.* **1990**, *25*, 572–577.
19. Jinzhong, Z. An Empirical Formula for Estimating Permeability Based on Well Logging Data and Its Physuical Background Analysis. *J. Xi'an Shiyou Univ.* **1991**, *6*, 16–20.
20. Xie, J.; Chen, W.; Zhang, D.; Zu, S. Application of principal component analysis in weighted stacking of seismic data. *IEEE Geosci. Remote Sens. Lett.* **2017**, *14*, 1213–1217. [CrossRef]
21. Inyang, N.J.; Agbasi, O.E.; Akpabio, G.T. Integrated analysis of well logs for productivity prediction in sand-shale sequence reservoirs of the Niger Delta—A case study. *Arab. J. Geosci.* **2021**, *14*, 1–9. [CrossRef]
22. Kuang, L.; He, L.; Yili, R.; Kai, L.; Mingyu, S.; Jian, S.; Xin, L. Application and development trend of artificial intelligence in petroleum exploration and development. *Pet. Explor. Dev.* **2021**, *48*, 1–14. [CrossRef]
23. Otchere, D.A.; Ganat, T.O.A.; Gholami, R.; Ridha, S. Application of supervised machine learning paradigms in the prediction of petroleum reservoir properties: Comparative analysis of ANN and SVM models. *J. Pet. Sci. Eng.* **2021**, *200*, 108182. [CrossRef]
24. Zhu, P.; Lin, C.; Wu, P.; Fan, R.; Zhang, H.; Pu, W. Permeability prediction of tight sandstone reservoirs using improved BP neural network. *Open Pet. Eng. J.* **2015**, *8*, 288–292. [CrossRef]
25. Mahdaviara, M.; Rostami, A.; Keivanimehr, F.; Shahbazi, K. Accurate determination of permeability in carbonate reservoirs using Gaussian Process Regression. *J. Pet. Sci. Eng.* **2021**, *196*, 107807. [CrossRef]

26. Gholami, R.; Shahraki, A.; Paghaleh, M.J. Prediction of Hydrocarbon Reservoirs Permeability Using Support Vector Machine. *Math. Probl. Eng.* **2012**, *2012*, 670723. [CrossRef]
27. Li, K.; Zhou, G.; Yang, Y.; Li, F.; Jiao, Z. A novel prediction method for favorable reservoir of oil field based on grey wolf optimizer and twin support vector machine. *J. Pet. Sci. Eng.* **2020**, *189*, 106952. [CrossRef]
28. Akande, K.O.; Owolabi, T.O.; Olatunji, S.O.; AbdulRaheem, A. A hybrid pARTICLE swarm optimization and support vector regression model for modelling permeability prediction of hydrocarbon reservoir. *J. Pet. Sci. Eng.* **2017**, *150*, 43–53. [CrossRef]
29. Okon, A.N.; Adewole, S.E.; Uguma, E.M. Artificial neural network model for reservoir petrophysical properties: Porosity, permeability and water saturation prediction. *Model. Earth Syst. Environ.* **2021**, *7*, 2373–2390. [CrossRef]
30. Hamidi, H.; Rafati, R. Prediction of oil reservoir porosity based on BP-ANN. In Proceedings of the 2012 International Conference on Innovation Management and Technology Research, Malacca, Malaysia, 21–22 May 2012; pp. 241–246.
31. Xiang, Y.F.; Kang, Z.H.; Hao, W.J.; Fu, K.N.; Wang, F.S. A Composite Method of Reservoir Parameter Prediction Based on Linear Regression and Neural Network. *Sci. Technol. Eng.* **2017**, *17*, 46–52.
32. Nasir, I.M.; Bibi, A.; Shah, J.H.; Khan, M.A.; Sharif, M.; Iqbal, K.; Nam, Y.; Kadry, S. Deep learning-based classification of fruit diseases: An application for precision agriculture. *CMC-Comput. Mater. Contin.* **2021**, *66*, 1949–1962. [CrossRef]
33. Ouahabi, A.; Taleb-Ahmed, A. Deep learning for real-time semantic segmentation: Application in ultrasound imaging. *Pattern Recognit. Lett.* **2021**, *144*, 27–34. [CrossRef]
34. Muhammad, A.N.; Aseere, A.M.; Chiroma, H.; Shah, H.; Gital, A.Y.; Hashem, I.A.T. Deep learning application in smart cities: Recent development, taxonomy, challenges and research prospects. *Neural Comput. Appl.* **2021**, *33*, 2973–3009. [CrossRef]
35. Yuyang, L.; Xinhua, M.; Xiaowei, Z.; Wei, G.; Lixia, K.; Rongze, Y.; Yuping, S. Shale gas well flowback rate prediction for Weiyuan field based on a deep learning algorithm. *J. Pet. Sci. Eng.* **2021**, *203*, 108637. [CrossRef]
36. Zhao, J.; Wang, F.; Cai, J. 3D tight sandstone digital rock reconstruction with deep learning. *J. Pet. Sci. Eng.* **2021**, *207*, 109020. [CrossRef]
37. An, P.; Cao, D.P.; Zhao, B.Y.; Yang, X.L.; Zhang, M. Reservoir physical parameters prediction based on LSTM recurrent neural network. *Prog. Geophys.* **2019**, *34*, 1849–1858.
38. Chen, W.; Yang, L.; Zha, B.; Zhang, M.; Chen, Y. Deep learning reservoir porosity prediction based on multilayer long short-term memory network. *Geophysics* **2020**, *85*, WA213–WA225. [CrossRef]
39. Zhang, Z.; Wang, Y.; Wang, P.; Dimitri, R. On a Deep Learning Method of Estimating Reservoir Porosity. *Math. Probl. Eng.* **2021**, *2021*, 6641678. [CrossRef]
40. Bai, S.; Kolter, J.Z.; Koltun, V. An Empirical Evaluation of Generic Convolutional and Recurrent Networks for Sequence Modeling. In Proceedings of the 6th International Conference on Learning Representations, Vancouver, BC, Canada, 30 April–3 May 2018.
41. Gehring, J.; Auli, M.; Grangier, D.; Yarats, D.; Dauphin, Y.N. Convolutional sequence to sequence learning. In Proceedings of the International Conference on Machine Learning, Sydney, Australia, 6–11 August 2017; pp. 1243–1252.
42. Tieleman, T.; Hinton, G. Divide the gradient by a running average of its recent magnitude. COURSERA Neural Netw. *Mach. Learn.* **2012**, *6*, 26–31.
43. Yu, F.; Koltun, V. Multi-scale context aggregation by dilated convolutions. In Proceedings of the International Conference on Learning Representations, San Juan, Puerto Rico, 2–4 May 2016.
44. He, K.; Zhang, X.; Ren, S.; Sun, J. Deep residual learning for image recognition. In Proceedings of the IEEE Conference on Computer Vision and Pattern Recognition, Las Vegas, NV, USA, 27–30 June 2016; pp. 770–778.
45. Lin, T.Y.; Dollár, P.; Girshick, R.; He, K.; Hariharan, B.; Belongie, S. Feature pyramid networks for object detection. In Proceedings of the IEEE Conference on Computer Vision and Pattern Recognition, Honolulu, HI, USA, 21–26 July 2017; pp. 2117–2125.
46. Huang, G.; Liu, Z.; Van Der Maaten, L.; Weinberger, K.Q. Densely connected convolutional networks. In Proceedings of the IEEE Conference on Computer Vision and Pattern Recognition, Honolulu, HI, USA, 21–26 July 2017; pp. 4700–4708.

MDPI
St. Alban-Anlage 66
4052 Basel
Switzerland
www.mdpi.com

*Energies* Editorial Office
E-mail: energies@mdpi.com
www.mdpi.com/journal/energies

www.ingramcontent.com/pod-product-compliance
Lightning Source LLC
LaVergne TN
LVHW071034170726
843515LV00006B/1282